"An engrossing book, a biography with a difference. . . Keynes gives an excellent overview of Victorian writings on medicine, religion, and science as they bore on the question of disease and death, especially the deaths of children, which raised with particular sharpness the question of God's intention and benevolence. . . . Above all, Keynes sets Darwin's lonely progress in his researches on species directly in the context of family love and loss." —*The Sunday Telegraph* (London)

"In this intimate portrait. . . the great-great-grandson of the scientist, Keynes uses published documents as well as family papers and artifacts to show how Darwin's thinking on evolution was influenced by his deep attachment to his wife and children. In particular, his anguish over his ten-year-old daughter Annie's death sharpened his conviction that the operation of natural laws had nothing to do with divine intervention or morality. Keynes shows that much of Darwin's intellectual struggle in writing *On the Origin of Species* and *The Descent of Man* arose from his efforts to understand the role of suffering and death in the natural order of the world. . . . A moving tribute to a thinker who, despite intimate acquaintance with the pain inflicted by the 'war of nature' could still marvel that, from this ruthless struggle, 'endless forms most beautiful and most wonderful have been, and are being, evolved.'" —*Publishers Weekly*

"Though there have been innumerable biographies of Darwin, there cannot have been any warmer portrayals of his humanity and his desire to discover meaning in human existence." —*The Irish Times*

"Above all this is a love story: the love of Charles and Emma . . . and the love of both for their children." —*The Times* (London)

"Fascinating . . . This is a wonderful portrait of a Victorian family in all its aspects . . . so extraordinary and so revealing."—*Birmingham Post* (UK)

*continued . . .*

"It is a rare biography that reveals the key emotional moment in its subject's personal and intellectual life so clearly as Randal Keynes does for Charles Darwin in *Darwin, His Daughter & Human Evolution*. . . . What makes this book so moving and illuminating is the way Keynes tracks his subject's emotional development and, more generally, shows how much his scientific thinking was influenced by his wife, Emma, and their ten children. . . . Besides its contribution to the intellectual history of Darwinism, *Darwin, His Daughter & Human Evolution* is a delightful portrayal of mid-nineteenth-century family life."

—*Financial Times* (London)

"The evolution of Darwin's theories played out against the evolution of his family life—in a graceful and insightful biography by the naturalist's great-great-grandson. . . . [The book] will do much to burnish Darwin's reputation as a husband and father; his scientific bona fides continue to stand tall." —*Kirkus Reviews*

"One of the most remarkable moving biographies of recent years."

—*The Scotsman*

"Sensitively told, this is a deeply human story in which Darwin, the caring father and husband, emerges with as much esteem as Darwin the great scientist." —*The Northern Echo*

"Rigorously, tenderly, Randal Keynes recounts the most emotional moments in Darwin's career. He opens up the sacred interiors of Darwin's marriage and family life to show how he drew on his heart-rending experiences to shed light on the evolution of human nature."

—James Moore, coauthor of *Darwin: The Life of a Tormented Evolutionist*

"*Darwin, His Daughter & Human Evolution* is . . . committed to humanizing a man whose personal life has inevitably become locked out of history. Keynes has reclaimed the piece of the past he aimed for. His bright and devoted biography makes the iconic, elevated figure of his great-great-grandfather seem entirely human [and] reminds us how difficult it is for any modern reader to even crudely imagine how much of a threat Darwin's views posed to the dominant beliefs of the day. Keynes absorbs himself into the period totally, and, in his dogged reconstruction of the Darwin family tree, creates an epic sense of lineage. His meticulous arrangement of notebook extracts supply *Darwin, His Daughter & Human Evolution* with both its structure and its intimacy. The extensive quotations from personal journals and letters—particularly the correspondence between Darwin and his wife in the build-up to Annie's death—for a surprising emotional weight at the book's heart."

—*Scotland on Sunday*

"Unique . . . a fascinating, detailed picture." —*Sunday Mail*

---

**Randal Keynes** is a great-great-grandson of Charles Darwin and a great-nephew of economist John Maynard Keynes. He lives and works in London. This is his first book.

*Annie Darwin in 1849*

# DARWIN, HIS DAUGHTER, *and* HUMAN EVOLUTION

RANDAL KEYNES

RIVERHEAD BOOKS
*New York*

Riverhead Books
Published by The Berkley Publishing Group
A division of Penguin Putnam Inc.
375 Hudson Street
New York, New York 10014


Book design by Michelle McMillian
Cover design by Honi Werner
Front cover images: Charles Darwin (upper left), Mohr Darwin Collection, The Huntington Library, Superstock; Annie Darwin (rectangular frame) © English Heritage Photo Library/Jeremy Richards; Darwin (oval frame), English Heritage Photo Library; evolution chart (lower left), The Grange Collection; quill and maps © Hans Neleman/The Image Bank



Previously published in England by Fourth Estate Limited as *Annie's Box: Charles Darwin, His Daughter and Human Evolution*.
First Riverhead hardcover edition: January 2002
First Riverhead trade paperback edition: November 2002
Riverhead trade paperback ISBN: 1-57322-955-5

Visit our website at www.penguinputnam.com

The Library of Congress has catalogued the Riverhead hardcover edition as follows:

Keynes, Randal.
Darwin, his daughter & human evolution / Randal Keynes.
p. cm.
ISBN 1-57322-192-9
1. Darwin, Charles, 1809–1882. 2. Darwin, Charles, 1809–1882—Family.
3. Darwin family. 4. Naturalists—England—Biography. 5. Evolution (Biology).
I. Title: Darwin, his daughter, and human evolution. II. Title.
QH31.D2 K48 2002 2001031852
576.8'092—dc21
[B]

Printed in the United States of America

10 9 8 7 6 5 4

For M.E.K.
1890–1974

*"Much love, much trial..."*

Charles Darwin to Joseph Hooker,
27 November 1863

# CONTENTS

# PLATES AND ILLUSTRATIONS

*Plates*

*Illustrations*

# FAMILY AND FRIENDS

**Darwins**

*Bernard* (1876–1961)—Francis's son, Charles and Emma's grandson. Brought up by Emma and Charles after his mother, Amy, died of puerperal fever. Essayist and journalist for *The Times.*

*Caroline* (1800–88)—Charles's elder sister. Brought up Charles after their mother died when he was eight. Married their cousin Josiah Wedgwood III in 1837. During Annie's childhood and afterwards, lived with her husband and their children at Leith Hill Place in Surrey.

*Catherine* (1810–66)—Charles's younger sister. Lived at the family home in Shrewsbury until 1863 when she married her cousin Charles Langton.

*Charles Waring* (1856–8)—Charles and Emma's last child. Died of scarlet fever.

*Elizabeth* (Betty in childhood, later Bessy) (1847–1928)—Charles and Emma's fourth daughter. Unmarried.

*Erasmus* (1731–1812)—Charles's grandfather. Physician and inventor.

*Erasmus Alvey* (Ras) (1804–81)—Charles's older brother. Unmarried. Lived in London.

*Francis* (1848–1925)—Charles and Emma's third son. Botanist. His first wife was Amy Ruck, who died after giving birth to their son Bernard.

*George* (1845–1912)—Charles and Emma's second son. Mathematician.

*Gwen* (1885–1957)—George's elder daughter, Charles and Emma's granddaughter. Married Jacques Raverat. Artist and writer of *Period Piece.*

*Henrietta* (Etty) (1843–1929)—Charles and Emma's third daughter. Married R. B. Litchfield. Writer of *Emma Darwin: A Century of Family Letters.*

*Horace* (1851–1928)—Charles and Emma's fifth son. Designer of scientific instruments.

*Leonard* (1850–1943)—Charles and Emma's fourth son. Soldier and MP.

*Margaret* (1890–1974)—George's younger daughter, Charles and Emma's granddaughter. Married Geoffrey Keynes.

*Robert* (1766–1848)—Charles's father. Physician in Shrewsbury.

*Susan* (1803–66)—Charles's elder sister. Unmarried. Lived at the family home in Shrewsbury until her death.

*William* (Willy) (1839–1914)—Charles and Emma's eldest son. Banker.

## Wedgwoods

*Alfred* (1842–92)—Son of Hensleigh and Fanny. Annie's first cousin.

*Allen* (1796–1882)—Son of Charles and Emma's uncle, John Wedgwood. Vicar of St. Peter's, Maer.

*Amy* (1835–1910)—Daughter of Francis. Annie's first cousin.

*Cecily* (1837–1917)—Daughter of Francis. Annie's first cousin.

*Charlotte* (1792–1862)—Emma's elder sister. Married Charles Langton as his first wife.

*Clement* (1840–89)—Son of Francis, Annie's first cousin.

*Elizabeth* (Bessy) (1764–1846)—Emma's mother, née Allen.

*Elizabeth* (1793–1880)—Emma's eldest sister. Unmarried. Lived with her parents at Maer while they were alive, then at Hartfield in Sussex, and spent her last years in Downe.

*Ernest* (Ernie) (1838–98)—Son of Hensleigh and Fanny. Annie's first cousin.

*Fanny* (1800–89)—Wife of Emma's brother Hensleigh. Daughter of Sir James Mackintosh, philosopher and historian.

*Fanny* (1806–32)—Emma's elder sister.

*Francis* (Frank) (1800–88)—Son of Josiah II. Emma's elder brother. Took over the firm after their father's retirement. Lived with his family at Barlaston in Staffordshire.

*Henry* (Harry) (1799–1885)—Son of Josiah II. Emma's elder brother. Barrister and writer of *The Bird Talisman*.

*Hensleigh* (1803–91)—Son of Josiah II. Emma's elder brother. Married Fanny Mackintosh and lived in London.

*Hope* (1844–1935)—Daughter of Hensleigh and Fanny. Annie's first cousin.

*James* ("Bro") (1834–64)—Son of Hensleigh and Fanny. Annie's first cousin.

*Josiah I* (1730–95)—The master potter. Established the firm of Josiah Wedgwood & Sons Ltd at Etruria in Staffordshire.

*Josiah II* ( Jos) (1769–1843)—Emma's father. Took over the company when his father died, and lived at Maer Hall in Staffordshire.

*Josiah III* ( Joe) (1795–1880)—Emma's eldest brother. Married Caroline Darwin and lived at Leith Hill Place with their children.

*Julia* ("Snow") (1833–1913)—Hensleigh and Fanny's eldest daughter, Annie's cousin. Unmarried. Essayist.

*Katherine Euphemia* (Effie) (1839–1934)—Daughter of Hensleigh and Fanny. Annie's first cousin.

*Lucy* (1846–?)—Daughter of Josiah II and Caroline. Annie's first cousin.

*Margaret* (Greata) (1843–1937)—Daughter of Josiah II and Caroline. Annie's first cousin. Married the Reverend Arthur Vaughan Williams. Ralph Vaughan Williams was their son.

*Sarah* (1776–1856)—Youngest daughter of Josiah I, Emma and Charles's aunt. Unmarried. Lived in Downe from 1847 until her death.

*Sophy* (1842–1911)—Daughter of Josiah II and Caroline, Annie's first cousin. Lived at Leith Hill Place.

*Susannah* (1765–1817)—Charles's mother. Josiah I's eldest daughter. Married Robert Darwin.

*Thomas* (Tom) (1771–1805)—Son of Josiah I. Friend of Samuel Taylor Coleridge.

## Other Relatives and Friends

*Fanny Allen* (1781–1875)—Emma's mother's sister. Unmarried.

*Robert FitzRoy* (1805–65)—Naval officer. Captain of HMS *Beagle* from 1828 to 1836.

*William Darwin Fox* (1805–80)—Charles's second cousin. With him at Cambridge University in the 1820s. Vicar of Delamere in Cheshire. His wife died in 1842 giving birth to their sixth child.

*Henry Holland* (1788–1873)—Emma and Charles's second cousin. Physician to Queen Victoria.

*Joseph Hooker* (1817–1911)—Botanist. Director of Royal Botanic Garden, Kew, from 1865 to 1885.

*Charles Lyell* (1797–1875)—Geologist.

*Jessie Sismondi* (1777–1853)—Emma's mother's sister. Married to the Swiss historian Jean Charles de Sismondi.

*Annie's writing case*

# INTRODUCTION

A CHILD'S WRITING CASE. The pale yellow ribbon curled inside is stitched with small glass beads. The goose-feather quills have dried ink on their tips, and the sealing wax has been melted over a candle flame. On the ribbon and the quills lies a fold of paper with a thick lock of fine brown hair. On the paper is written "April 23rd 1851." And on a leaf torn from a pocketbook is a map of a churchyard: "Annie Darwin's grave at Malvern."

The writing case was Annie's, and is filled with her things. She was Charles and Emma Darwin's first daughter. She died when she was ten. Charles wrote a "memorial" of her, and Emma kept the case to remember her by. It was passed down to my father, one of their great-grandsons.

I came across the case one day when I was looking through a box of family odds and ends. I was struck by a note in Charles's untidy scrawl. He had headed it "Anne" and wrote how she felt every day and night during her last months. She was often well but he noted when she was distressed. "Late evening tired and cry." "Early morning cry." "Poorly in morning." It was haunting to sense how he had been watching her day after day, night after restless night.

I found other traces of Annie's life in Charles and Emma's notebooks and letters. In the pages that follow I piece together a jigsaw of her childhood, and tease out some of Charles and Emma's feelings and ideas through the years after her death. I draw links with Charles's thinking about human nature, both before and after her short life. He learnt from his feelings for her about the lasting strength of the affections, the paradox of pain, the value of memory and the limits of human understanding.

There is one idea at the heart of my account. Charles's life and his science were all of a piece. Working at home on things he could study there, spending every day with his wife, children and servants, living at a time when science meant knowledge and understanding in the broadest view, and dwelling on issues that bear directly on the deepest questions about what it is to be human, he could not keep his thinking about the natural world apart from feelings and ideas that were important to him in the rest of his life.

This book explores Darwin's life with his family and his thinking about human nature in the interweavings around Annie and her memory.

CHAPTER ONE

# MACAW COTTAGE

*Marriage—First home in London—First child—*
*Annie's birth—Infancy*

When at twenty-nine Charles Darwin thought about marrying, he took a piece of paper and wrote: "This is the question." Under "Not Marry" he jotted down: "Freedom to go where one liked—choice of society and *little of it.* Conversation of clever men at clubs. Not forced to visit relatives and to bend in every trifle—to have the expense and anxiety of children—perhaps quarrelling—*loss of time* . . . How should I manage all my business if I were obliged to go every day walking with my wife. Eheu! I never should know French, or see the Continent, or go to America, or go up in a Balloon." Under "Marry" he noted: "Children (if it please God), constant companion (and friend in old age) who will feel interested in one." He weighed all the points for and against, and made up his mind. "My God, it is intolerable to think of spending one's whole life, like a neuter bee, working, working, and nothing after all. No, no, won't do. Imagine living all one's day solitarily in smoky dirty London house. Only picture to yourself a nice soft wife on a sofa with good fire, and books and music perhaps . . . Marry—Marry—Marry. Q.E.D."

A few days later, in July 1838, he visited his uncle, Josiah Wedgwood II, at his home, Maer Hall, near the Wedgwood factory in Staffordshire. Josiah's daughter Emma was there. She was a year older than Charles and had been a companion to him since childhood. She was lively and attractive and had been courted by many young men, but she was now looking after her elderly mother who had lost her mind, and faced the prospect of remaining single. Charles had met her in London earlier in the month, and they now had a long talk together by the fire in the library. He decided that he wanted her to be his wife. She was very happy in his company, and felt tentatively that if he saw more of her, he might really like her. When he proposed three months later, she accepted him eagerly. She went straight to her Sunday school for the village children after the "important interview," but "found I was turning into an idiot, and so came away." She wrote to her Aunt Jessie Sismondi: "He is the most open transparent man I ever saw and every word expresses his real thoughts." He was "the most affectionate person possible." Like many of the Wedgwood family, she often found it difficult to show her feelings. She felt it was a great advantage to have the power of expressing affection, and was sure that he would "make his children very fond of him."

Charles and Emma lost no time in planning for the future. They agreed to live in London while Charles was tied there by his scientific work, and the next month he was back at his lodgings in Great Marlborough Street, house-hunting anxiously. Emma wrote to him from Maer: "It is very well I am coming to look after you, my poor old man, for it is quite evident that you are on the verge of insanity and we should have had to advertise you—'Lost in the vicinity of Bloomsbury, a tall thin gentleman &c. &c., quite harmless. Whoever will bring him back shall be handsomely rewarded.'"

After the five years from 1831 that he had spent on HMS *Beagle,* sailing round the world as ship's naturalist, and his two years back in London since then working on his collections and findings from the voyage, Charles was looking forward to this change in his life. A few days before their wedding he wrote to Emma: "I was thinking this morning how on earth it came that I, who am fond of talking and am scarcely ever out of spirits, should so entirely rest my notions of happiness on quietness and a good deal of solitude; but I believe the explanation is very simple, and I mention it, because it will give you hopes that I shall gradually grow less of a *brute*." During the voyage, "the whole of my pleasure was derived from what passed in my mind, whilst admiring views by myself, travelling across the wild deserts or glorious forests, or pacing the deck of the poor little *Beagle* at night. Excuse this much egotism. I give it to you, because I think you will humanise me, and soon teach me there is greater happiness, than building theories and accumulating facts in silence and solitude." Charles had been thinking about matters of great importance to him. The theories he had been building were parts of the idea he was forming about the origin of species. He was having to work "in silence and solitude" because he recognised how fiercely his ideas would be attacked as soon as he revealed them to anyone, and he could not risk an argument until he was sure of his ground. His hope that Emma would humanise him was a deep wish that she could draw him out of his lonely work into the company and care of a close family circle.

But he could joke about the difficulties she would have. After spending a morning with his friend, the geologist Charles Lyell, he wrote to her: "I was quite ashamed of myself today; for we talked for half an hour unsophisticated geology, with poor Mrs Lyell sitting by, a monument of patience. I want *practice* in ill-treating the female sex. I did not observe Lyell had any compunction. I hope to harden my conscience in time: few husbands seem to find it difficult to effect this."

He found a house for them in Upper Gower Street, a long terrace on what was then the northern edge of London. The street led to the recently founded University College with its teaching hospital across the road, and a school whose pupils played in the grounds in front of the main building. University College was known as "the Godless College." Under the guidance of the Whig politician Lord Brougham

*Upper Gower Street by George Scharf in 1840*

and other progressive reformers, it gave literary and scientific education to students of all backgrounds and denominations. The main building with its grand ten-column portico was modelled on a temple in Athens. Together with the monumental Euston Arch (now sadly destroyed), St. Pancras New Church, the Royal College of Surgeons and the great colonnade of the British Museum, University College gave the neighbourhood the distinctive high-minded tone of the "Greek" revival. The imposing new buildings stood for progressive enterprise, free inquiry and the life of the mind. The people who commissioned them felt they were constructing a "brave new world." In his "Ode to Liberty," Shelley had written of Athens with its "crest of columns" set on the will of man "as on a mount of diamond."

Many of the hospital's patients were poor people from the crowded slums immediately behind the smart terraces and public buildings of the neighbourhood. Charles Dickens said at a fund-raising dinner that the hospital represented "the largest liberality of opinion. It excludes no one patient, student, doctor, surgeon or nurse because of religious creed. It represents the complete relinquishment of claims to coerce the judgement or the conscience of any human being." Like his namesake, Charles Darwin made regular donations.

The Darwins' neighbours in the terrace were well-to-do professionals—surgeons, lawyers, artists, a publisher and a famous Shakespearean clown. Emma's brother Hensleigh Wedgwood lived a few doors away with his wife Fanny, daughter of Sir James Mackintosh, who was one of the governors of University College and known as the "Whig Cicero." The mews at the back of the long narrow gardens were tenanted by coachmen, stable keepers and their families.

The house Charles found had a kitchen and a room for the manservant in the basement, the dining room and a study for Charles on the ground floor, the main drawing room on the first floor and a small back room with a bay window looking out over the garden. The family bedrooms were on the second floor, and the cook and maids slept in the attic rooms. Charles planned to move in before the wedding, and jotted in his notebook: "Remnants of carpets; mat for hall . . . white curtains washed; two easy chairs; blinds in red rooms washed." The yellow curtains in the drawing room clashed with the blue paintwork and the furniture, and there was a dead dog in the garden. Charles kept the yellow curtains but had the dog removed, and looked forward to walking in the garden. He was grateful for the plants and the open air, but the atmosphere was poisoned by the smoke of the city. A physician who lived nearby wrote that the trees and bushes in the squares and gardens were stunted and often died. "If you pluck a branch from one of them, your fingers are smeared with soot . . . By the time a person has been in the streets two or three hours, the glory of the laundress and the clear-starcher is laid, not in the dust, but in smoke, which forms itself into myriads of flocculi, designated 'blacks,' and the blacks are by no means capricious, for they stick most assiduously to ladies' and gentlemen's dresses, if the weather be more than ordinarily dense."

Charles looked forward to Emma's arrival. "Is it not *our* house?" he wrote to her. "What is there, from me the geologist to the black sparrows in the garden, which is not your own property?" Thinking back in later years, he often laughed over the house's ugliness. Remembering splendours in the tropical forests of South America, he called it "Macaw Cottage" because the furniture in the drawing room combined all the macaw's colours "in hideous discord."

Charles moved in on the last day of December. His servant Syms Covington, who had been his assistant on HMS *Beagle,* helped him load two large vans with his "specimens of natural history." A few dozen drawers of shells were carried by hand. He wrote to Emma that one of the front attics was quite filled, and was to be called the Museum. "I wish I could make the drawing room look as comfortable as my own studio; but I dare say a fire and a little disorder will temporarily make things better."

Charles and Emma were married at Maer in January 1839. Emma wrote that on their first Thursday together in London they "went slopping through the melted snow to Broadwood's," where they tried a pianoforte and asked if it could be delivered to their home. On Saturday they walked out again and as they came back met "a pianoforte van in Gower Street, to which Charles shouted to know whether it was coming to No. 12, and learnt to our great satisfaction that it was. Besides its own merits, it makes the room look so much more comfortable . . . I have given Charles a large dose of music every evening."

When the schoolchildren were not playing in the college grounds, the neighbourhood was quiet. There were no shops or pubs on the street, and the road between University College and the hospital was a private right-of-way with gates that were often closed at night. Charles and Emma heard a strange "wailing whistle" from time to time, a sound of the new railway age from Euston Station. Locomotives approaching on the London & Birmingham Railway did not run down the last falling mile to the terminus because they could not manage the steep return climb. The carriages were uncoupled and rolled down on their own. They were hauled back up on a continuous chain drawn by two stationary steam engines at the top of the long incline, and staff at the station would signal that they were on their way by blasts on a great organ pipe operated by compressed air.

Charles might also have heard, or thought he heard, or sensed, cries from the operating theatre in the hospital across the road. The surgeon, Robert Liston, had been a well-known figure at the Royal Infirmary in Edinburgh when Charles studied medicine there twelve years before. Charles had given up his medical studies, partly because he was distressed by patients' suffering during operations without anaesthetic. At University College Hospital, Liston continued to improve the methods that had gained him his reputation at Edinburgh. His great skill was speed, essential for any major surgery because of the trauma of pain and loss of blood. He could amputate a leg in under thirty seconds. A casebook for his operations, with doodles of cut-throat razors on the cover, is still kept in the medical school. In January 1840, a country girl aged nineteen was admitted with a tumour of the right lower jaw. "The patient being seated in a chair, Mr Liston extracted the lateral incisor tooth . . . The jaw was then partly sawn through and its division completed with the cutting pliers . . . The operation lasted eight or nine minutes and was borne with the most heroic fortitude by the patient." Six years later, effective anaesthetics came into use. In 1846, Liston performed the first operation under ether in London, and a newspaper proclaimed: "We have conquered pain!"

But the wailing whistle and patients' cries were some way away. Charles found the stillness in his new study a welcome contrast to the many noises he had had to put up with in Great Marlborough Street. He wrote to his cousin and close friend William Darwin Fox, a clergyman in Cheshire: "If one is quiet in London, there is nothing like its quietness—there is a grandeur about its smoky fogs, and the dull distant sounds of cabs and coaches." Emma, though, remembered "how the passage of a rattling cab seemed in the night a matter of eternity."

Two days after Charles and Emma's wedding, his elder sister Caroline lost her six-week-old baby. Emma's sister Elizabeth wrote: "She does her utmost not to yield, but she is very unwell, and I never felt greater pity for anyone in my life." Caroline's husband Josiah Wedgwood III was now running the factory for his father with some reluctance. Elizabeth wrote bitterly that the loss of their child would make him "not so unwilling to go as usual to his employment, but what poor Caroline will find to do I cannot think; for the last so many months the thoughts of this precious child and the preparations for it have occupied her in an intense way that I never saw in anyone else."

In April, Emma found that she was pregnant. In August she noted in her diary: "Half way now, I think, from symptoms." She and Charles were living a full life, visiting and receiving many friends including their cousin Dr. Henry Holland and his wife, the Lyells, Thomas and Jane Carlyle, the mathematician Charles Babbage (known in the family as Baggage), Professor Richard Owen of the Royal College of Surgeons and Harriet Martineau the writer. Charles and Emma went together to the Zoological Gardens, to Handel's *Messiah* and Bellini's *La Sonnambula*. They sampled sermons at a number of churches, and attended the Unitarian Chapel in Little Portland Street, another new "Greek" building, where Hensleigh and Fanny Wedgwood worshipped. The minister, James Tagart, preached the future triumph of Unitarian Christianity, rejecting the doctrine of the depravity of human nature and emphasising social concord, domestic piety and fraternal union.

When Emma was due to give birth, her sister Elizabeth came to be with her, and Charles engaged a doctor to attend. There were different views at the time about childbirth and the pain of delivery. Some considered "the endurance of pain during delivery essential to the fulfilment of the primaeval curse, consequent upon the temptation and fall of our first mother, Eve." But one obstetrician wrote that childbirth was a natural process and suggested that "no sentiment is more pregnant with mischief than the opinion which almost universally prevails, that this process is inevitably one of difficulty and danger."

Another obstetrician suggested how the doctor should cope with the shyness of a young lady having her first child. "In the case of a woman who has been long married, and has borne children before, there is no difficulty or delay on the score of delicacy. The nurse brings you towels and hog's lard at once . . . But a newly-married woman dislikes and dreads the examination; and, therefore, you sit down by the bedside, and talk to her about other things. Presently the nurse asks how the baby is lying; and this makes the lady anxious about it. A pain comes on, and you relieve it by putting your hand on the sacrum. When the next pain comes on, introduce one or two fingers of the other hand into the vagina; ascertain that the passages are all right; and the arch of the pubis, and the outlet of the pelvis, natural. Then feel for the os uteri . . ." After the delivery, the doctor should tie the umbilical cord, place a cap on the baby's head and hand it to the companion who would have a flannel or woollen shawl to wrap the child in. "Do not stay to nurse your patient . . . for it alarms her, and you get bothered."

Charles and Emma's first child was born on 29 December. It was a boy, and he was christened William Erasmus, both Darwin family names. They called him Doddy first, then Willy as he grew into childhood. Charles wrote to his cousin Fox: "What an awful affair a confinement is; it knocked me up, almost as much as it did Emma herself." He mentioned in another letter that he had become a father. "The event occurred last Friday week: it is a little prince." On 10 February, Emma wrote in her diary: "Baby smiled for the first time." And the next day, as she thought of spring flowers in the dirt and cold of the London winter, "Baby made little noises. Got the hyacinths." Charles wrote to Robert FitzRoy, his captain on HMS *Beagle,* about "my little animalcule of a son." This word was a naturalist's term for living organisms so small that they could not be seen by the naked eye; FitzRoy might have thought of the creatures that Charles had fished eagerly from the sea onto the deck of HMS *Beagle,* and studied so intently with his microscope in the poop cabin.

Charles was surprised by his absorption in his son. He wrote to Fox in June: "He is a charming little fellow, and I had not the smallest conception there was so much in a five-month baby. You will perceive by this, that I have a fine degree of paternal fervour." "He is a prodigy of beauty and intellect. He is so charming that I cannot pretend to any modesty. I defy anybody to flatter us on our baby, for I defy anyone to say anything in its praise, of which we are not fully conscious."

Emma was to have eight more children in the next twelve years. Her life was a treadmill of pregnancy, delivery, suckling, weaning and waiting for the next conception. After bearing her fifth child, she wondered if she might have "the luck to escape having another soon," but Charles does not seem to have appreciated her feelings. Shortly afterwards she wrote about the possibility of "having a respite" for another year, but her sixth child was conceived five months later.

After the experience of her first pregnancy, Emma wanted to be prepared for the next delivery. There were four signs to watch for: missing a period, which was referred to in polite conversation as "ceasing to be unwell," morning sickness, changes in the breasts, and feeling the baby move, which was known as "quickening." After one or more of the signs, the due date was calculated by "the reckoning"—counting forty weeks from three days after the last menstruation. Emma marked her periods in her diary with a special cross, and when she missed one just seven months after Willy was born, she numbered forty weeks forward from three days after the last cross.

She suffered morning sickness a month later, but by then she had Charles to care for as well as herself. He became ill while they were staying with her parents at Maer in August, and from then until November she noted his symptoms in her diary every day. This "Maer illness," as he later called it, was the first long and serious attack of the disorder which was to dog him for the rest of his life, and for Emma it eclipsed her own discomfort. Charles had been healthy and lively as a young man. He had suffered acutely from sea-sickness on HMS *Beagle* and was laid low by fever a few times, but otherwise was one of the hardiest and most energetic members of the ship's company. By the time he married Emma he was already showing signs of his later illness, and she wrote: "I shall scold you into health." But he could not recover by an effort of will. Among his recurring symptoms were a state of languor and discomfort in which he found he could not work, swimming of the head, dying sensations and black spots before the eyes, spasmodic stomach pains, wind and vomiting, bouts of eczema and boils.

While at Maer, Charles was unhappy to be so ill and weak, and spent many hours in the nursery with his baby son. Emma's daily notes record a rich diet for Charles's delicate digestion. "Pulse 60, oysters and artichokes . . . pulse 52, partridge and pudding . . . very good day, hare, oysters, pulse about 54." One day, she wrote: "Turtle did not agree."

In the twenty-first week of her second pregnancy, Emma felt the child move in her womb, and wrote "quicken" in her diary. A week later, as Charles was recovering from his illness, the family returned home to Macaw Cottage. The novelist Maria Edgeworth, who knew the Wedgwoods and Darwins through her father, Richard Lovell Edgeworth, paid a call on Emma and Charles after Christmas. She wrote to a friend: "Mrs Darwin is the youngest daughter of Jos Wedgwood, and is worthy of both father and mother, affectionate and unaffected and, young as she is, full of old times. She has her mother's radiantly cheerful countenance even now, debarred from all London gaieties, and all gaiety but that of her own mind, by close attendance on her sick husband."

When Emma was nearly eight months pregnant, a fifteen-year-old girl, Bessy Harding, came from Maer to be Willy's nursemaid. Emma was in discomfort and preoccupied, and found it difficult to look after her child. She wrote later that in the weeks before her second confinement, "I could take so little notice of the little boy that he got not to care a pin for me, and it used to make me rather dismal sometimes."

As the days passed, Emma continued to jot symptoms in her diary. "Very languid . . . Great lassitude . . ." But the words "At his work" show that once again it was Charles she was watching, not herself.

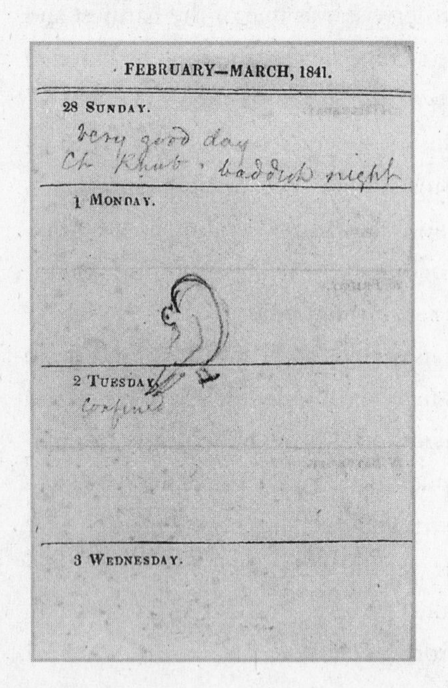


Emma's sister Elizabeth had come again to be with her for the birth. In her thirty-eighth week, Emma made a small sign of how she felt in her diary. On the first of March she drew a pencil doodle of a fancy pigeon, a "pouter" with a huge inflated crop. The next day, she wrote "confined," and Annie was born.

It seems to have been a difficult

birth. Emma and Charles's cousin Dr. Holland attended. Emma was ill afterwards and a nurse, almost certainly from University College Hospital, came to help. A wet-nurse was also engaged. She was probably chosen from the daily advertisements in *The Times.* The day after Annie was born, a notice appeared in the "Want Places" column. "As wet-nurse, a young woman from the country, with her first child, who has a good breast of milk, and has been confined a week." Most who advertised had their own infant at the breast, but some did not. A few days later, a notice appeared: "As wet-nurse to take an infant to nurse to whom every attention will be paid, a respectable female who has just lost her own child."

There was a strong feeling at the time that it was natural and right for a mother to breastfeed her own child, and women in society who chose to avoid the bother were criticised harshly. Mothers who were unable to breastfeed may have felt the reproach in obscure ways. A doctor wrote: "It may be called a fixed law of nature that a healthy woman should suckle her offspring." Not to comply with this "arrangement of Providence" was to forgo the first reward after the pain of childbirth. "It is plainly intended to cherish and increase the love of the parent herself, and to establish in the dependent and helpless infant from the first hours of its existence those associations on which its affections and confidence afterwards will be most securely founded. The evidence of design is manifest. So long as the child is unborn, no milk is secreted in the mother's breast, but no sooner does she give it birth, than this fluid is prepared and poured forth, admirably fitted in its qualities for the rapid growth of its delicate organism." The doctor made a link with animals to drive his point home. "Animals, even those of the most ferocious character, show affection for their young; they do not forsake or neglect them, but yield them their milk and watch over them with the tenderest care. Woman, who is possessed of reason as well as instinct, must not manifest a love below that of the brute creature."

Emma had some difficulties caring for Annie during the first weeks after the birth, but after two months, she felt quite well and able to nurse her. "The baby too, which began by being a very poor little thing, is now thriving and smiling very sweetly. I believe Elizabeth thought me a very unnatural mother while she was here, and I think she did care more for it than I did, but I like its company very much now." Emma was returning to normal life, playing again on the piano and enjoying being able to "play with the little boy and walk about and do what I like, without always thinking about oneself which is very tiresome."

She jotted down in the back of her diary some piano music to buy: sonatas by Clementi and Beethoven's *Diabelli Variations.* She had learnt the piano as a child, and when George IV's wife Mrs. Fitzherbert visited her school in the 1820s, she was chosen to play a piece as the best pupil. At one time she had lessons from Chopin. Charles now paid for her to receive lessons from Ignaz Moscheles, the Czech virtuoso who had taught Mendelssohn and was one of the foremost pianists of the day. His style was incisive and he disliked the flamboyant romanticism of Chopin and Liszt. Emma had a crisp and fine touch, and it was said that she played always with intelligence and simplicity. But "she could endure nothing sentimental, and 'slow movements' were occasionally under her treatment somewhat too 'allegro.'"

The Darwins and Wedgwoods all looked to Charles's father, Robert, a wealthy and successful physician in Shrewsbury, for medical judgements and prescriptions. Dr. Darwin gave Charles robust advice about his own ailments, and provided "receipts" for Willy and Annie. "In all inflammatory ailments of very young children, three drops of Antimonial wine repeated twice a day, is usually sufficient, but in decided fever a grain of Calomel with a little Chalk may be safely given." "A drop of Sal volatile will sometimes compose an infant to sleep, given at night." For a baby with a constant cough and soreness in the mouth, he suggested "one grain of chalk with Opium, with three drops of Antimonial wine to be dropped on brown sugar." Antimonial wine was a solution of tartar emetic and sherry wine. Calomel was mercurous chloride. Sal volatile was carbonate of ammonia. These compounds had almost no value in treating the conditions for which they were used, and some could be very harmful.

With one in five infants dying in their first year, most children were baptised within a few weeks after their birth. But Charles and Emma were in no hurry, and took Annie to Maer in late May to be christened with her cousin Sophy, daughter of Caroline and Josiah Wedgwood. As it happened, the government's new General Register Office took its first National Census while they were there. Enumerators visited every household in the country and listed every person present on the census night. There were twenty-one people in the return for Maer Hall—Emma's father Josiah and her mother Bessy, her brother Josiah, Caroline, Charles, Emma, Willy and Annie, and thirteen servants.

The house was a Jacobean mansion in a small park with a lake. Dr. Holland's wife described the Wedgwood family's life there. "They have freedom in their actions in this house as in their principles. Doors and windows stand open. You are nowhere confined. You may do what you like. You are surrounded by books that all look most tempting to read. You will always find some pleasant topic of conversation or may start one, as all things are talked of in the general family."

The parish church was in the grounds, and Emma's father had appointed his lame and eccentric nephew Allen Wedgwood as vicar. Charles privately considered Allen "half idiotic in some respects," though "with a store of accurate and even profound knowledge." Allen baptised the two children Anne Elizabeth and Ann Sophy. Annie's names were, like Willy's, "proper *family* names." Charles's great-great-grandmother had been Anne Waring; she had brought an estate in Nottinghamshire into the family, and her tablet in the parish church of Elston commemorated her as "daughter, wife, mother, mistress, neighbour answering Solomon's character of a good woman." The name Elizabeth was chosen for Emma's mother in her sad and slowly deepening dementia.

After the baptism, Charles went to stay with his father and sisters in Shrewsbury while Emma stayed with Annie at Maer. Bessy the nursemaid had taken Willy ahead and Charles was touched by his pleasure at seeing him. "He sat on my knee for nearly a quarter of an hour . . . and looked at my face and pointing, told everyone I was Pappa . . . When I had had him for about five minutes, I asked him where was Mama, and he repeated your name twice in so low and plaintive a tone, I declare it almost made me burst out crying. He is full of admiration at this new house and is friends with everyone and sits on Grandpapa's knees. He shows me the different things in the house. Dear old Doddy—one could write for ever about him." Charles looked forward to hearing from Emma about herself and Annie who, "as I have several times remarked to myself, is not so bad a girl, as might be expected of Doddy's rival." But Charles feared his son was a coward. "A frog jumped near him and he danced and screamed with horror at the dangerous monster, and I had a deal of kissing at his open bellowing mouth to comfort him. He threw my stick over the terrace wall, looked at it as it went, and cried 'Tatta' with the greatest sangfroid and walked away."

A few days later, Charles warned Emma: "A thunder storm is preparing to break on your head, and which has already deluged me, about Bessy not having a cap." Emma was not particular about their maid's appearance, but Charles's sisters said that she looked "like a grocer's maidservant," and his father added angrily: "The men will take liberties with her, if she is dressed differently from every other lady's maid!" Charles told Emma that he had taken half the blame on himself, and "never betrayed that I had beseeched you several times on that score. If they open on you, pray do not defend yourself, for they are very hot on the subject."

When the family was back in Macaw Cottage, Charles wrote to his cousin Fox: "We are all well here . . . our two babies are, I think, strong healthy ones, and it is an unspeakable comfort, this." At the end of the year, Emma was breastfeeding Annie, but she had little milk and the doctor told her it would not matter if she stopped. When Annie was nine months old, Emma felt she was "very ugly, poor body, with a broken out ear just like mine."

Charles was a doting father to Willy and Annie, and was eager for their attention. Willy, just two in January 1842, sat with his parents at table and behaved "with great decorum." But Annie was "very naughty" about her father and would not go to him. So, Emma wrote one day, "he has given her up and devoted himself to Doddy." That month, Emma became pregnant with her third child. She numbered the weeks ahead to forty-one, writing at the sixth week, "Taken ill at this stage last time," and at the tenth week, "I got better at about this time last time."

A few weeks later, William Darwin Fox's wife died giving birth to their sixth child. With Emma now expecting her third, Charles must have had the dangers of her forthcoming confinement in mind when he wrote to his cousin: "What a comfort it must be to you; that is, I think I should find it the greatest, the having children. It must make the separation appear less entire. The unspeakable tenderness of young children must soothe the heart and recall the tenderest, however mournful remembrances." He told Fox that Emma was "uncomfortable enough all day long and seldom leaves the house, this being her usual state before her babies come into the world." But, he wrote, "my two dear little children are very well and very fat."

Emma was more frank in a letter to her aunt. "My little Annie has taken to walking and talking for the last fortnight. She is thirteen months old and very healthy, fat and round, but no beauty."

CHAPTER TWO

# PTERODACTYL PIE

*Charles's secret—Ideas at Cambridge—HMS* Beagle*—Species—Human nature—Man and ape—Jenny the orang*

When he completed his almanac for 1841, Charles wrote "Annie born," and on the next line, "Sorted papers on Species theory." Annie was a little child like a hundred others in the neighbourhood, but her father was living with a secret idea that put him in a place of his own. For four years he had been working quietly on his theory to explain how new species came into being. He had not revealed it to anyone yet, because it was a direct challenge to accepted ideas about the Creation and Man's place in Nature, and he could guess all too easily how most people would react. The theory was eventually to shock the world and change our understanding of natural life. It was bound up with his strongest interests, his ideas about religion and human values, his ambition as a scientist, his relationship with Emma and his fascination with his children.

As a young boy in Shrewsbury, Charles had loved long walks on his own, collecting insects and bird-nesting. Dr. Darwin thought his son a very ordinary child and said: "You care for nothing but shooting, dogs, and rat-catching, and you will be a disgrace to yourself and all your family." At sixteen he was sent to study medicine at Edinburgh University. Some years later he thought back to the teaching with a shudder. "I shall ever hate the name of Materia Medica, since hearing Duncan's lectures at 8 o'clock in a winter's morning—a whole, cold, breakfastless hour on the properties of rhubarb!" He often went to see patients at the hospital but could not bear to see them in pain and remembered with horror for the rest of his life one operation on a poor street boy who cursed and screamed as the surgeon worked without anaesthetic. He turned to natural history, taking careful notes as he walked round the mineral collection in the college museum, and spending days wandering along the shores of the Firth of Forth looking for small sea creatures.

In his second year at Edinburgh, Charles persuaded his father that medicine was not for him, and they agreed that he would instead go to Cambridge to read for holy orders. His idea was to become a country parson, caring for his parishioners but living for natural history. When he arrived at Christ's College in 1828, he joined his cousin William Darwin Fox and a group of friends who shared a passion for collecting beetles in the countryside and fens around Cambridge. He attended Professor Adam Sedgwick's lectures on geology and learnt botany from Professor John Henslow.

Charles and his friends read William Paley's *Natural Theology,* which explained the accepted way of understanding the natural world as a grand array of evidence of God's creative power and goodness towards man. Paley argued that if you found a watch on the ground and someone asked how it came to be there, you would guess and reply that the watch had a maker who had built it for the purpose which it served; and the artificer must have "comprehended its construction, and designed its use." Every organ of every living creature was clearly designed for a purpose in the same way, and gave evidence of God's power and wisdom as the Creator of all things.

Natural theology was part of the thinking of the age. In the late 1850s, Mrs. Beeton, author of the best-selling *Book of Household Management,* was to open her section on cooking fish with a chapter on their natural history. She reminded her readers that "In studying the conformation of fishes, we naturally conclude that they are, in every respect, well adapted to the element in which they have their existence." Using one of Paley's examples, she explained how a fish's air bladder worked, and exclaimed: "How simply, yet how wonderfully, has the Supreme Being adapted certain means to the attainment of certain ends!'

It was a premise of natural theology that each kind of plant and animal had been first brought into being by God in a separate act of creation. Once created, species were fixed. The idea that one species might change through time into another was rejected because it clashed with the Book of Genesis and put into question both the need for a Divine Creator and the wisdom of his designs.

The task of natural theology was to interpret "the Book of Nature" and draw people up to "Nature's God." A clergyman writing on marine life drew a lesson from the myriads of organisms that could be seen with a magnifying glass on a frond of seaweed. "Their name was legion . . . a legion of God's creatures doing good, actively employed in doing his will, and consequently happy. Though they had never been seen by man, God would have not lost his praise, for he gave them life, and rendered that life uninterruptedly happy . . . If one frond is the habitation of a million happy creatures, how great must be the amount of happiness which God is giving every moment to the utterly uncountable myriads of his creatures that inhabit the deep!"

Other writers recognised the suffering involved in the unending struggle for life, and found ways to justify it according to God's benign purpose. Taking up a suggestion by Paley that killing was a mercy when an animal grew old and infirm, the Reverend William Buckland, Professor of Geology at Oxford, declared that "the appointment of death by the agency of carnivora" was "a dispensation of benevolence" as it eliminated "the misery of disease, and accidental injuries, and lingering decay."

Theologians and men of science, many themselves clergymen, saw humans as creatures in the natural world but not of it. They recognised our physical similarities with animals and suggested that our bodily organs showed the wisdom of the Creator in providing for our physical needs. But they also believed that our inner being was entirely different. Animals and plants lived and died within the natural order but man had an immortal soul and a spiritual destiny beyond the present life. One naturalist suggested that man "is placed upon this earth, not as necessary to its well-being, or to perform a part in its regulation, but as one who is undergoing a state of probation; who is journeying, indeed, as a stranger and a pilgrim."

When Charles read Paley's *Natural Theology* at Cambridge, he did not question his arguments, but the whole structure was coming under strain with new findings and ideas in geology and zoology. There were questions in the air, challenges and fears. The extraordinary variety of natural life was one puzzle. Professor Robert Grant of University College London, whom Charles had known in Edinburgh, told his students in 1828 that "Nine hundred species of intestinal worms have already been extracted from the bodies of animals, although comparatively little of the attention of naturalists has hitherto been directed to this extraordinary tribe of beings." Taxonomists found it difficult to decide on the differences between some species, or to make ordered sense of the similarities and discrepancies that close comparison revealed.

More critically for people outside the narrow world of expert knowledge, geology and palaeontology were revealing how long and strange the history of the earth and its inhabitants had been before man's first appearance. By the 1820s many geologists had abandoned a literal interpretation of the biblical account of the Creation. The aeons of prehistory revealed by the careful study of geological formations prompted questions about man's significance in the scheme of things.

As the fossil record became clearer, the extinction of kingdom after kingdom of monstrous prehistoric animals before the appearance of man raised issues about God's purpose in Creation. To Alfred Tennyson writing in the 1840s, Nature "lent evil dreams," and he faltered when he looked for her "secret meaning in her deeds." Some had suggested that if Nature was indifferent to individual creatures, she was "careful of the type" or species. But no.

*From scarped cliff and quarried stone,*
*She cries, "A thousand types are gone:*
*I care for nothing, all shall go."*

Man trusted that God was love and that love was "Creation's final law," but "Nature red in tooth and claw with ravine, shriek'd against his creed."

Many, though, found excitement in the geologists' new longer view. In 1826, a character in Benjamin Disraeli's novel *Vivian Grey* asked: "Can you conceive anything sublimer than the gigantic shadows and the grim wreck of an antediluvian world? Can you devise any plan which will more brace our powers, and develop our mental energies, than the formation of a perfect chain of inductive reasoning to account for these phenomena? What is the boasted communion which the vain poet holds with nature compared with the conversation which the geologist perpetually carries on with the elemental world?"

Charles was thrilled by the challenge. A friend who sat with him at Professor Sedgwick's lectures in Cambridge remembered many years later how he talked eagerly about "the enlarged views both of time and space which geology could give." When Professor Sedgwick explained his speculations about the probable age of the earth, Charles exclaimed: "What a capital hand is Sedgwick for drawing large cheques upon the Bank of Time!"

In 1830 he read John Herschel's *Preliminary Discourse on the Study of Natural Philosophy.* By "natural philosophy" Herschel meant the physical sciences in their broadest sense. His title was dry and discursive, but he gave an account of the aims of scientific understanding, and a view of what the future held, which seized Charles's imagination. The book gave him a "burning zeal" to make his own contribution.

Although Herschel did not challenge natural theology directly, he suggested a radically different approach. While Paley had simply celebrated every instance of design in nature as another piece of evidence of the Creator's goodness and creative power, Herschel wrote of man as a "speculative being" who "walks in the midst of wonders," intrigued by the hints of underlying patterns in the infinite variety of the living world and searching for grand principles to explain them. Newton's theory of universal gravitation and his laws of motion were a triumph of the scientific approach. After Laplace had consolidated Newton's achievement by demonstrating that the solar system was stable, Herschel claimed that the laws of nature were "not only permanent, but consistent, intelligible and discoverable." The way forward was to embark on an inductive inquiry into natural phenomena, searching for the underlying patterns and inferring the grand causes. On the vexed question of the link between the truths of religion and of science, the word of the Bible and the evidence of nature, Herschel argued that science should claim its own authority, separate from that of Scripture. "We must take care that the testimony afforded by science to religion, be its extent or value what it may, shall be at least independent, unbiased and spontaneous."

Many at the time felt with Herschel that science should approach the natural world with awe and an openness to possibilities beyond the limits of current factual knowledge. Science was not seen as a narrow method of knowing; the word itself stood for understanding, broadly conceived. Science shared concerns with metaphysics, religion and art, and connections were made. In 1833, Thomas Carlyle wrote in *Sartor Resartus* of the view of the universe as "one huge, dead, immeasurable steam-engine," a machine "fixed to move by unalterable rules." He suggested that "to the wisest man the system of Nature . . . remains of quite infinite depth, of quite infinite expansion." Although he was often critical of their work, many men of science shared his view.

Professor Henslow befriended Charles; he recognised his promise as a naturalist and encouraged him to take up geology as a serious pursuit. Professor Sedgwick was planning a field trip in North Wales to examine some problems in the geological history of the area, and Professor Henslow persuaded him to take Charles along as his apprentice. Charles mastered the use of his "clinometer" for measuring slopes with careful practice on his bedroom furniture, and spent a week in August walking with Professor Sedgwick in the Welsh mountains. Sedgwick taught him how to read a landscape for its ancient history, and how to develop guesses into a theory by testing them with other observations. The two men took parallel lines across the country; Charles collected specimens and, when he rejoined Sedgwick, he explained his ideas about the stratigraphy. Sedgwick discussed his suggestions, and Charles wrote later that "this of course encouraged me greatly, and made me exceedingly proud."

Sedgwick also showed Charles the imaginative approach to science. A bluff Yorkshireman, he had met the poet William Wordsworth when he was geologising in the Lake District in the early 1820s. They went on many long walks together, talking freely about their shared love for nature and poetry. When Sedgwick became President of the Geological Society of London in 1830, he told the society that their science could never be exact like astronomy because of the infinite complexity of the material facts. "There is an intense and poetic interest in the very uncertainty and boundlessness of our speculations." Some years later Wordsworth asked Sedgwick to write an account of the geology of the Lake District to add to his well-known *Guide to the Lakes.* The man of science was glad to make his contribution to the poet's book, writing that "No one has put forward nobler views of the universality of Nature's kingdom than yourself." The nobility lay in Wordsworth's metaphysical themes. Sedgwick believed that geological understanding complemented the insights about the natural world, the human mind and ultimate truths that Wordsworth had drawn from the landscape he loved. The reach of geology was still not yet clear, "and there is still so much of wild untamed nature about it, that it is almost as well fitted to inflame the imagination, as to inform the reason." There was room in Sedgwick's science for intuitions and insights that could not be conveyed in objective description, including perhaps those "questionings of sense and outward things" that Wordsworth valued from his early experiences. Sedgwick would say to his students: "I cannot promise to teach you all geology; I can only fire your imaginations."

When Charles was seized with the ambition to make a mark as a man of science, he was twenty-one and recognised that one way to gain experience and attention would be to travel abroad as a naturalist. The opportunity came in 1831, when he eagerly accepted an invitation to join HMS *Beagle* on a two-year expedition in the southern oceans. Captain FitzRoy planned to survey the coast of South America and determine longitudes for the Admiralty. He wanted a young gentleman naturalist to accompany him on the voyage, and Charles's name was suggested, as much for his good manners as for his knowledge and promise as a naturalist.

As HMS *Beagle* sailed through the mid-Atlantic to South America, Charles fished with a net for marine invertebrates and caught huge numbers to examine under his microscope. "Many of these creatures so low in the scale of nature are most exquisite in their forms and rich colours. It creates a feeling of wonder that so much beauty should be apparently created for such little purpose." He first saw the "glory of tropical vegetation" on the Cape Verde Islands. "Tamarinds, bananas and palms were flourishing at my feet . . . I was afraid of disappointments; how utterly vain such fear is, none can tell but those who have experienced what I today have . . . I returned to the shore, treading on volcanic rocks, hearing the notes of unknown birds, and seeing new insects fluttering about still newer flowers. It has been for me a glorious day, like giving to a blind man eyes." When, weeks later, he walked into a Brazilian forest, he wrote: "The delight one experiences in such times bewilders the mind. If the eye attempts to follow the flight of a gaudy butterfly, it is arrested by some strange tree or fruit; if watching an insect, one forgets it in the strange flower it is crawling over . . . The mind is a chaos of delight." He wrote from Rio de Janeiro to Professor Henslow: "I am at present red-hot with spiders; they are very interesting, and if I am not mistaken, I have already taken some new genera."

Before Charles embarked, Professor Henslow had suggested he read Charles Lyell's *Principles of Geology*, and FitzRoy gave him a copy of the first volume which had just appeared. Lyell stated in the title that his aim was to "explain the former changes of the earth's surface by reference to causes now in operation." This simple notion had been developed by other geologists before Lyell but he gave it definitive form. It came to be known as "uniformitarianism"; the label is dull but two elements are fundamental to our present understanding of the history of natural life. First, the approach focused on natural processes like sedimentation and erosion that could be observed as they happened, and it avoided explanations that depended on supposed events outside the experience of living people—global floods, volcanic cataclysms and other catastrophes. Learning from Lyell's writings about the ways in which observable processes could be used to explain how geological strata were formed, folded and worn away through time, Charles studied plants and animals and sensed how much he might be able to explain about their forms as well from natural processes that he could watch as they worked around him.

The second element of the approach was the length of time available for the geological processes to work through. The few thousand years set by Archbishop Ussher in the standard chronology of world history were nowhere near enough to accommodate any gradual trends. But when the geologists argued from known rates of change and other pointers, that each geological era must have lasted for many millions of years, Charles was able to use these periods of time to allow for the slow evolution of living things as well as geological change. He was to write in *The Origin of Species* that Lyell's *Principles of Geology* produced a "revolution in natural science." Charles himself took the revolution into the study of plant and animal life.

During his years on HMS *Beagle* Charles collected thousands of specimens and filled many hundreds of pages with observations on geology and zoology. Thinking boldly about what he saw, he built a theory of the raising of the South American continent, and another about the formation of coral islands. He wrote about a walk through a forest in Brazil among "numberless species of ferns and mimosas," that "it is nearly impossible to give an adequate idea of the higher feelings which are excited; wonder, astonishment and sublime devotion fill and elevate the mind." And some time later, "Among the scenes which are deeply impressed on my mind, none exceed in sublimity the primeval forests undefaced by the hand of man; whether those of Brazil, where the powers of Life are predominant, or those of Tierra del Fuego, where Death and Decay prevail. Both are temples filled with the varied productions of the God of Nature: no one can stand unmoved in these solitudes without feeling that there is more in man than the mere breath of his body."

Charles responded to the richness and variety by ranging freely, and with the same wonder, from the minute particulars of a single plant or insect to the widest view of a whole forest, plain or mountainscape. Every day he observed, collected and made notes on his detailed findings. When he reached the crest of the Andean Cordillera, "the profound valleys, the wild broken forms, the heaps of ruins piled up during the lapse of ages, the bright coloured rocks contrasted with the quiet mountains of snow, together produced a scene I never could have imagined. Neither plant or bird, excepting a few condors wheeling around the higher pinnacles, distracted the attention from the inanimate mass. I felt glad I was by myself; it was like watching a thunderstorm, or hearing in the full orchestra a chorus of the *Messiah*." He was puzzled by his haunting memories of the empty wastes of Patagonia. "Without habitations, without water, without trees, without mountains, they support merely a few dwarf plants. Why then . . . do these arid wastes take so firm possession of the memory? . . . I can scarcely analyse these feelings: but it must be partly owing to the free scope given to the imagination." The plains of Patagonia are "unknown: they bear the stamp of having thus lasted for ages, and there appears no limit to their duration through future time."

In the last months of the voyage, after he had visited the Galapagos Islands in the Pacific off the coast of Ecuador, Charles noticed some strange patterns of likeness and variation between tortoises and mockingbirds on different islands. He noted that if islands close to each other were inhabited by distinct species which were nevertheless closely linked and not found elsewhere, it would be worth making a special examination of the "zoology of archipelagoes . . . for such facts would undermine the stability of species." That was the start of his thinking about change and evolution. At the time, though, he still accepted the idea that God had brought all forms of life into existence by separate acts of creation. At that point he had no idea where his conjectures would lead him.

• • •

During the voyage, two experiences which had nothing to do with his work as a naturalist shook Charles deeply. Both were encounters with people—black slaves in Brazil and "savages" in Tierra del Fuego. As with the creatures of the Galapagos, Charles saw likenesses and differences; they triggered strong feelings, and were to come to mind again and again in later years as Charles developed his ideas about species and human life. When he eventually tackled the question of man's place in nature, slaves and savages were close to the centre of his thinking.

Charles had first learnt about the tropics from a freed black slave in Edinburgh. John Edmonston earned his living as a "bird-stuffer"; Charles took lessons from him and enjoyed their conversations; he commented later how like Edmonston's mind was to the minds of Europeans. At the time that was an exceptional view among people in Charles's social position, but his grandfather Josiah Wedgwood I had been a leading member of the Committee for the Abolition of the Slave Trade in the 1780s, and produced at his pottery the well-known cameo of the chained slave with the question "Am I not a man and a brother?"

When HMS *Beagle* landed at Bahia and Rio de Janeiro, Charles saw many slaves and met slave-owners. Captain FitzRoy saw no evil in the institution, and Charles, quite out of character, quarrelled with him. One day FitzRoy told Charles that he had just visited a rich slave-owner who had summoned many of his slaves and asked them whether they wished to be free. "All answered 'No.' I then asked him, perhaps with a sneer, whether he thought that the answers of slaves in the presence of their master were worth anything. This made him excessively angry, and he said that as I doubted his word, we could not live any longer together." Charles thought he would have to leave the ship but, after a few hours, FitzRoy sent an apology and asked Charles to continue to share his table.

A few weeks later, Charles and his companions met an Irish trader who took them to a plantation he had cleared from the forest six days' ride into the interior. The trader had a violent quarrel with his agent in which he threatened to sell at auction an illegitimate mulatto child of whom the agent was very fond. He also said he would take all the women and children from their husbands and sell them separately at the slave market in Rio. Charles wrote in his diary: "Can two more flagrant and horrible instances be imagined?" Faced by such cruelty, he rejected out of hand the argument that slavery was "a tolerable evil." He was angry at some English writers who showed sympathy towards slave-owners, but gave no thought to the feelings of their slaves. "Picture to yourself the chance, ever hanging over you, of your wife and your little children—those objects which nature urges even the slave to call his own—being torn from you and sold like beasts to the first bidder."

With his liberal upbringing and values, Charles was able to recognise his common nature with negro slaves, but the inhabitants of Tierra del Fuego were the greatest challenge for civilised people, as they were reckoned to be among the "lowest Barbarians" known. The first Fuegians that Charles met were on board HMS *Beagle* when he joined the ship, and had been trained in European ways. When Captain FitzRoy had visited the Straits of Magellan on a previous voyage in 1830, he took three young men and a small girl from their families, just as a slave-owner might have taken one of his slaves' children. He brought them back to England for a Christian education. One died of smallpox but the others, Jemmy Button, York Minster and the nine-year-old Fuegia Basket, were clothed and schooled in a village on the outskirts of London. They were so pliant and took their instruction so well that Captain FitzRoy was able to arrange for them to be presented to King William and Queen Adelaide at the Court of St. James's. When Charles embarked on HMS *Beagle* the three were on board for their return to their people together with a missionary to make a Christian settlement among them. As the ship sailed south, Charles got to know Jemmy Button and York Minster well. He found Fuegia Basket "a nice, modest, reserved young girl, with a rather pleasing but sometimes sullen expression, and very quick in learning anything." As he had been with John Edmonston the freed slave, Charles was "incessantly struck" while living with the Fuegians on board ship, "how similar their minds were to ours."

When HMS *Beagle* reached Tierra del Fuego, Charles found that the Fuegians in their own surroundings were "without exception the most curious and interesting spectacle I ever beheld." "Four or five men suddenly appeared on a cliff near to us. They were absolutely naked and with long streaming hair; springing from the ground and waving their arms around their heads, they sent forth most hideous yells. Their appearance was so strange, that it was scarcely like that of earthly inhabitants." Charles felt they were "man in his lowest and most savage state." Seeing a Fuegian in his native surroundings was like watching "the lion in his desert, the tiger tearing his prey in the jungle, the rhinoceros on the wide plain, or the hippopotamus wallowing in the mud of some African river." And "the reflection at once rushed into my mind—such were our ancestors."

Here it was the difference that struck Charles deeply. He knew York Minster, Jemmy Button and Fuegia Basket as quiet and well-mannered young people who could follow the etiquette of an audience with the King and Queen of England. But their fellow tribesmen, yelling and waving on the rocks, looked "scarcely like earthly inhabitants." The gulf between savagery and civilisation was enormous, and yet the three young Fuegians had stepped across it.

Seeing how savages could be brought so close to civilisation, Charles also recognised an element of the savage in his own being. He wrote of his feelings when hunting game on the vast empty steppes of Patagonia, that the love of the chase was said to be "an inherent delight in man, a

*Fuegia Basket*

relic of an instinctive passion." If so, he felt that living on the steppes as he had done, "with the sky for a roof, and the ground for a table," was part of the same feeling. "It is the savage returning to his wild and native habits." He always looked back to that way of life in "unfrequented countries, with a kind of extreme delight, which no scenes of civilisation could create." The young English gentleman, shocked by the unearthly savagery of the Fuegians on their wild and rocky shores and yet seeing how similar their minds were to ours, recognising that "such were our ancestors" and feeling himself to be "the savage returning to his wild and native habits," was ready to step further beyond accepted boundaries in seeking to understand mankind's place in the natural world.

When Charles arrived back in England in October 1836, he took lodgings in London with his servant Syms Covington and two pet tortoises they had brought back from the Galapagos. He met the geologist Charles Lyell and was welcomed into his circle, as Lyell recognised his promise and felt he was a "glorious addition" to his group of friends and supporters. Charles set to work on his geological notes but gave his zoological and botanical specimens to experts who could describe and classify them for him. Living in Great Marlborough Street, he was depressed to see "nothing but the same odious house on the opposite side as often as one looks out," but he dined at the Athenaeum Club, and felt "just like a Duke" as he sat reading on a sofa in the great drawing room.

The *Beagle* specimens were found to be remarkable in a number of ways. They posed some taxing questions and set Charles thinking again about the nature of species and the disturbing possibility that they might change. On his expeditions in Patagonia, he had made an impressive assembly of fragments of giant fossil mammals, and Professor Owen at the Royal College of Surgeons now examined them for him. Among them were a giant ground-sloth, an enormous armadillo and a capybara the size of a rhinoceros. All were long extinct, but it was striking how similar they were to the living species, all much smaller, which were unique to the South American continent. Owen drew attention to a remarkable "persistence of type" which Lyell suggested amounted to a "law of succession." The ornithologist John Gould pointed out that a number of bird specimens from the Galapagos which Charles had shown to a meeting of the Zoological Society as finches, wrens, "Gross-beaks" and blackbirds, were in fact all finches of a special kind, so peculiar in their likenesses and differences as to form an entirely new genus and three subgenera. Charles's earlier hunch about the "zoology of archipelagoes" was thus borne out.

Charles kept notebooks for jotting down ideas as they occurred to him. He had developed the habit while observing in the wild and then working in his cabin on HMS *Beagle*. In 1837, he wrote down his first gnomic suggestion on the creation of new species: "Speculate on neutral ground of two ostriches." Shortly afterwards, he opened a special notebook for his notes on the "transmutation of species," and from then on he recorded his ideas and questions as they came to mind. Between July 1837 and the end of 1839, the years of his boldest and most far-reaching thinking, he filled six small volumes. They are a unique record of the free flow of creative thought, strictly private, utterly frank and fascinating in what they reveal of his wayward progress towards the conclusions that eventually became his theories of evolution and the descent of man. He had no fellow naturalist to confide in at the time; no one he could trust to consider his ideas without writing him off as a heretic or crank.

He started his first notebook on transmutation with a string of comments on the processes of sexual reproduction and generation, finding in them clues to the critical factors of fixity and change between parents and offspring. He took up the theme of inheritance and variation, and formed the idea that species might be related by branching descent from a common ancestor. On the thirty-sixth page he sketched a tree of life to help him envisage what he had in mind.

Charles used his *Beagle* specimens and notes as a store of facts for his conjectures, but Herschel's emphasis on the universal laws of nature was part of the framework for his thinking. Herschel had himself suggested to Lyell in 1836 that the replacement of species by others was the "mystery

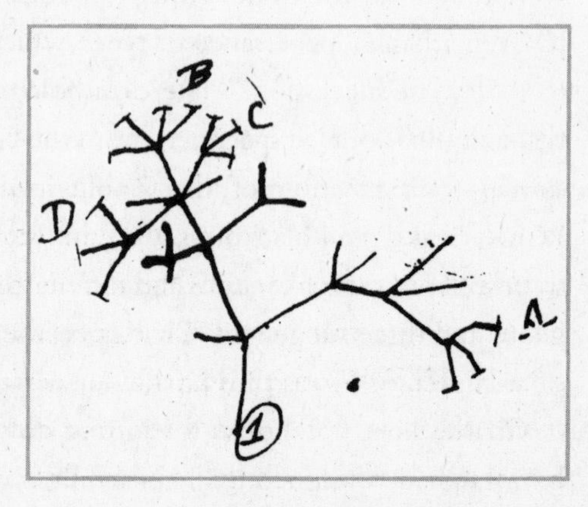


*Evolution by branching descent*

of mysteries" for natural science. He challenged the idea of separate creation, and suggested that God might work at one remove through laws of nature which somehow brought new kinds of creature into being. If, then, we could ever observe the origin of new species, he suggested that it "would be found to be a natural in contradistinction to a miraculous process." Charles now made it his aim to find that process.

In the first months Charles did not say clearly in his notes how he saw mankind in the scheme of things, or what possibilities he was considering, but he quickly came to believe that humans were a species of animal like any other and had evolved naturally from animal ancestors just as the others had. This drew him at once to metaphysics as it put into question the idea that God governed events in human lives according to a "particular Providence" or special moral purpose for each person, and that individuals' suffering had a moral meaning for them or people close to them.

Radical thinkers were playing with materialist ideas, and Charles toyed with an organic view of mind and brain in order to make sense of the inheritance of instincts and other mental capabilities. He also, though, held on for some time to the possibility that humans had immaterial souls, and an afterlife in which they enjoyed some form of reward or punishment for their conduct in this world.

Charles showed his interest in human nature in his reading. While Milton's *Paradise Lost* had been his favourite book on HMS *Beagle,* he now explored Wordsworth's writings and found many passages that struck chords with his thinking. His moments of intense awareness during the voyage—the thrill he felt on the crest of the Cordillera and his haunting memories of the empty wastes of Patagonia—could be seen in the light of Wordsworth's experiences in his early years. He remembered his pleasure as a ten-year-old child "in the evening or on blowy days, walking along the beach by myself and seeing the gulls and cormorants wending their way home in a wild and irregular course."

He was surprised to realise how early in his life he had first experienced "such poetic pleasures, felt so keenly in after years."

Charles noticed Wordsworth's comments on the links between poetry and science in his "Preface" to the *Lyrical Ballads.* Wordsworth had written that poetry was "the breath and finer spirit of all knowledge"; it was the "impassioned expression" which was in "the countenance of all science." Thinking perhaps of Wordsworth's remarks in the "Preface" about the ways in which poetry gave pleasure, Charles noted the pleasures of imagination with which, for him, it was linked. "I, a geologist, have ill-defined notion of land covered with ocean, former animals, slow force cracking surface &c.—truly poetical." A botanist might also view plants and trees in the same way. "I am sure I remember my pleasure in Kensington Gardens has often been greatly excited by looking at trees as great compound animals, united by wonderful and mysterious manner." Wordsworth had suggested that the "remotest discoveries" of science were "as proper objects of the poet's art" as any on which it could be employed. Charles understood him to mean that science might one day become "sufficiently habitual to become poetical," and welcomed the idea.

At the same time, though, the secrecy with which he felt he had to work on his "species theory" gave Charles a special understanding of one contrast which Wordsworth drew between the poetic and scientific approaches. Wordsworth had suggested that the knowledge that poets gain in their work was shared by everyone as "our natural and unalienable inheritance." Giving voice to a song in which all human beings join with him, the poet "rejoices in the presence of truth as our visible friend and hourly companion." The knowledge of the man of science, on the other hand was a "personal and individual acquisition, slow to come to us," and not shared with our fellow beings by any "habitual and direct sympathy." "The man of science seeks truth as a remote and unknown benefactor; he cherishes it and loves it in his solitude."

Charles read Wordsworth's long poem *The Excursion* twice, and found links with his own interests and ideas in the poem's rhetoric. Wordsworth's underlying theme was always human nature, rather than nature on its own. In his preface to the poem he offered a passage from another poem as a "kind of *Prospectus.*" He had found his inspiration while musing in solitude "on Man, on Nature, and on Human Life."

> *. . . Not Chaos, not*
> *The darkest pit of lowest Erebus,*
> *Nor aught of blinder vacancy, scooped out*
> *By help of dreams—can breed such fear and awe*
> *As fall upon us often when we look*
> *Into our Minds, into the Mind of Man,*
> *My haunt, and the main region of my song.*

Charles may have recognised a kindred spirit in Wordsworth's figure of the Wanderer. Growing up in wild country, the Wanderer "had felt the power of Nature." In one of the poem's climaxes, he spoke of the wisdom he had gained in words which Charles could have read as voicing his own wish to find scientific laws to explain the pattern of natural species.

> *Happy is he who lives to understand,*
> *Not human nature only, but explores*
> *All natures,—to the end that he may find*
> *The law that governs each; and where begins*
> *The union, the partition where, that makes*
> *Kind and degree, among all visible Beings;*
> . . .
> *Up from the creeping plant to sovereign Man.*

Wordsworth looked forward to a time when science would teach humane values from all things. Then her heart would "kindle" and she would find her "most noble use . . . in furnishing clear guidance . . . to the mind's *excursive* power."

> *So build we up the Being that we are;*
> *Thus deeply drinking-in the soul of things,*
> *We shall be wise perforce . . .*

Charles saw that if one wanted to understand human nature, one must look first into the natural history of mankind. It was widely believed that humans were part of a "vast chain of being" which was a fundamental element in the Divine plan of Creation. Man was the head of the chain of terrestrial beings, and "from him all the other links descend by almost imperceptible gradations." Next below were the great apes. The likenesses were remarkable, but they posed no challenge to our pride as long as each animal in the sequence was seen as a separate creation, fixed in its position in the ascending scale. A number of thinkers before Charles had suggested that humans might be tied more closely with the great apes. Jean-Jacques Rousseau and the Scottish philosopher Lord Monboddo both believed that chimpanzees and orang-utans were human beings in a more primitive state. Each had different points to make about the nature of the link between man and animal, but Charles found little of value in the ideas of either.

Most naturalists in London in the 1830s ridiculed the suggestion of shared ancestry, and saw mankind as completely separate from the "brute creation." They recognised that according to the best principles of comparative anatomy, humans were close to apes and monkeys in almost all anatomical details, but the human mind was so utterly superior that the anatomists refused to group us with our animal cousins. The naturalist William Swainson spoke for many when he urged the "innate repugnance, disgust, and abhorrence, in every human being, ignorant or enlightened, savage or civilised" against the admission of any relationship between humans and other mammals. The French comparative anatomist Baron Cuvier and most of his successors put mankind into a taxonomic order of its own. Professor Owen and others were quick to seize on small anatomical differences and claim that they were critical.

Nevertheless, the similarities continued to puzzle many people, and some who looked carefully into human and animal thinking found that the boundaries between them were unclear. Lord Brougham noted how the customary distinction between human "reason" and animal "instinct" broke down when it was seen that humans had some instincts and some animals reasoned. When a new periodical, the *Magazine of Natural History,* was launched in January 1837, the first article in the first issue was "On the psychological distinctions between man and all other animals." It opened: "There is not, within the wide range of philosophical inquiry, a subject more intensely interesting to all who thirst for knowledge, than the precise nature of that important mental superiority which elevates the human being above the brute."

One reason for the special interest during those years was that people were only then coming to know the man-like apes for the first time, and discovering just how disconcertingly close to humans they were. Before the 1830s a few travellers had written vague accounts of orang-utans, gibbons and chimpanzees but every one that had been brought from Africa or the Far East had died on board ship or shortly after arrival. Skeletons and skins had been exhibited, and taxonomists had put their specimens in various places in their different classifications, but people had almost no idea how the living animals looked and behaved. The first chimpanzee to be exhibited in London arrived at the Zoological Gardens in 1835. He was called Tommy and lived for a few months before dying of tuberculosis.

All who saw Tommy were fascinated, and he was treated in a remarkable way which reflected the obsession with the difference between man and brute. The pattern was repeated with the young orang-utan called Jenny who came to the zoo in 1837 and survived until 1839, and a second orang also called Jenny who came in 1841. People insisted on the gulf between ape and man, but they dressed Tommy and the two Jennys in children's clothes; they taught them to eat and drink at table with spoons, dishes and cups, to understand what their keepers said to them, and to recognise things they were allowed to do and others that were forbidden. Visitors were eager to watch them behaving like human children, and yet almost all found the spectacle disquieting.

Mrs. Lyell saw Tommy in 1835 and was struck by his "painfully humanlike expression." The Zoological Society's veterinary surgeon, William Youatt, looked after him, and wrote a detailed account of his patient. Tommy wore a Guernsey frock and a little sailor's hat "and as he sat within his cage gazing composedly around, he looked like an old weather-beaten sailor." Youatt had to inspect all the animals every day, and wrote that "it was long before he could get rid of a feeling of dislike, and almost of loathing, when he paid him his usual morning visit." However, they became friends and Tommy would give him his hand every time he came. When Tommy fell ill, Youatt and the other staff nursed him like a child. Youatt described his last hours as if he were human. After Tommy had screamed with pain for a while, "the screams became less violent, and assumed a resemblance, painful to hear, to the cries of a sick and sinking infant." At the end "he flung his arms around the keeper's neck and clenched his hands for firmer hold—he threw back his head a little, and brought it before that of the keeper, gazed intensely on his face, with an expression which the man says he never shall forget; and so he continued for one or two minutes, when his hold gradually loosened, his arms fell, and he had died without a struggle."

Youatt's account, which appeared in a veterinary periodical, was of such interest to others that it was reprinted only weeks later in two leading medical journals.

Jenny the orang was brought to London in 1837 and purchased by the Zoological Society from her importer. She appeared in her child's clothes on the front page of the widely-read *Penny Magazine of the Society for the Diffusion of Useful Knowledge.* The question she posed about animal and human nature was clearly recognised to be important for the magazine's many readers, and a long article comparing her with Tommy concluded with apparent relief that "extraordinary as the Orang may be compared with its fellows of the brute creation, still in nothing does it trench upon the moral or mental provinces of man." But William Broderip, an acquaintance of Charles who was helping him with the *Beagle* specimens, struck a slightly disquieting note in his account of Jenny in the *New Monthly Magazine.* "The personage who has lately arrived at the gardens of the Zoological Society in the Regent's Park, and is now 'the observed of all observers,' is of the softer sex, and very young. She receives company in the Giraffe-house, and appears amiable, though of a gravity and sage deportment far beyond what is usual at her years." Her keeper watched as a carpenter worked on her cage. He said: "Come, Jenny, you must leave the carpenter alone," and gently led her away. " 'Dear me!' said a lady; 'Dear me! does she know what is said to her?' 'Yes, she knows her name, Ma'am,' was the cautious reply: upon which the lady said 'Dear me!' again."

Queen Victoria saw the second Jenny on a visit to the Zoological Gardens with Prince Albert in 1842. She wrote in her diary that Jenny was "too wonderful," preparing and drinking her tea and "doing everything by word of command." She was "frightful, and painfully and disagreeably human."

• • •

In early 1838, while Jenny was holding court in the giraffe house, the Geological Society persuaded Charles to become their secretary. The president was the Reverend William Whewell, a Cambridge polymath and close friend of Herschel, Henslow and Sedgwick. The Cambridge men of science were greatly impressed by Charles's geological findings on HMS *Beagle,* and his ambitious theories about coral atolls and the elevation and subsidence of continents.

The first meeting that Charles attended as secretary was the anniversary meeting in February at which Whewell presented the society's medal to Professor Owen for his descriptions of Charles's Patagonian fossils. The fellows sat round a grand horseshoe table in the society's rooms in Somerset House, as an icy wind rattled the windows and deep snow drifted in the streets outside. It was a red-letter day for Charles; the realisation of his dream of making a contribution to the "noble structure of natural science." As he sat at the table with Lyell, Professor Sedgwick and many of the other men of science whom he most admired, Whewell praised Owen's achievements, and then spoke of Charles. Thinking of his work on the succession of fossil forms in South America, Whewell referred to "the profound and enlarged speculations on the diffusion, preservation, and extinction of races of animals to which Mr Darwin has been led by the remains which he has brought home." He hoped that Professor Owen and Mr. Darwin, "so fitted by their endowments and character to advance the progress of science, may long go on achieving new triumphs," and that they would succeed in throwing light "upon the darkest and widest of the vast problems which they have proposed to themselves." Charles alone in the room knew the darkest and widest problem he was working on. He was ready to carry forward his "profound and enlarged speculations." But he also knew that his conclusions would be rejected out of hand by almost everyone in the room.

Moving on to survey the geological discoveries of the year, Whewell described one of the most remarkable, the first find of fossil primates in geological strata long predating man. With his branching diagram in mind, Charles could see at once how the fragments pointed to the possibility of a common ancestry for man and the man-like apes. Whewell also recognised the extraordinary possibility, but saw the dangers lurking behind it and quickly called a halt to speculation. "The origin and end of man's being" was clearly at issue, but the question must be tackled by focusing on "the most remarkable facts in his nature," his "intellectual, moral and religious constitution," and "civilisation, art, government, writing and speech." Such matters were beyond the natural sciences. "The geologist may well be content to close his own volume, and open one which has man's moral and religious nature for its subject."

After the meeting, Whewell and Owen went with Lyell, Buckland and others for a meal of "pterodactyl pie" (woodcock, as Lyell explained to a friend) and bumpers of cognac in a friend's apartment in Mayfair. Charles did not join them, but returned alone to his lodgings to carry on his secret thinking. He was struck by Whewell's evasive dismissal of the clear pointers to man's common ancestry with the apes. Taking up Whewell's suggestion that "the most remarkable point in man's nature" was the mental and moral capabilities that distinguished him from animals, he noted that people "often talk" about the "wonderful event of intellectual Man appearing" for the first time. He did not agree that the point was the most remarkable one. "The appearance of insects with other senses is more wonderful," and the emergence of man was "nothing compared to the first thinking being."

Charles noted in his almanac that during the following days, he "speculated much about existence of species, and read more than usual." His thinking about animal and human nature now came together—man and brute, slave-owner and slave, savage and civilised, fossils and living beings, differences and likenesses. He was developing a theory of how new species might be formed continually through descent and physical isolation. He saw its power to explain the development of both animals and humans, bodily forms and the human mind. He noted how the theory would lead to a study of instincts and mind as inherited traits and capabilities, and from there how it might lead to a "whole metaphysics." The theory could be used to examine the causes of change in the natural world "in order to know what we have come from and to what we tend." Charles was now racing forward with his ideas. "Animals whom we have made our slaves, we do not like to consider our equals." Do not slave-owners wish to make the black man a separate kind of being from themselves? "If we choose to let conjecture run wild," as animals are "our fellow brethren in pain, disease, death and suffering and famine, our slaves in the most laborious work, our companions in our amusements," we may all "partake, from our origin in one common ancestor, we may all be netted together." His grammar broke down in the last wild sentence, but from his use of the word "netted" elsewhere, he clearly meant not "caught" but "meshed" or "interlaced." Common descent meant shared nature, and that idea was one key to understanding man's place in the living world.

Charles was now challenging the thinking of the age on human nature. One of his guiding lights was David Hume, the Scottish philosopher of the previous century. Hume's name was not one to bandy about at the time because his sceptical treatment of religion was reckoned to be atheistical and "obnoxious," but his writings were in the library of the Athenaeum Club and at Maer Hall, and during 1838 and 1839 Charles became familiar with all the main strands of his thinking. The unique way in which Hume combined the sharpest critical reasoning about received views with the aim of creating a "science of man" based on common sense and empirical observation was helpful.

In his *Treatise of Human Nature* Hume had proposed that the science of man should take account of the links between human reason and the mental powers of animals, and put the reasoning faculty in its proper place in the life of sentient beings. "Reason is, and ought only to be, the slave of the passions, and can never pretend to any other office than to serve and obey them." In his *Enquiry Concerning the Principles of Morals* he had developed that idea, basing his whole theory of the moral sense and its compelling power on sources among the natural human affections and sympathy rather than Divine instruction, abstract principles or Utilitarian self-interest. In his *Dialogues Concerning Natural Religion* he had tackled arguments for the existence of God and pointed to serious weaknesses in the so-called argument from design in natural theology, and his *Natural History of Religion* suggested that human notions of God had developed naturally from our understanding of our own actions and our anxiety about events we could not control. Hume's success in covering all these issues together from a single naturalistic point of view encouraged Charles as he tackled the high questions about human origins. He recognised the value of the way in which Hume had tackled philosophical riddles. "I suspect the endless round of doubts and scepticism might be solved by considering the origin of reason as gradually developed. See Hume on sceptical philosophy." He followed Hume in speculating that the idea of God might have developed from features of the human mind, and then dealt with the nature and workings of our moral sense in the same way.

On a warm spring day a few weeks after the meeting of the Geological Society of London, Charles went with his idea to the Zoological Gardens. He wrote to his sisters in Shrewsbury that when the animals were turned out, "Such a sight has seldom been seen, as to behold the rhinoceros kicking and rearing (though neither end reached any great height) out of joy . . . The elephant was in the adjoining yard and was greatly amazed at seeing the rhinoceros so frisky. He came close to the palings and after looking very intently, set off trotting himself, with his tail sticking out at one end and his trunk at the other, squealing and braying like half a dozen broken trumpets."

Charles was able to see Jenny the orang "in great perfection." The keeper showed her an apple, but would not give it to her, "whereupon she threw herself on her back, kicked and cried, precisely like a naughty child. She then looked very sulky and after two or three fits of passion, the keeper said 'Jenny if you will stop bawling and be a good girl, I will give you the apple.' She certainly understood every word of this, and though, like a child, she had great work to stop whining, she at last succeeded, and then got the apple, with which she jumped into an arm chair and began eating it, with the most contented countenance imaginable."

A few weeks later, Charles toyed obliquely in his notes with the possibility that man had evolved from an animal ancestor, and saw that once it was granted that one species might change into another, "the whole fabric totters and falls." He took man down from his pedestal, comparing the wild savage in Tierra del Fuego with the tame orang in the Zoological Gardens. "Let man visit Ourang-outang in domestication, hear expressive whine; see its intelligence when spoken [to], as if it understood every word said; see its affection to those it knew; see its passion and rage, sulkiness and very actions of despair; let him look at savage . . . and then let him dare to boast of his proud pre-eminence." The links here were as strange and disturbing to him as to any of the others who had watched Tommy or Jenny, but he wanted to explore them. He concluded: "Man in his arrogance thinks himself a great work, worthy the interposition of a deity. More humble and I believe true to consider him created from animals."

CHAPTER THREE

# NATURAL HISTORY OF BABIES

*Mind and memory—Religion—Human nature—Emma's faith—Observations on Willy and Annie—Daguerreotype*

IT WAS NOW IN MID-1838, while Charles's mind was full of apes, philosophers and savages and his conjectures were running wild, that he decided "Marry Marry Marry." He went to stay with his father in Shrewsbury and opened another notebook for his "metaphysical inquiries." His mind was also drawn to religion, and as he faced new possibilities in every direction, idea led to idea.

He asked his father about patients of his with mental disorders, and looked for clues to the workings of the human mind in what his father told him about their breakdowns. Was the mind purely spiritual as most Christians believed, or might it have some material basis in the brain, and be affected by the condition of the body? His father told him that one of his patients, after a paralytic stroke, could not remember a neighbour's name but could recall it and talk about him if an "early association," say from their schooldays together, was called up. Might that be due to a physical factor which had different effects on the power to remember in childhood and in old age?

Charles read about the science of mind and found a book which dwelt at length on memory as one of the most revealing and intriguing of our mental faculties. John Abercrombie, First Physician to Queen Victoria in Scotland, had published his *Inquiries Concerning the Intellectual Powers* to encourage young doctors to take an interest in the subject. He distinguished between memory, which was involuntary, and recollection, which was a conscious act. Control of your thoughts was the aim, and it was best understood by the contrast with the mind's lack of control in dreaming, insanity and other conditions. Charles noted how a "vivid thought" could not be dismissed even by the strongest will, and asked himself: "Is insanity an unhealthy vividness of thought?" Abercrombie mentioned the ways in which associations influence memory, and commented on the inexplicable power of "local" or "incidental" associations to revive deep feelings. He wrote that "the accidental discovery of some trifling memorial" "produces a freshness and intensity of emotion known only to those who have experienced it."

Charles thought carefully about the way his own memory worked, and wrote a seven-page note on his recollections between the ages of four and eleven to see whether he could trace any patterns. Did he remember happenings or only pictures from them? What feelings came back with the thoughts? He distinguished between true recollections of things themselves and memories of memories; he was intrigued by his recollections of the feelings of fear, pride and shame, and he was fascinated to find that he seemed to have acquired his own power of memory quite abruptly, as he could remember the earliest things "quite as clearly as others very much later in life which were equally impressed on me." He asked his sisters Catherine and Susan what they could call to mind, and gathered that Catherine had a better memory for ideas than images. For her, "a vivid thought is repeated, a vivid impression forgotten."

Charles thought also about Emma's mother at Maer in her dementia. He had gathered that her affections had "failed" more than her memory. He wrote cryptically about the sick old lady, who was both his aunt and soon possibly to be his mother-in-law, "Therefore affections effect of organisation, which can hardly be doubted when seeing Nina with her puppy." Nina was his sisters' dog at Shrewsbury. What he meant by the comment was that since Aunt Bessy's affections were breaking down together with her mind and body, they must stem, like a bitch's instinctive mothering, from her organic makeup rather than any spiritual element of her being. In the language of the day, this idea was "materialism," heretical and subversive. Charles thought at once of the danger it presented, and worked out a way to conceal the direction of his thinking. "To avoid stating how far I believe in materialism, say only that emotions, instincts, degrees of talent which are hereditary, are so because brain of child resemble[s] parent stock."

Charles talked at the Athenaeum with a physiologist, Herbert Mayo, who took a particular interest in the power of memory and its relation to our sense of personal identity. Charles read his book *The Philosophy of Living* and noted his accounts of dreaming and "double consciousness" in which a person switched from one identity to another with no memory of either while in the other. Charles wrote: "The possibility of the brain having whole trains of thoughts, feeling and perception separate from the ordinary state of mind is probably analogous to the double individuality implied by habit . . . These facts showing what a train of thought, action &c will arise from physical action on the brain, render much less wonderful the instincts of animals."

Charles thought about his religious beliefs. By his own later admission, "the religious sentiment" was never "strongly developed" in him, and he laid no store by soul-searching or prayer; but he wanted to be clear about the articles of faith. Many points were being argued about at the time. Three main elements of Holy Scripture were in question—the Genesis account of the creation of the world and the Fall of Man, the wrathful character of the God of the Old Testament, and the New Testament Revelation with Christ's promise of eternal life and salvation by faith. Underlying these doctrines was the issue of belief in an ever-present God, wise, purposeful and beneficent. And below that lay the ultimate question of the "First Cause," the mystery behind the creation of the universe and whatever gave rise to sentient life. The grounds of religion were also a concern. Many at the time based their beliefs on reasoning from the "evidences" of Scripture and natural theology. For others, the feeling of unworthiness before God and the act of faith were the key.

A note Charles made at the time reveals how down to earth his own approach was. "It is an argument for materialism that cold water brings on suddenly in head a frame of mind analogous to those feelings which may be considered as truly spiritual." After rejecting a literal reading of the Genesis account of the Creation as he learnt about the vastness of geological time, Charles questioned other historical parts of the Hebrew Bible, and found that he could not accept the God of the Old Testament because he was described as a vengeful tyrant. In rejecting that figure, he was following many others at the time who based their faith on the loving God of the Christian Revelation. But he did not stop there. He was also puzzled by the way in which the message of salvation was revealed in the New Testament. Having found flaws in the Old Testament account of the Creation and God's nature, he could not believe that God expected us to accept Christ's message on the authority of the New Testament, because the Gospels placed such emphasis on the fulfilment of the Old Testament prophecies. If God had meant us to accept Christ's message, he would surely have given it a more credible and persuasive foundation.

While he was in Shrewsbury, and thinking perhaps of Emma, Charles talked with his father about marriage and religious difficulties that might follow. Dr. Darwin suggested firmly that he should conceal his doubts from a future wife. He had known "extreme misery thus caused with married persons." "Things went on pretty well until the wife or husband became out of health, and then some women suffered miserably by doubting about the salvation of their husbands, thus making them likewise to suffer."

. . .

Charles returned to his lodgings in London, and worked hard through the high summer on a problem in Scottish geology. But he kept coming back to his new metaphysical themes. He wrote page after page of notes in a bout of what he called "mental rioting," free play with guesses and intuitions in which soundness of reasoning counted for little. He got into difficulties, for example, with free will and oysters, but got out of the tangle with disarming ease. Seeing young mammals play, he noted, it could not be doubted that they had free will. If they had, all animals must have. "Then an oyster has." Charles could imagine that the free will of an oyster was a direct effect of its physical makeup. "If so, free will is to mind what chance is to matter. The free will . . . makes change in bodily organisation of oyster. So may free will make change in man." He then stopped, writing simply: "Probably some error in argument. Should be grateful if it were pointed out."

But Charles was confident about other intuitions. He looked into the sources of the moral sense, which many like Whewell in his speech to the Geological Society saw as an exclusive and defining feature of mankind. Whewell had written that the moral sense was "an impress stamped upon the human mind by the Deity himself; a trace of His nature, an indication of His will, an announcement of His purpose, a promise of His favour." Charles saw that his own radically different theory of human origins would stand up only if he could show that morality could have been derived by a natural process from animal life. Harriet Martineau, his brother Erasmus's friend, had suggested in her book *How to Observe: Morals and Manners* that there were some universal feelings of right and wrong. Charles took up Hume's idea that the affections we share with other sentient beings, like "tenderness to our offspring," are fundamental to our nature, and instinctive. They have "no manner of dependence" on self-love, as some philosophers have argued. Charles guessed that they might be linked with the social instincts, and might have arisen because humans, "like deer" and other mammals, had become "social animals" by an evolutionary process.

Fresh ideas came to Charles day by day. One short string of casual comments has entered the history of ideas. "Origin of man now proved. Metaphysic[s] must flourish. He who understands baboon would do more towards metaphysics than Locke." A few jottings later, he sketched out an example of what he meant. Humans have instincts like anger and revenge which experience shows they must restrain if they wish to achieve happiness. Those instincts must at one time have been helpful in preserving the species, but external conditions changed and humans have become more cooperative. The conflict between aggression and restraint is not odd, because the older fierce instincts were once necessary but are now slowly vanishing. The tension is part of the layered history of human instincts and awareness, with elements from past adaptations surviving below other more recent ones, just as every species carries bodily traces of its ancestors in tell-tale features like the remains of a tail or limbs. The human mind has been shaped gradually from our animal past and is no more perfectly designed for current needs than any animal's body. "Our descent, then, is the origin of our evil passions!! The Devil under form of Baboon is our grandfather!!"

While Charles was exploring these metaphysical issues, he was also looking for possible links between animal and human feelings which could be checked by direct observation. He asked himself whether animals had likes and dislikes like humans. Jenny the orang came to mind, and he wrote in the back of his notebook: "Do the Ourang Outang like smells, peppermint and music?" He went to the Zoological Gardens in early September with a mouth organ, some peppermints and a sprig of verbena. He played the mouth organ to Jenny and she listened with great attention. He gave it to her and she "readily put it when guided to her mouth." She "seemed to relish the smell of verbena . . . and liked the taste of peppermint." A mirror was handed to Jenny. Charles noted that she was "astonished beyond measure" at the looking glass, and "looked at it every way, sidelong, and with most steady surprise." He was also fascinated to see Jenny "take bread from a visitor, and before eating every time, look up to [her] keeper to see whether this was permitted and eat it." She understood commands. "Jenny understands, when told door open, to give up anything and to do what she is told." But she would also "often do a thing, which she has been told not to do. When she thinks keeper will not see her, then knows she has done wrong and will hide herself." Charles did not know whether this was fear or shame. "When she thinks she is going to be whipped, will cover herself with straw or a blanket." Using objects for a purpose in this way was, he realised, a step in the progression from animal to man.

He went to see Jenny again a few weeks later. She was "amusing herself by getting out ears of corn with her teeth from the straw, and just like [a] child not knowing what to do with them, came several times and opened my hand, and put them in." A male orang who had joined her was unwell and Charles observed that he had an "expression of languor and suffering." He remembered the expression many years later in *The Expression of the Emotions in Man and Animals.* "The appearance of dejection in young orangs and chimpanzees, when out of health, is as plain and almost as pathetic as in the case of our own children. This state of mind and body is shown by their listless movements, fallen countenances, dull eyes, and changed complexions."

Charles was intrigued by the likenesses between some of Jenny's expressions in fear and anger, and the ways in which humans showed those feelings. He decided to examine the link between animal and human nature by looking at emotions and their facial expressions. He would see how the expressions were formed by the facial muscles, and the comparison with humans would be particularly telling if it could be shown that animals shared any of our expressions and might therefore share the feelings that went with them. Charles jotted inside the back cover of his notebook: "Natural history of babies. Do babies start (i.e. useless sudden movement of muscle) very early in life? Do they wink when anything placed before their eyes, very young, before experience can have taught them to avoid danger? Do they frown when they first see it?"

Charles found the key to the idea of natural selection as the mechanism for adaptation in a flash of insight at the end of September. Reading Thomas Malthus's *Essay on the Principle of Population,* he realised that the unending need for animals to compete for the means of life provided a continual pressure for selection that might work with the facts of variation and inheritance to create new species as circumstances changed. The three factors could be seen as "laws of life," and the process might be able to explain the endless diversity of natural forms in the grand and simple way that Herschel had suggested was the aim of scientific achievement.

Charles had a worry about his new argument, but made almost nothing of it at the time. Natural theology celebrated God's wisdom in the Creation, seeing every feature of the natural world as part of a Divine contrivance showing evidence of an all-powerful mind working with infinite benevolence. People recognised that animals fought, killed and fed on each other, but found arguments to justify the violence and suffering involved as a factor in the perfect balance of life which God maintained for the good of all. Charles believed he had identified a natural process which could explain the development of all forms of life without Divine intervention; the process was ridden with suffering and death and it seemed to be quite heartless and mechanical. The way in which the endless struggle for existence worked to produce the infinite richness and variety of animal and human life was a riddle to which

Charles had no answer. He was clear only that the arguments of natural theology were no solution. He wrote a few words charged with irony: "Pain and disease in world, and yet talk of perfection."

Through September and October 1838 Charles developed his new natural approach to metaphysics, moving back from the feelings of anger and revenge to the roots of the moral sense. His comments were carelessly worded, but he was in no doubt about his underlying aim. Might not our sense of right and wrong stem from reflection with our growing mental powers on our actions as they were bound up with our instinctive feelings of love and concern for others? He recognised that he needed to "analyse this out" and that it was important to bear in mind how the power of speech might have been a factor in the development. If any animal with affections and social instincts developed the power of reflection, it must have a conscience. He commented, "This is capital view," and applied the idea at once to a familiar animal. "Dog's conscience would not have been same with man's, because original instincts different. Man . . . who reasons much on his actions, makes his conscience far more sensitive." But that difference was one of degree, not of kind. Charles decided that conscience might be "an hereditary compound passion" and believed he had found a way in which the human sense of right and wrong could be shown to have developed gradually and naturally from feelings that were part of the life of social animals.

One feature of his emerging picture prompted a thought about mental activity that took place below the threshold of awareness. Herschel had emphasised in his *Preliminary Discourse on the Study of Natural Philosophy* that man could enter "only very imperfectly" into the "recesses" of his mind, while Abercrombie and Mayo had pointed in their books to aspects of behaviour which suggested that the mind had hidden depths. Charles noted that we assume we have free will but often find it difficult to analyse our motives for action. If our motives were "originally mostly instinctive," it would make sense that we now had to make "great effort of reason to discover them." He felt that this was an "important explanation."

Charles found support in Wordsworth's writings for his belief in the primary importance of the affections. The poet had declared in the "Preface" to his *Lyrical Ballads* that his "principal object" in the poems was to make themes from common life interesting by tracing in them "the primary laws of our nature." He aimed "to follow the fluxes and refluxes of the mind when agitated by the great and simple affections." Poems like "Michael" about the strength and constancy of a man's love for his son, and "The Ruined Cottage" about a woman's love for her husband, set a value on the natural feelings of ordinary people which Charles reflected in his scientific ideas. In "The Old Cumberland Beggar," Wordsworth wrote that "the poorest poor"

> *. . . have been kind to such*
> *as needed kindness, for this single cause,*
> *That we have all of us one human heart.*

Some years later he suggested to Coleridge that *The Excursion* was an attempt to treat "the commonplace truths of the human affections." It did not aim to convey "recondite or refined truths" but "rather to remind men of their knowledge, as it lurks inoperative and unvalued in their own minds." Wordsworth also emphasised the central importance of sympathy in his view of poetry and the Romantic imagination. It is "the wish of the poet to bring his feelings near to those of the persons whose feelings he describes." The object of poetry is truth "carried alive into the heart by passion," and through his writings, the poet "binds together by passion and knowledge the vast empire of human society."

In November Charles combined his new thinking about human

morality with some ideas he was exploring about the role of sexual reproduction in the development of species. He suggested that if animals produced offspring on their own rather than by pairing, inherited traits would never be pooled and species would never be formed with social instincts. The idea was important to him because he now "hoped to show" that the social instincts were "probably the foundation of all that is most beautiful in the moral sentiments of the animated beings." These tentative conclusions brought him to a staging point in his argument. He had found an answer to the challenge posed by human morality to his theory about mankind as a natural species. The moral sense was founded in the affections, and was a natural development from reflection on them just as they were a natural development for survival in some species. Charles's ideas were loose and rough, and a few were incoherent, but he now had a clear sense of the power of his species theory to explain central elements in human nature.

As he explored these themes, Charles thought again about religion. After his engagement to Emma at Maer in November, they had another long talk by the fire in the library. Frank and forthcoming as he was by nature, he went against his father's advice and told Emma about his "honest and conscientious doubts" about the Christian Revelation. He explained that he fully accepted the Christian morality, and wanted to believe in the afterlife and the promise of salvation, but could not persuade himself with the arguments others found satisfactory.

Emma's Christian beliefs differed in one way from Charles's critical and reasoning approach. They shared the family roots in the eighteenth-century tradition of Rational Dissent which placed a strong emphasis on the believer's personal responsibility for right conduct, a wide and humane concern for others, common sense and avoidance of doctrinal dispute. Emma combined her liberal Unitarian views with a deep commitment to Christ's promise of salvation through faith, which was also the message of the Evangelical movement. Unitarians emphasised the value of reason in supporting religious views, but insisted that religion was "an affair of the heart, not the intellect." Emma did not insist on reasons to justify the promise of salvation, as there might not be any compelling ones. "For now we see through a glass, darkly; but then face to face: now I know in part; but then shall I know even as also I am known."

Emma did not mention matters of faith in her diary; she seldom referred to them in her letters, and unlike many pious ladies of the time, she had no wish to make a display of her religion. But traces survive of her quiet and intent concerns in her interleaved Bible, reading lists and notes for prayers. The afterlife was especially important for Unitarians like the Wedgwoods, and the hope of reunion with loved ones after death was a mainstay of their faith. Emma had read *A View of the Scripture Revelations Concerning a Future State* by an Anglican clergyman, Richard Whately. He argued that the "perfection of friendship" with loved ones would be a great part of "the future happiness of the blest." When Emma's beloved elder sister Fanny had died suddenly in 1832, Emma wrote that she was trying to keep her mind "fixed upon the hope of being with her again . . . Such a separation as this seems to make the next world feel such a reality—it seems to bring it so much nearer to one's mind and gives one such a desire to be found worthy of being with her."

Just as Dr. Darwin had feared might happen between a devout wife and a husband with doubts about religion, Emma was unhappy about Charles's uncertainties about his faith and salvation. She was deeply afraid that she might lose him in the next world which was so important to her. When Whately had tackled the problem in his book, he dealt with it quite ruthlessly. A person who went to heaven might be supposed to feel grief for "the total and final loss of some who may have been dear to him on earth," but a wise and good man would try as far as possible "to *withdraw* his thoughts from evil which he cannot lessen." The eternal suffering of former friends might "often cloud his mind," but it was reasonable to suppose that in the world to come the blest "will occupy their minds entirely with the thought of things agreeable . . . and will be able, by an effort of the will, completely to banish and exclude every idea that might alloy their happiness." If that advice was any help to anyone, it was difficult for Emma to accept, loving Charles as she did.

After Charles had returned to London for the weeks before their marriage, Emma wrote to him that while they were together, "I think all melancholy thoughts keep out of my head, but since you are gone, some sad ones have forced themselves in, of fear that our opinions on the most important subject should differ widely. My reason tells me that honest and conscientious doubts cannot be a sin, but I feel it would be a painful void between us." She offered no arguments to overcome his doubts, but asked him simply to read Christ's farewell to his disciples in St. John's Gospel. "It is so full of love to them and devotion and every beautiful feeling. It is the part of the New Testament I love best. This is a whim of mine. It would give me great pleasure." She ended with painful diffidence, "Though I can hardly tell why, I don't wish you to give me your opinion about it."

Charles read the passage and said what he could to allay Emma's fears, but could not give her full assurance. She replied: "Thank you, dear Charles, for complying with my fancy. To see you in earnest on the subject will be my greatest comfort and that I am sure you are. I believe I agree with every word you say, and it pleased me that you should have felt inclined to enter a little more on the subject." In another letter Emma hoped that "though our opinions may not agree upon all points of religion, we may sympathise a good deal in our *feelings* on the subject." The two were eager to find a faith to share for their life together, but honest as they were with each other, they could not overcome all their differences.

. . .

Emma became aware of Charles's interest in facial expressions and how he was keeping a notebook for points that struck him. He explained his ideas to her, and she joked in a letter that this might affect the way he treated her. "I believe from your account of your own mind that you will only consider me a specimen of the genus (I don't know what, simia I believe). You will be forming theories about me, and if I am cross or out of temper, you will only consider 'What does that prove?' Which will be a very grand and philosophical way of considering it." The simia were the apes. Her joking made light of her anxiety about her differences with Charles.

Meanwhile, in his last weeks on his own in London, Charles made jottings in his metaphysical notebook about blushing, human, chameleon and octopus. He thought about the affections. "Why does joy and other emotion make grown up people cry? What is emotion?" And "What passes in a man's mind when he says he loves a person? Do not the features pass before him marked with the habitual express emotions which make us love him or her? It is blind feeling, something like sexual feelings. Love being an emotion, does it regard—is it influenced by—other emotions?"

Three months after the marriage, Emma and Charles returned to Maer for a three-week stay. While they were there, he wrote a long note on the moral sense. He traced its development from three essentials: sympathy, memory and reflection. David Hume had put sympathy at the centre of his thinking about the natural sources of moral principles. He saw it as a natural feeling rather than an attitude based on reasoning from some abstract notion. "There is some benevolence, however small, infused into our bosom; some spark of friendship for human kind; some particle of the dove kneaded into our frame, along with the element of the wolf and the serpent." Charles now developed this idea and speculated how our moral sense might also grow naturally from that feeling. "Looking at Man, as a Naturalist would at any other mammiferous animal, it may be concluded that he has parental, conjugal and social instincts, and perhaps others . . . These instincts consist of a feeling of love or benevolence to the object in question. Without regarding their origin, we see in other animals they consist in such active sympathy that the individual forgets itself, and aids and defends and acts for others at its own expense."

Charles believed that any action in accordance with the social instincts would give great pleasure, and any action that interfered with them would involve pain. He saw the genesis of sympathy in "a very slight change in association" whereby a person observing another's action would feel part of that pleasure or pain. Moral approval and disapproval would then develop from that feeling whenever man, "from his memory and mental capacity of calling up past sensations," reflected on the choice between following a social instinct which would give lasting pleasure, and yielding to a personal appetite, the pleasure of satisfying which would fade. The strength of our feeling that a social instinct ought to be followed showed that it was part of our nature, while selfish emotions, although equally natural, did not have lasting effects. In this way, Charles saw memories, both of pleasure and pain, as things of value, to nurture and cherish.

Sometime during the year, perhaps when Emma realised she was pregnant and had to think of the possibility that she might die in childbirth as Fox's wife had done, she was troubled again by her worries about Charles's faith, and wrote a note to him about a difficult matter she found she could not discuss with him face to face. Having to write in that way even though they were now living together was a painful admission of her difficulty. It is not known precisely what had happened between them because she chose her words with discretion, but he had apparently explained to her that he was carrying on with what she called his "discoveries"; he was still uncertain about the Christian Revelation, but his opinion was not formed. He had suggested to her that "luckily there were no doubts as to how one ought to act." He had claimed that some important point of Christian faith was unprovable and she felt that he was disregarding the essential point that if it was not provable in the scientific way he was familiar with, that was likely to be because it was "above our comprehension." Emma believed he was "acting conscientiously, and sincerely wishing and trying to learn the truth," and she wanted to feel he could not be wrong, but she was afraid that he might indeed be. She revealed a deep concern; there were dangers in "giving up revelation," that is Christ's offer of eternal life, and also in the sin of ingratitude for his suffering, "casting off what has been done for your benefit as well as for that of all the world."

Emma explained: "I do not wish for any answer to all this—it is a satisfaction to me to write it, and when I talk to you about it I cannot say exactly what I wish to say, and I know you will have patience with your own dear wife." Her fear was again for the afterlife. "Everything that concerns you concerns me and I should be most unhappy if I thought we did not belong to each other for ever." She declared that she could not tell him how happy he made her and how dearly she loved him, and she thanked him for all his affection "which makes the happiness of my life more and more every day." But her fear that they might not "belong to each other for ever" reached beyond the happiness of her life into eternity.

Emma's reference to Charles's "discoveries" and his comment that "there were no doubts as to how one ought to act" strongly suggest that he had told her about his theory of the natural origin of our moral sense. But, as a devout Christian, she could not accept his ideas. Forty-five years later, after his death, she made the point firmly to their son Francis. "Your father's opinion that all morality has grown up by evolution is painful to me."

Charles kept her note for the rest of his life. At some point, perhaps many years later, he wrote to her on the outer fold: "When I am dead, know that many times I have kissed and cried over this. C. D." He felt deeply for her, but could only think of breaking his silence to her after his death. The "painful void" between them was there before they married, and despite their deep devotion to each other, could never be bridged. From their shared family background in the Unitarian faith, Charles knew full well what rejection of Christian salvation would mean to Emma in her hope for eternal life with him after death.

They could, though, joke about other people's religion. On one occasion after a railway journey, Charles wrote to Emma: "In my carriage there was rather an elegant female . . . so virtuous that I did not venture to open my mouth to her. She came with some female friend, also a lady, and talked at the door of the carriage in so loud a voice that we all listened with silent admiration. It was chiefly about family prayers, and how she always had them at half past ten, [so as] not to keep the servants up. She then charged her friend to write to her either on Saturday night or Monday morning, Sunday being omitted in the most marked manner. Our companion answered in the most pious tone, 'Yes, Eliza, I will write either on Saturday night or Monday morning.' As soon as we started, our virtuous female pulled out of her pocket a religious tract in a black cover and a very thick pencil. She then took off her gloves and commenced reading with great earnestness, and marking the best passages with the aforesaid thick lead-pencil. Her next neighbour was an old gentleman with a portentously purple nose, who was studying a number of the *Christian Herald,* and his next neighbour was the primmest she-Quaker I have ever seen. Was not I in good company? I never opened my mouth and therefore enjoyed my journey."

Charles was able to start his "natural history of babies" when his own son Willy was born at the end of the year. Charles took a used notebook with a white vellum cover and wrote his observations on the blank pages. "W. Erasmus Darwin born Dec 27th 1839. During first week yawned, stretched himself just like old person—chiefly upper extremities; hiccupped, sneezes, sucked. Surface of warm hand placed to face seemed immediately to give wish of sucking, either instinctive or associated knowledge of warm smooth surface of bosom. Cried and squalled, but no tears. Touching sole of foot with spill of paper (when exactly one week old), it jerked it away very suddenly and curled its toes, like person tickled, evidently subject to tickling. I think also body under arms more sensitive than other parts of surface. What can be origin of movement from tickling?"

Charles tried to maintain his analytical detachment, and was, for example, always "anxious to observe accurately the expression of a crying child," but he found often that "his sympathy with the grief spoiled his observation."

In the first months, Charles was interested in Willy's senses and his responses to sights, sounds and sensations. "At his eighth day he frowned much . . . At his ninth day . . . he appeared to follow a candle with his eyes . . . When one month and one day old, perceived bosom when three or four inches from it, as was shown by protrusion of lips and eyes becoming fixed. Was it by smell or sight? It was *not,* certainly, by *touch.*" Charles believed that he had observed Willy's first reasoned action, noting that in his fourth month he "took my finger to his mouth, and as usual could not get it in, on account of his own hand being in the way; then slipped his own back and so got my finger in."

Charles watched for the earliest signs of each emotion. Fear came first with the child starting and crying at any sudden sound. In Willy's fifth month, "I had been accustomed to make close to him many strange and loud noises, which were all taken as excellent jokes, but at this period I one day made a loud snoring noise which I had never done before; he instantly looked grave and then burst out crying. Two or three days afterwards, I made through forgetfulness the same noise with the same result." Pleasure and affection followed, and then anger. "During last week has got several times in passion with his playthings . . . When in a passion he beats and pushes away the offending object."

Around the same time, Charles showed Willy a looking glass to see himself in, just as he had given the pocket mirror to Jenny the orang. Charles wrote that Willy "smiled at himself in glass." "How does he know his reflection is that of [a] human being? That he smiles with this idea, I feel pretty sure." In the seventh month, when Willy was looking in the mirror, he "was aware that the image of the person behind was not real, and therefore when any odd motion or face was made, turned round to look at the person behind."

Following his experiment with Jenny and the mouth organ, Charles watched carefully for Willy's first signs of musical taste. At four months he believed he had "shown decided pleasure in music—his whole expression appearing pleased." But Charles was not sure. Six months later he was still watching. Willy "cried when Emma left off playing the pianoforte." He cried so often and showed "such decided pleasure as soon as she turned round to go back" to the piano, that Charles was "certain there was no mistake."

Charles was also eager to see when and how Willy picked up the sense of right and wrong. When Willy was just over a year old, "I repeated several times in reproving voice, 'Doddy won't give poor Papa a kiss,—naughty Doddy.' He unquestionably was made slightly uncomfortable by this." Charles felt that this "showed something like first shade of moral sense." As for the dark feelings, "Jealousy was plainly exhibited when I fondled a large doll, and when I weighed his infant sister, he being then fifteen and a half months old." Charles must have been in the children's room on this occasion, helping to look after the month-old Annie.

In August 1842, when Willy was two and a half, Charles met him coming out of the dining room, "with an unnatural brightness of eye and an odd affected manner, so odd that I turned back to see if any one was hiding behind the door. I then found out by marks that he had been taking the powdered sugar which he had once or twice been told he ought not to do." A fortnight later Charles met him again "with his pinafore folded up carefully and he eyeing it. I asked him what he had got there. He said 'Nothing,' looking all the while to see that his pinafore was well folded, and as I came nearer he cried 'Go away; Doddy going to send; go away.' From his odd manner I determined to see what was concealed, when I found he had stained with yellow pickle his pinafore when taking pickle, like he had done sugar. Here was natural acting and deceit."

Around the time when Annie was born in March 1841, Charles bought a set of Wordsworth's *Poetical Works* in six small volumes. In the two weeks after her birth, he read the first volume, which opened with a section of "poems referring to the period of childhood," and the second, which included a poem Wordsworth had written about his daughter Dora when she was a month old. On her face smiles were beginning

*. . . like the beams of dawn,*
*To shoot and circulate; smiles have there been seen,*
*Tranquil assurances that Heaven supports*
*The feeble motions of thy life, and cheers*
*Thy loneliness: or shall those smiles be called*
*Feelers of love, put forth as if to explore*
*This untried world, and to prepare thy way*
*Through a strait passage intricate and dim?*
*Such are they; and the same are tokens, signs,*
*Which, when the appointed season hath arrived,*
*Joy, as her holiest language, shall adopt;*
*And Reason's godlike power be proud to own.*

In Annie's first weeks, Charles watched for the points he had observed in Willy, to see whether they emerged in a regular sequence during the infant's growth, according to a law of nature. Annie's first smile, "as if to explore this untried world," came on her forty-sixth day.

Whenever Charles was struck by a similarity with Willy, a difference or some other feature, he wrote it down. When Annie was almost a year old, she was "rather amused, at a wafer sticking first to one hand and then to the other as she tried to disengage it." He remembered how Willy at about the same age "grew frantic with passion at a bit of wet paper thus sticking to his fingers." He found that Annie had "infinitely less observation and animation" than Willy at the same age. He showed her the reflection of her face in the polished case of his watch, but "She now hardly understands a person coming up behind her when she is standing in front of the glass and looking at her image."

A fortnight after her first birthday, Annie "walked loose about four feet, walked well and says 'Goat.'" Charles watched her first simple reasoning by association of ideas. She was "accustomed to see the keys taken out to go to cupboard and tea-chest for good things, and the keys being given her today to play with in farther part of room, she immediately led Emma by the hand towards the tea-chest." Another observation bore on the philosophical puzzle of awareness of self and others and its link with consciousness. "Younger children, such as Annie, now a year old, look at people with a degree of fixedness which always strikes me as odd. It is like the manner older people only look at inanimate objects. I believe it is because there is no trace of consciousness in very young children—they do not think whether the person they are looking at is thinking of them." Charles watched Annie's expressions carefully, and wrote: "I have some months remarked how much Annie wrinkles up her nose in smiling or half laughing. This she did very early, more so than Doddy, and certainly with her is a more essential part of a smile than the play of the muscles round the corners of her mouth." When she was fourteen months old, it was "curious to see how neatly Annie takes hold in proper way of pens, pencils and keys. Willy to present time with equal or greater practice cannot handle anything so neatly as Annie does, often in exact manner of grown up person."

One day Willy "was running to give Annie a punch with a little wooden candlestick, when I called sharply to him and he wheeled round and instantly sent the candlestick whirling over my head. He then stood resolute in the middle of the room as if ready to oppose the whole world." He "peremptorily refused to kiss Annie, but in short time, when I said 'Doddy won't throw a candlestick at Papa's head.' He said 'No, won't; kiss Papa.' " Charles commented: "I shall be curious to observe whether our little girls take so kindly to throwing things when so very young. If they do not, I shall believe it is hereditary in male sex, in the same manner as the S. American colts naturally amble from their parents having been trained." A day before Annie's second birthday, Charles noticed that she showed "no signs of skill in throwing things either as an amusement, or as an offensive act, in [the] same ready way as Willy did: nor does she so readily give slaps." He watched her as she grew, by inheritance of a difference between the sexes, into a little girl.

Many parents wanted "likenesses" of their children in infancy, to remember them by when they grew older or if they died young. In 1841, Hensleigh and Fanny Wedgwood had commissioned the artist George Richmond to do a watercolour of their three-year-old son Ernie. Charles and Emma had sat for Richmond in 1840, but they did not ask him to paint their children. Instead they had "sun pictures" made by the new method of photography which was the latest wonder of technical progress.

The first portrait studio for daguerreotypes in London had been opened in 1841 by Richard Beard on the roof of the Polytechnic Institution in Regent Street, in order to catch the sunlight; Paul Claudet followed suit with a studio above the Adelaide Gallery off the Strand later in the year. The Polytechnic and the Adelaide were both exhibition halls for popular science; they were locked in competition through the 1840s with ever-varying displays and entertainments. Each had working models, lectures and a "gas microscope" projecting images of minute objects onto a large screen. The Polytechnic had magical "dissolving views" which children loved, but the Adelaide Gallery had a steam cannon and was just next to the Lowther Arcade, where every kind of cheap toy could be bought.

In the first years, both daguerreotype studios were very popular and a visit was an intriguing experience. Maria Edgeworth wrote to a friend in 1841 about having her likeness taken. "It is a wonderful, mysterious operation. You are taken from one room into another, up stairs and down, and you see various people whispering, and hear them in neighbouring passages and rooms unseen, and the whole apparatus and stool on high platform under a glass dome casting a *snap-dragon blue* light making all look like spectres, and the men in black gliding about . . ."

Many were disconcerted by the images the apparatus caught, but they had to accept them as objective, and truthful in that new way. One commentator wrote: "The common remark upon showing your sun picture to friends is, 'Well, it isn't a flattering portrait, but it must be like, you know!' " And George Cruikshank, the *Punch* cartoonist, wrote in a poem, "Photographic Phenomena": " 'Well, I *never!*' all cry; 'it is cruelly like you!' But truth is unpleasant to prince and to peasant." The commentator warned ladies not to "make up a face for the occasion" because the result was often disastrous. "If ladies, however, *must* study for a bit of effect, we will give them a recipe for a pretty expression of mouth. Let them place it as if they were going to say *prunes.*"

Charles first took Willy to Claudet's studio in August 1842 around the time of Willy's secret raid on the pickle jar. Annie was only a year and five months old at the time and could not have been kept still for the full minute required for the exposure. Ernie Wedgwood had appeared in Richmond's watercolour as an angelic little child elegantly posed in a rocky landscape. Charles had to hold Willy firmly on his lap in front of Claudet's painted backcloth, and the camera caught them with the ruthlessness for which it was feared. Charles avoided the camera's eye, as if to say "This picture is of my son, not me." Willy was wearing his best frock; after walking past the toy stalls, climbing the stairs and the long queue, the loading of the camera and being posed by the studio assistant under the mysterious blue light, he sat without moving but looked warily at the photographer, waiting for the next strange thing to happen.

The contrast between Richmond's watercolour of Ernie and Claudet's daguerreotype of Willy shows clearly the change from art to photography which people were coming to understand and accept in the early 1840s. The disquiet people had felt when they first saw Tommy and Jenny at the Zoological Gardens in the late 1830s was like the effect of a cruel sun picture. The man-like apes were recognisable and yet different, and the young man watching his small children in Macaw Cottage was drawing a truth about human nature out of the likeness.

CHAPTER FOUR

# YOUNG CROCODILES

*Down House—The Big Woods—Servants—Brodie and Parslow—*
*Charles's theory—Life for the young children*

IN JULY 1842, on a gloomy day under a cold northeast wind, Charles and Emma found Down House in Kent. After a year in Macaw Cottage, they knew they wanted to live in the country. Life in their London terrace was confined and unhealthy for growing children; Charles and Emma wanted to escape from the social round of calls and dinners, and they dreamt of a garden like the ones they had been brought up in at Maer and in Shrewsbury. At the same time, though, Charles wanted to be able to visit friends in London to talk about natural history, and be home with Emma for supper. As the new railway lines snaked out from the metropolis through the home counties, Charles and Emma looked along the routes for a house from which Charles could get to London and back the same day. In Annie's first months, Emma started noting properties and their acreages in her diary—forty acres at Langley, six near Reading and twenty-five in Harrow Weald. They wanted to buy a house in Woking but missed it, and looked at Down House as a second best.

Charles thought the house ugly and Emma was disappointed with its surroundings, but it was situated just as she liked "for retirement, not too near or too far from other houses." The village of Downe (spelt with an "e" from the 1850s, while the Darwins have always referred to the house as "Down House" or "Down") was a cluster of about forty households round an old flint church with an ancient yew and a gnarled walnut tree. There was a butcher, a baker and a post office. Charles wrote to his sister: "The little pot-house where we slept is a grocer's shop and the landlord is the carpenter, so you may guess style of village." The villagers touched their hats "as in Wales," and sat at their open doors in the evening. The manor was "a most beautiful old farm-house with great thatched barns and old stumps of oak trees," and there were a number of so-called "gentry houses," villas with a few acres each for people with independent means.

Downe was sixteen miles from London, a two-hour journey by train and carriage from London Bridge Station. But Charles felt that he and Emma were "at the extreme verge of the world." "It is the quietest country I ever lived in . . . To the east and west there are impassable valleys, to the south only one very narrow lane, and to the north, through the village, only two other lanes." The village is situated high on the chalk uplands of the Kentish North Downs. The landscape has been patterned through the centuries with settlements, fields and lanes on the high ground, and thick woods beneath, in which cattle, sheep and pigs were fattened on acorns and beechmast. Emma wrote that the valleys were "crowned at the top with old hedges and hedgerows, very disorderly and picturesque, and with enormous clusters of clematis and blackberries, and a great variety of yews and other trees."

Down House was at the time a gaunt villa a short way out of the village; Charles felt that it had "somewhat of a desolate air." Around the house there were many old trees—cherry, walnut, yew, spanish chestnut, pear, larch and mulberry. There were quinces, medlars, plums and a purple magnolia. The field was a good hay-meadow and the previous owner had kept three cows, a horse and a donkey. Emma, now six months pregnant with her third child, agreed with Charles that they could make a home for their family there, and he set about buying the property at once so that they could move from Macaw Cottage and be settled in good time before the birth.

The soil over the chalk was heavy red clay, "abounding" as Charles noted "with great irregularly shaped, unrolled flints, often with the colour and appearance of huge bones." He wrote: "The charm of the place to me is that almost every field is intersected . . . by one or more footpaths. I never saw so many walks in any other country. The country is extraordinarily rural and quiet, with narrow lanes and high hedges." Reading the landscape with his geologist's eye, he decided that the valleys, now dry, were "in all probability ancient sea-bays," and wondered "whether the sudden steepening of the sides does not mark the edge of vertical cliffs formed when these valleys were filled with sea-water." Three miles to the south, the ground fell away sharply to the Kentish Weald, thickly wooded and rolling over rock formations with a long history that Charles knew well from the writings and talk of friends in the Geological Society.

In 1825, Dr. Gideon Mantell had found fossil fragments of what he thought was a monstrous lizard in a quarry in the Weald near Lewes in Sussex, thirty miles to the south. He called it "iguanodon" and in 1841 Professor Owen coined the word "dinosaur" for the beast and its cousins in the remote geological past. Dr. Mantell gave a vivid account of the world of the iguanodon in his popular book *The Wonders of Geology.* He wrote that the country had then been "diversified by hill and dale, by streams and torrents, the tributaries of its mighty river. Arborescent ferns, palms, and yuccas, constituted its groves and forests; delicate ferns and grasses, the vegetable clothing of its soil; and in its marshes, equiseta [horsetails], and plants of a like nature, prevailed. It was peopled by enormous reptiles, among which the colossal Iguanodon and the Megalosaurus were the chief. Crocodiles and turtles, flying reptiles and birds, frequented its fens and rivers, and deposited their eggs on the banks and shoals." Charles Lyell speculated in his *Principles of Geology* about what might happen if the warmth of the ancient climate were to be restored. "Then might those genera of animals return, of which the memorials are preserved in the ancient rocks of our continents. The huge iguanodon might reappear in the woods, and the ichthyosaur in the sea, while the pterodactyl might flit again through umbrageous groves of tree-ferns."

A view far out over the Weald from a point on the escarpment three miles south of Down House came to have a special importance for Charles, and he later explained it to his children when he told them about geology and the history of the landscape around their home. He was to write in *The Origin of Species* how his theory of evolution needed aeons of time for the long sequence of branching development, "all changes having been effected very slowly through natural selection." Past ages had been "incomprehensibly vast," and their extent impressed his mind "almost in the same manner as does the vain endeavour to grapple with the idea of eternity." One way to grasp the vastness was "to stand on the North Downs and look at the distant South Downs." Charles pictured "the great dome of rocks" which must have filled and risen over the wide valley in recent geological time. It had been 1,100 feet thick on one reckoning, and he calculated that it might have taken three hundred million years to eat away the rocks. "I have made these few remarks because it is highly important for us to gain some notion, however imperfect, of the lapse of years. During each of these years, over the whole world, the land and the water have been peopled by hosts of living forms. What an infinite number of generations, which the mind cannot grasp, must have succeeded each other in the long roll of years!"

In August, when Charles and Emma were back in London preparing for the move, labourers in the "manufacturing districts" of the north and west went on strike for the Chartists' demands for electoral reform, and a sudden wave of fear spread among the well-off. Emma had read Thomas Carlyle's *Chartism* with its urgent plea for serious attention to the bitter discontent of the "Working Classes." She had found his arguments "full of compassion and good feeling, but utterly unreasonable." On the fifteenth of the month she wrote in her diary, "Riots in Potteries," the only public event she ever recorded. Violence had broken out very close to her family home near the Wedgwood factory in Staffordshire. *The Times* had reported a confrontation between strikers and armed soldiers near the house of a Wedgwood cousin in Burslem, in which three rioters were killed. "The fatal charge on the mob yesterday morning has had a salutary effect, and probably saved two churches from the destruction threatened. It was made about half-past ten . . . opposite the corner of Mr Wedgwood's house. The town was menaced by the advance of two columns of the mob . . . The Riot Act had been previously read, and the commanding officer gave the order to charge and fire." A clergyman's house nearby was a smouldering ruin. "It is remarkable that the mob gave the old clergyman ten minutes . . . to get away, being lame and only able to walk with two sticks. They made a capture of his surplice, put it on one of the mob in mockery, who paraded in front of the house when on fire; while others tore his Bible into tatters, and scattered the leaves in the street. Several Dissenters were calm and smiling witnesses of this gross outrage!" In the 1840s, an "age of revolutions," respectable people feared all kinds of nonconformity. The Unitarian Wedgwoods counted as Dissenters, and Unitarians were reckoned to be all the more dangerous for their claimed faith in reason.

When serious disturbances broke out in Manchester, the government decided that troops should be sent from London to impose order, and trains were prepared at Euston Station to take them north. Crowds gathered to shout support for the strikers, and thousands of people

*Soldiers and Chartists at Euston Station*

milled in the streets around Macaw Cottage waiting for the soldiers to march past on their way to the station. Policemen cleared the way for a troop of a hundred and fifty cavalry and four artillery pieces, but when seven hundred foot guards approached with three wagons of ammunition, the crowd shouted, "Don't go and slaughter your starving fellow-countrymen," and groaned loudly as they entered the railway yard. The next day a crowd of five thousand gathered. They listened to the "harangues of several Chartist demagogues who were speaking the most seditious language in the hearing of the police," and yelled discordantly when another six hundred soldiers came.

The disturbances died down after a few days and *The Illustrated London News* reported: "The semi-revolutionary movements that have been spreading so much alarm over all parts of the Kingdom are now happily subsiding into peace; the law has vindicated its majesty, and order and tranquillity are once more beginning to fold their wings over the land."

. . .

On the first Sunday in September Charles went to Down House. Emma, now eight months pregnant, followed a week later with Willy, Annie and Bessy the nursemaid. Two days after arriving, she paid a courtesy call on their closest neighbour, Mrs. Mary Price at Down Lodge. Charles wrote to his cousin Fox in December that the family's move from London had "answered very well; our two little souls are better and happier, which likewise applies to me and to my good old wife." But he had left out a sharp pain. A week after visiting Mrs. Price, Emma had been feverish and had a violent headache. The next day, in the thirty-sixth week of her pregnancy by the careful reckoning she had made in her diary, she gave birth to a second daughter. The child was baptised Mary Eleanor in the village church at the beginning of October, and Charles wrote to a friend two days later that Emma was making a quicker recovery "owing, I think, to country air, than she has ever done before." But Mary died within three weeks. Charles chose a burial plot for the family close to the west door of the church, and Mary was buried there. Emma wrote to her sister-in-law Fanny Wedgwood: "Our sorrow is nothing to what it would have been if she had lived longer and suffered more. Charles is well today and the funeral is over, which he dreaded very much." Emma felt she regretted the loss of her baby more from her "likeness to Mamma, which I had often pleased myself with fancying might run through her mind as well as face . . . With our two other dear little things, you need not fear that our sorrow will last long, though it will be long indeed before we either of us forget that poor little face."

Emma was soon taking walks, and noticed the wild climber, traveller's joy, which was covering the autumn hedgerows with its seedheads. Willy approved strongly of "country 'ouse," and Annie at eighteen months tottered here and there over the unaccustomed space of floor and grass. Emma's chief impression of the first year at Down was that "we enjoyed the intense quiet, as we knew no one."

In her early childhood at Down, Annie's world was bounded by the deep woods in the valleys around the village. During the first winter, when Emma's brother Hensleigh was ill, three of his children, Julia or "Snow" aged nine, James or "Bro" aged eight and Ernie aged five, came to stay. Living as they did in a London terrace, they were eager to play out of doors, in the fields, woods and clay-pits. The weather was sharp and cold and there was snow on the ground, but Emma let Bessy the nursemaid take them with Willy and Annie into the Big Woods in the wide valley to the south. The woods, where Charles once saw a polecat, were a mass of hazel-copse with occasional oaks and crossed by narrow footpaths. The children lost their way and wandered helplessly for hours. Bessy cried and Annie was distressed. Snow and Willy found themselves on their own and managed to make their way home. Charles met them as they came back, and gathered what he could from Snow about the other four. He asked some neighbours to help look for them, and they eventually found them near a farmhouse in the valley, after Bessy had been carrying Annie for three hours. Charles noted in his book of accounts that he had paid seven shillings to "People hunting children."

Bessy was ill for a year afterwards, and though Annie cannot have been able to understand anything at the time, Bessy must have told her about the day when she could talk with her. Two years later, Emma found that Annie was upset by one of her books. It was called *Little Robert and the Owl* and was about a boy who got lost in a wood as he went to visit his grandmother one winter evening. He crept inside a hollow tree, driving out the owl, and fell asleep while thinking confidently of a verse he remembered from the Prophet Ezekiel: "And I will make with them a covenant of peace, and will cause the evil beasts to cease out of the land: and they shall dwell safely in the wilderness, and sleep in the woods." His dog Faithful found him next morning and all was well. Annie knew a hollow ash tree in the Darwins' meadow, and the children heard the owls in the woods during the winter nights. Annie could not stand hearing the story but insisted that Little Robert "must have somebody to take care of him."

When spring came, Charles walked along the narrow lanes and found the banks clothed with primroses and pale blue violets. "A few days later some of the copses were beautifully enlivened by *Ranunculus auricomus,* wood anemone and a white *Stellaria.*" In June 1843, he wrote that "the sainfoin fields are now of the most beautiful pink, and from the number of hive bees frequenting them, the humming noise is quite extraordinary." He was also struck by the number of different kinds of bush in the hedgerows, all entwined by traveller's joy and bryony. The neighbourhood he was exploring was to be the range of his daily walks and observations as a naturalist for the remaining forty years of his life.

As the family settled in, Charles and Emma got the measure of their new home. They wanted more rooms, an orchard, a good kitchen garden, dry paths, a view over the valley to the west and shelter from the fierce north wind. They were also most anxious for privacy from people walking along the lane to the next village, as Charles found their gaze "intolerable." He employed a builder and filled every afternoon with his schemes for the grounds, planting bushes, levelling banks, digging the lane down by two feet and raising mounds in the garden with the earth.

The building works began in the spring of 1843. Charles took advice from the builder's foreman, "a sort of Jack of all knowledge." "From the effects of a £1 present and the hopes of another" the foreman was very helpful. Charles commented: "I suspect he is an old rogue, but he is a useful one." The house had been badly laid out, and one of the bricklayers told Charles with a gloomy shake of his head, "A most deceptious property to buy, sir." Charles felt that the alterations made it a quite substantial home. On the ground floor the family had three rooms—a large and open drawing room with a new bay window looking southwest into the garden, a dining room and Charles's study looking over the lane. In the hall, a wide staircase led to the first floor. The bedroom above the drawing room shared the newly built bay and the windows opened out over the garden, the meadow and the valley to the west. Charles and Emma had separate bedrooms. It appears that the one with the new bay window was to be for her; according to Charles, "Emma's bedroom will be truly magnificent; I quite grudge it her."

In the passage between the drawing room and the dining room was the door to the "offices," the kitchen, pantry and scullery, and the back staircase to the servants' bedrooms on the second floor. During the first two years, the offices were cramped and inconvenient for the many servants who worked for the young family, and when Charles and Emma needed to add a schoolroom on the first floor for Willy and Annie, the servants asked if they also could be given more room. Charles felt it was "so selfish making the house so luxurious for ourselves and not comfortable for our servants, that I was determined if possible to effect their wishes." He engaged the village builder, John Lewis, to rebuild the offices with the schoolroom above and a new passage and back door into the yard.

While the building works were going on, Emma took Willy and Annie to stay with her Allen cousins in Wales. Charles reported to her that Mr. Lewis had quarrelled with his carpenters and laid them all off. Lucy, one of the servants, "was very good natured and took keen interest about one man whose wife had come from a distance with a baby and is taken very ill. The poor man was crying with misery, but we have persuaded Lewis to take him back again." These were the "hungry forties" when many labourers had to take their families from place to place for any casual work they could find.

Charles's cousin Fox wrote to him in the weeks before the first anniversary of his wife's death in childbirth. The passing of time had not softened Fox's grief. Charles wrote back that he had thought "time *inevitably* would have done you more good than it seems to have done." He continued humbly, "I had hoped (for experience I have none) that the mind would have refused to dwell so long and so intently on any object, although the most cherished one." He thought about the value of feelings and the roots of human nature. "Strong affections have always appeared to me the most noble part of a man's character . . . you ought to console yourself with thinking that your grief is the necessary price for having been born with (for I am convinced they are not to be acquired) such feelings." He ended on another note of deference to Fox's understanding of grief. "But I am writing away without really being able to put myself in your position. You have my sincerest sympathy and respect in your sorrow."

In September Emma and Charles's next child was born. She was christened Henrietta, and was known as Etty. George (my great-grandfather) followed in 1845, Elizabeth, known in her childhood as Betty, in 1847, Francis in 1848, Leonard in 1850 and Horace in 1851. All except Betty and Horace wrote accounts of their childhoods which are the sources of much that follows. Thinking back on her early years, Etty's impression was that her mother was almost entirely wrapped up in her husband and the children. "The life of watching and nursing which was to be my mother's for so long . . . cut her off from the world." When Charles became ill, "intercourse with our neighbours almost ceased, and we children had a rather desolate feeling that we were aliens. But I think that my mother never felt this as any loss. She was not essentially sociable as he was." Emma was often unwell with recurring headaches and fevers, and she noted her ailments in her diaries alongside Charles's "bilious vomitings" and fever. Her jottings were all brief and matter-of-fact, but they show how she cared for Charles every day while coping with her own illness.

Annie was brought up in a household of servants. During the 1840s Charles was spending around £1,500 each year from his father and family trusts. That was equivalent to around £67,000 today according to the Bank of England's index of values, but the comparison cannot be precise. A manual called *The Complete Servant* suggested that a gentleman and lady with children and an income of £1,500 would normally employ a cook, two house-maids, a nursery maid, a coachman, a footman and a man to help in the stable and garden. During Annie's early childhood, Charles had a butler, a footman and two gardeners. Emma had a cook, a kitchen maid, a laundry maid, a housemaid, a nurse for the children and one or two nursery maids. The butler had the cottage next to the coachhouse in the yard, and the two gardeners lived in the village.

There was an open market for domestic staff in *The Times* and other newspapers, but the Darwins preferred to engage people they knew or whom sisters, brothers, cousins, close friends or neighbours could vouch for. According to Etty, her mother managed the household with a light hand and played down domestic difficulties. "It was remarkable how she infused this spirit into the household and made the servants ready to co-operate with her, often even at great inconvenience to themselves . . . She would take any trouble to help them or their relations, and in return there was nothing they would not do to please her. In an emergency they would cheerfully work like horses; or anyone would change their work; the cook would nurse an invalid, the butler would drive to the station, and anybody would go an errand anywhere or be ready to help in looking after the poor people."

Francis wrote that his father was loved and respected as a master of servants; "he always spoke to them with politeness, using the expression 'Would you be so good' in asking for anything." Francis remembered being reproved by him for using more spoons than he needed to, because it meant more work in cleaning. His father tried to avoid harsh words with the staff. When Francis was a small boy, he heard the head gardener being scolded for abusing the second gardener, and "my father saying angrily 'Get out of the room, you ought to be ashamed of yourself.' It impressed me as an appalling circumstance, and I remember running upstairs out of a general sense of awe."

According to a family story, once when it was time for supper, Charles went into the kitchen and played the cook's hand at whist for her, so that she could get on with preparing the meal. The cook was Jane Davies from Wales; she was known as Mrs. Davies although she was unmarried. Annie knew her as "Daydy" and Francis remembered her kindness in spite of constant threats of "tying a dish-cloth to your tail" which he never understood. Annie was at home in the kitchen; she often ate meals there, and could generally extract gingerbread and other good things from Daydy when she wanted.

There was no bathroom in the house, nor any hot water except in the kitchen, but the housemaids carried bath-cans here and there. Bessy Harding was the first of a succession of young maids who helped with the children and did other menial tasks. Emma used to let the children's mess accumulate till the room "becomes unbearable," and then called Bessy in to tidy up. Once when Bessy was looking after Etty and the small child was being particularly trying, Emma wrote that Bessy was "so tender and sweet-tempered, she is a jewel."

There were also a housemaid, a nursery maid and a laundry maid, all girls from villages nearby. With the children and the young servants, the

Darwin household at that time had only four people over twenty-five, and twelve or more under. A book for children, *Little Servant Maids,* gave an account of a servant's life. It was a great favourite of Etty's, and she always remembered how it "described in the most delightful way all the activities of the little maids, their sins and the sins of their mistresses." It declared with merciless severity the assumptions of respectable people who expected subservience from their staff. Children going into service "should consider what sort of behaviour is most likely to gain the good-will of ladies and gentlemen . . . Let poor boys and girls . . . be quiet and orderly in their manners, and civil in their mode of answering. A servant . . . will find it necessary to be constantly subduing her own will, and giving up her inclinations at the direction and command of others."

Emma put the book into a lending library she ran for the village children. Every Sunday afternoon they came to the village schoolroom near the church to borrow and return their books. Etty watched jealously as her favourite books were lent out, and pointed out to her mother when they were not returned. She believed that "if a book was much enjoyed, the proof was that it was stolen however often it was replaced. This was the fate of my beloved *Little Servant Maids.*" It seems remarkable now that the village children wanted to read the book, but many girls had to seek a "place" because it was the only way to earn money for the family or escape from drudgery at home. As the book made clear, the ways of a genteel household were quite different from those that poor girls were used to, and the book gave them an idea of what to expect. The Darwins' maids probably had an easier time than many, but it must have been a very hard life for any who were attached to their family and friends, and unhappy to submit to the discipline of quiet obedience to strangers.

. . .

The servant to whom Annie was closest from her earliest years was the children's nurse, Jessie Brodie. She was known to everyone as Brodie, not Miss like the governess or Mrs. like the cook because she had a lower position than either. When she came to the household in 1842 she was forty-nine, a tall, erect woman with carroty hair, china-blue eyes and marked features deeply pitted with smallpox. She looked severe in a photograph taken some years later, but Etty remembered that she had a "most delightful smile." She came from a small fishing port on the northeast coast of Scotland. Her father had been a ship's master who had been kept prisoner by the French in the Napoleonic Wars, and she had heard nothing of him for ten years.

In 1839, before coming to the Darwins, she had worked for the novelist William Makepeace Thackeray as nurse to his two young daughters, Anny aged two and the newborn Jane. In Brodie's first months with the Thackerays, Jane fell ill with a chest infection and died. Brodie helped at the birth of Thackeray's third daughter, Minnie, in May 1840, and Anny Thackeray later wrote that she remembered first seeing her little sister wrapped in flannel on Brodie's lap. "My nurse said 'Come here, Missy, and look at your little sister.' And I said 'I can't see her, Brodie,' and Brodie said 'Look at her kicking her little feet.'"

Shortly after Minnie's birth, Thackeray's wife became severely disturbed and after apparently trying to drown the three-year-old Anny in the sea at Margate, tried to drown herself while the family was travelling on a steamer from London to Cork. Thackeray wrote to his mother that he never saw anything more beautiful than Brodie's care for the children on board the steamer. "She was sick every quarter of an hour, but up again immediately staggering after the little ones, feeding one and fondling another. Indeed a woman's heart is the most beautiful thing that God has created, and I feel I can't tell you what respect for her (I have)." In the following weeks while Thackeray was having to cope with his wife's breakdown, there was "only poor Brodie of whom I can make a friend; and indeed her steadfastness and affection for the little ones deserves the best feelings I can give her. The poor thing has been very unwell, but never flinched for a minute, and without her, I don't know what would have become of us all."

*Minnie Thackeray by her father*

At the time Brodie had plans to marry a well-to-do man, but she decided that she would not go until Mr. Thackeray could arrange something for his wife and the children. She stayed with them; the man went to Australia, and she lost her chance. Thackeray could not afford to keep her on, and she went to the Darwins in 1842. Early the next year, when his daughters were being looked after in Paris, he wrote from London to his Anny: "How glad I shall be to see all my darlings well; and there is somebody else who wants to see them again too, and that is Brodie who longs to come back to them." Anny Thackeray knew that Brodie had gone to work for Mr. and Mrs. Darwin "who had a little girl of my own age called Anny too."

In the early months at Down House, Emma had to support Brodie when Bessy Harding was "pert" to her, but after a while things went smoothly. Brodie also kept in touch with Mr. Thackeray and saw his two children often, but on one occasion he did not tell her that they were in London. He wrote to his sister: "She is so prodigiously fond of the children that to see them for an hour would give her more pain than pleasure, and we have not had the heart to send to her." Anny Thackeray always remembered that Brodie had "a genius for loving."

A story has been passed down in the Darwin family that Brodie once said it was a pity Mr. Darwin had not something to do like Mr. Thackeray. She had seen him watching an ant-heap for a whole hour.

Etty in her old age could still remember Brodie "almost as if she was before my eyes, sitting in the little summer-house at the end of the Sand-walk, and hear the constant click-click of her knitting needles. She did not need to look at her stocking, knitting in the Scotch fashion with one of the needles stuck into a bunch of cock's feathers, tied to her waist, to steady it. There she sat hour after hour patiently and benevolently looking on, whilst we rushed about and messed our clothes as much as we liked."

We can hear Brodie's voice and sense her passionate affection for the children she looked after, in a letter she wrote to congratulate Etty on her marriage, twenty years after leaving the Darwins' service. She was then in her late seventies, living on her own in a stone tenement in Aberdeen. "Think how happy I was when I got your very kind letter . . . It was so very kind of you My owen Sweet pet to Send it. I Cannot Describe how Much I estame it . . . I not going to writ much more as Present it is very Dark & Dismall to Day . . . Now My Sweet Child I Am just going to finish . . . I hoop you will Both excuse Me for My Shortcummings—Godd by My Derest one. May the blessings of the Lord Rest uppon you and your Dear Husband. I am My Sweet Child your ever affect Miss Old Brodie."

Annie played with Joseph Parslow, the butler, who came from Gloucestershire and had been Charles's manservant in London. Emma's Aunt Jessie found him "the most amiable, obliging, active, serviceable servant that ever breathed." Once when he went with Charles to Shrewsbury,

Dr. Darwin was shocked by his "long greasy hair" and asked him in the hearing of the other servants "whether he was training to turn into my Lord Judge with a long wig." Parslow married Emma's maid Eliza in the early 1840s and she then set up as a dressmaker in a cottage in the village, where she taught needlework to girls apprenticed to her by Emma.

According to Francis, "no man ever had a truer affection for the whole family than Parslow." However, "he was a curiously simple-minded man and, if sent to buy a cow, would say the seller, 'a most respectable man, assured him it was a good cow.'" The children all remembered him as a kind friend. Francis did not remember ever being checked by him "except in being turned out of the dining-room when he wanted to lay the table for luncheon, or being stopped in some game which threatened the polish of the sideboard, of which he spoke as though it was his private property." "He had what may be called a baronial nature: he idealised everything about our modest household, and would draw a glass of beer for the postman with the air of a seneschal bestowing a cup of Malvoisie on a troubadour . . . It was good to see him on Christmas Day—with how great an air would he enter the breakfast-room and address us: 'Ladies and Gentlemen, I wish you a happy Christmas, etc. etc.' I am afraid he got but a sheepish response from us." Parslow enjoyed music; he sang tenor in the parish church choir and helped the village children with their musical performances in the schoolroom. He also won prizes for potatoes, carrots, beans and onions at the village vegetable and flower shows.

When Annie was two and three-quarters, Emma put two little combs in her hair, and Charles felt that it made her look "quite a beauty." Annie and Willy liked to play with pictures, and were allowed to cut them out from *Punch* and *The Illustrated London News,* and from other magazines, almanacs and penny chapbooks which were bought from the street-sellers who came out from London to hawk their wares from door to door. The children coloured these "scraps" with their paints and stuck them onto the stitched cloth pages of their scrapbooks. A roughly sewn album, which may have been Annie's, has a picture from *March's Penny Book of Sports* and others from an alphabet book, *Park's Cock Robin.* Charles noted when Annie was three and a half years old that he had seen her "looking at a print of a girl weeping at her mother's grave. I heard Willy say 'You are crying.' She burst out laughing and said 'No I aren't; it is only the water coming out of my eyes.' Her face was red and eyes full of tears. She seemed to wish to excite the emotion again, and went on saying 'Poor Mamma, poor Mammy.' Willy then seemed to find it rather melancholy and said 'Is her Mamma really dead? Has she got no nurse?' "

The nursery rhymes that Charles and Emma spoke and sang to the two young children had come to them by word of mouth in their own childhood, from parents, nurses and playmates. In 1842, James Halliwell published the first printed collection, *The Nursery Rhymes of England,* and Emma bought a copy. The simple timeless worth of the rhymes was not obvious to some. Halliwell had to defend them against claims that they were immoral, and did so by suggesting that they were "harmless and euphonious nonsense." But Charles was happy to acknowledge their value. Thinking about a botanical point, the distribution of plants through the dispersal of seeds, he once wrote to a friend about how a seed could be eaten, and the eater itself be eaten in turn before the seed eventually sprouted somewhere far away. "I find fish will greedily eat seeds of aquatic grasses, and that millet seed put into fish and given to stork and then voided, will germinate. So this is the nursery rhyme of 'This is the stick that beat the pig' &c &c."

Etty remembered several nursery songs her mother used to sing to the children when they were very young. "When Good King Arthur

Ruled This Land" was one, and "There was an Old Woman as I've Heard Tell" was another. Emma also had "a particular lilt for the babies when they were being joggled on her knee." She had a special trick of cutting out animals in paper. She could do bears and lions, but Etty remembered "pigs as being her *chefs d'œuvre.*"

In the family's first years at Down, Charles worked quietly in his study on his theory of the origin of species. In 1843, he met the young botanist Joseph Hooker in London, when it was arranged for Hooker to identify and describe all the plants which Charles had collected in South America and on the Galapagos. At the time, Hooker's father, Sir William, was director of the Royal Botanic Gardens at Kew. Joseph had been the naturalist on Sir James Ross's expedition to the Antarctic between 1839 and 1843, and was working on his *Flora Antarctica.* Charles found his work on plants in different habitats and locations very helpful for his own ideas about plant distribution and the flora of islands, and their bearing on the species question. He referred to Hooker as "the first authority in Europe on that grand subject, that almost key-stone of the laws of creation, geographical distribution," and asked for his views on every point on which he might be able to help. Hooker quickly became Charles's closest and most trusted friend, and his most penetrating, helpful and encouraging critic. He was to remain both for the rest of Charles's life.

In January 1844, Charles let Hooker into the secret he had been living with since coming back from the *Beagle* voyage. He wrote to him that he had for some time been engaged in "a very presumptuous work" which everyone he knew would say was "a very foolish one." He explained how he had started with his Galapagos specimens and the Patagonian fossils, and that he was now "almost convinced (quite contrary to opinion I started with) that species are not (it is like confessing

*Joseph Hooker by T. H. Maguire*

a murder) immutable . . . I think I have found out (here's presumption!) the simple way by which species become exquisitely adapted to various ends. You will now groan, and think to yourself 'On what a man have I been wasting my time in writing to.' I should, five years ago, have thought so." Hooker wrote back that he could believe there might have been "a gradual change of species" and would be interested to hear how Charles thought it might have taken place "as no presently conceived opinions satisfy me on the subject."

As his young children played around the house, Charles worked up his outline of the case for evolution by natural selection into an essay, arguing each step carefully, identifying all the strongest arguments against it, and dealing with each as best he could. Using the idea of natural laws in the way Herschel had suggested, focusing on the never-ending struggle for existence, and proposing natural selection as the mechanism for developing new species, he sketched out a simple process which explained the infinite variety and richness of the natural world and allowed for endless further developments in any direction. "There is much grandeur in looking at every existing organic being either as the lineal successor of some form now buried under thousands of feet of solid rock, or as being the co-descendant of that buried form of some more ancient and utterly lost inhabitant of this world." He suggested that it accorded with "what we know of the laws impressed by the Creator" on the physical world, that species should be developed through the operation of natural causes. We feel at first that they must each have been separately designed and created, but he argued that "there is a simple grandeur in this view of life . . . that from so simple an origin, through the selection of infinitesimal varieties, endless forms most beautiful and most wonderful have been evolved."

The range and depth of this conjecture together with the simplicity of the process which led to change were the essence of Charles's view of natural and human life, and he saw the primary laws he was hoping to work out as universal principles that could take their place in a philosophy of nature to be considered alongside Scriptural Revelation. He spoke of God's "most magnificent laws," of everything flowing "from some grand and simple laws," and suggested that "the existence of such laws should exalt our notion of the omniscient Creator."

He had still to deal with his concern about the problem of pain and suffering. In one book he had read, the Whig politician Lord Brougham had declared that he could not explain why a benevolent Creator had made a world full of misery and death, as it was "repugnant" to all our feelings and reason to suppose that God desired "the misery of all sentient beings." Henry Hallam touched on the question in his survey of European thought which Charles also looked through. Hallam wrote that he had found no answer, and concluded that "the creation of a world so full of evil must ever remain the most inscrutable of mysteries."

Now, paradoxically, by placing the struggle for existence at the heart of his theory, Charles believed he had part of an answer. Cruelty and pain were not a moral issue because they were the outcome of purely natural processes. By 1839, Charles had abandoned the idea that God had a particular providence for humans. He had then noted: "Man acts on and is acted on by the organic and inorganic agents of this earth, like every other animal." He now argued that what held for individuals held also for all species. "To marvel at the extermination of a species appears to me to be the same thing as . . . to look at illness as an ordinary event, [but] nevertheless to conclude, when the sick man dies, that his death has been caused by some unknown and violent agency." He distinguished between the natural laws ordained by God and the processes that resulted from them; he believed that God had no particular concerns about any of the individual consequences of those processes in the infinite elaboration of events. "We cease to be astonished that a group of animals should have been formed to lay their eggs in the bowels and flesh of other sensitive beings; that some animals should live by and even delight in cruelty . . . that annually there should be an incalculable waste of the pollen, eggs, and immature beings; for we see in all this the inevitable consequences of one great law, of the multiplication of organic beings not created immutable."

In taking this view, Charles implicitly rejected one of his favourite poet's main tenets, which had become one of the presumptions of the age, that nature was at heart benign. Wordsworth had claimed in his "Lines Composed a Few Miles above Tintern Abbey" that

*Nature never did betray*
*The heart that loved her; 'tis her privilege,*
*Through all the years of this our life, to lead*
*From joy to joy.*

There was clearly value in the idea for all who were able to find comfort or inspiration in experiences of the natural world, but it was not true of nature in all its workings. Charles now saw what Matthew Arnold was to write five years later:

*Nature is cruel, man is sick of blood;*
*Nature is stubborn, man would fain adore;*

*Nature is fickle, man hath need of rest;*
*Nature forgives no debt, and fears no grave;*
*Man would be mild, and with safe conscience blest.*

*Man must begin—know this—where Nature ends;*
*Nature and man can never be fast friends.*

In July 1844, when Annie was three, Charles wrote a note for Emma: "I have just finished my sketch of my species theory. If, as I believe . . . my theory is true, and if it be accepted even by one competent judge, it will be a considerable step in science. I therefore write this, in case of my sudden death, as my most solemn and last request, which I am sure that you will consider the same as if legally entered in my will, that you will devote £400 to its publication." He showed the essay to Emma and she found time to read it carefully, noting a few places where she did not understand his train of thought. She also questioned one important point in his argument. Paley in his *Natural Theology* and generations of naturalists before him had found the structure and functioning of the eye the most persuasive of all the proofs of the existence of God "from design." It was clearly the contrivance of a creative mind, and Paley claimed it was impossible to imagine how such a complex mechanism could have developed by a chance succession of small steps from an organ with another function. Charles recognised that if he was to persuade others to accept his theory, he must be able to show that the structure could have evolved in this way by a purely natural process. He wrote in the essay that this was "the greatest difficulty to the whole theory" and offered an ingenious suggestion as to how the development might have occurred. But Emma was not persuaded by his argument, and wrote in the margin "A great assumption—E.D." Charles's suggestion was, indeed, a "great assumption," and Emma knew that it was a key part of the argument for the theory which he hoped would be a major contribution to science. It was only ever a conjecture; the evidence for it was widely scattered, indirect and fragmentary. For Emma to question the point in the direct way that she did was to strike at the heart of the theory.

Charles put the essay away while he carried on with other work. He was now confident that the theory would prove to be sound, but was reluctant to publish until he was ready to cope with the close and fierce criticism to which his argument was certain to be subjected. He had to find convincing answers to the various difficulties he had identified; he needed to gather all the evidence that would be required to support his most challenging claims, and he wanted to build his scientific reputation so that when the time came, his argument would be given the careful attention it would need for any hope of a fair judgement. He planned to build up his case carefully and looked forward to announcing it in due course, but he saw no reason to hurry. Some years later he wrote to Hooker: "How awfully flat I shall feel, if when I get my notes together on species &c &c, the whole thing explodes like an empty puff-ball."

. . .

Watching Annie at three, Charles could appreciate a poem Wordsworth had written about his daughter Catherine at the same age.

> *Loving she is, and tractable, though wild;*
> *And innocence hath privilege in her*
> *To dignify arch looks and laughing eyes;*
> *And feats of cunning; and the pretty round*
> *Of trespasses, affected to provoke*
> *Mock-chastisement and partnership in play.*

In the next year the time came for Annie and Willy to learn to read. When Emma had taught at the Sunday school for the villagers at Maer, she had written some stories and had them printed in large letters as a reading book for her pupils. She now read the book with her own children, and they enjoyed the stories. One was about a boy who ate a plum pie and then lied to his mother about it. He was very sad all the next day, and at last admitted what he had done. "'But I am very sorry now, and I wish I had not taken the pie, for I knew how wrong it was at the time.' 'Well my dear boy,' said his mother, 'I am glad you have told me, and I hope you will try never to say what is not true again. I hope you will ask God to forgive you, for a lie is a sin in the sight of God.'"

Annie may also have known a story book, *Cobwebs to Catch Flies,* with which her aunt Caroline taught her cousins Sophy, Lucy and Greata to read. One theme which many books for children touched on was kindness to animals, or "humanity to the brute creation." In one story a maid let fall a drop of honey as she mixed milk for the little boy she was serving. A fly landed on the edge of the boy's bowl to suck the honey. "The good child laid aside his spoon to avoid frightening the poor fly. His mamma asked: 'What is the matter, William? Are you not hungry?' 'Yes, mamma; but I would not hinder this little fly from getting his breakfast.' 'Good child,' said his mamma, rising from her tea, 'we will look at him as he eats. See how he sucks through his long tube. How pleased he is.'" Later, she told the boy about insects which disguise themselves to escape the dangers which they meet. "She picked up a wood-louse, and laid it gently on his little hand. 'There,' said she, 'you see the wood-louse roll itself into a little ball, like a pea: let it lie awhile, and when it thinks you do not observe it—' 'Ah! mamma, it unrolls. Oh! it will run away; shall I not hold it?' 'No, my dear, you would hurt it.' 'I would not hurt any creature, mamma.' 'No, surely. He who made you, made all creatures to be happy.'"

When Emma was away staying with her parents, Charles wrote to her about each day's happenings. One morning he did hardly any work and was "much overcome" by the children. "The day was so thick and wet a fog, that none of them went out." John Lewis, the village builder, came to mend the plumbing and paper the water closet. Willy, Annie and Bessy the nursemaid were there, and the one-year-old Etty "insisted on going in, I dare say, greatly to the disturbance of Bessy's delicacy."

Willy was good-natured as an elder brother, but Annie got caught in the rough and tumble. "Poor Annie has had a baddish knock by Willy's ball in her eye—it is swelled a bit, but not otherwise bad." A few days later, "Willy told me to tell you that he had been very good and had given Annie only one tiny knock." Charles felt "It is really wonderful how good and quiet the children have been; sitting quite still during two or three visits, conversing about everything, and much about you and your return. When I said 'I shall jump for joy, when I hear the dinner bell,' Willy said 'I know when you will jump much more—when Mamma come home.' 'And so shall I' responded many times Annie." Charles ended one letter: "Played with children till 6 o'clock; read again and now have nothing to do, but most heartily wish you back again."

Charles told Emma that Etty had been neglecting him "and would not play. She could not eat any jam, because she had eaten so much at tea . . . She was rather fidgety, going in and out of the room, and Brodie declares she was looking for you. I did not believe it, but when she was sitting on my knee afterwards and was looking eagerly at pictures, I said, 'Where is poor Mama?' She *instantaneously* pushed herself off, trotted straight to the door, and then to the green door, saying 'Kitch'; and then Brodie let her through, when she trotted in, looked all around her and began to cry; but some coffee-grains quite comforted her."

Emma was always easygoing with her young children. The Darwins' neighbour, Louisa Nash, remembered her saying that she had never thwarted them needlessly, but would say: "You seem to care very much about so and so, and I don't care at all, and when you are older you won't care for it either, so you may have it now." She would often use small bribes to get over "small childish difficulties." According to her grandson Bernard, "Nobody could have been firmer on what she deemed to be matters of principle, and the most bold-faced child could never have taken a liberty with her; but she had a wide-minded sense of proportion, and in what she deemed immaterial things, she believed in corruption rather than coercion." Willy, Annie and Etty were obedient children, "and anything like deliberate disobedience may be said to have never entered our heads. The rules of life were very simple, and when anything could be explained to us it was, and even when it could not, we never questioned the absoluteness of a definite command."

Emma cared little about neatness or order in the household. When she was young, she had been "Little Miss Slip-slop" to her elder sister's "Mrs Pedigree." Her Aunt Jessie had expected her to "lark it through life" as she had a "pretty gaiety" about her, "always ready to answer to any liveliness and sometimes to throw it out herself." But Etty remembered her as "serene but somewhat grave. The jokes and the merriment would all come from my father." Emma always disliked displays of emotion but at some time in the 1830s or 1840s, perhaps with the strain of looking after her sick mother, then Charles and then her children, she developed the deep restraint and determined self-sufficiency which she kept for the rest of her life. There was "a certain reserved gravity in her expression" which acquaintances sometimes "strangely misunderstood" before she spoke. She talked herself once of her "foolish habit of not looking in people's faces." She avoided exuberant expression; Etty wrote that "Simplicity, even bareness of manner, was more to her taste." She was also "calm over music, deeply as she enjoyed it."

Charles played often with the children when they were small. He liked to hold them, and he liked them to be close and affectionate to him. He also, though, had to cope with their furies, and comparisons came to mind when he did so. He wrote many years later in *The Expression of the Emotions in Man and Animals*: "Everyone who has had much to do with young children must have seen how naturally they take to biting when in a passion. It seems as instinctive in them as in young crocodiles who snap their little jaws as soon as they emerge from the egg."

Etty remembered her father as "the most delightful play-fellow" to Annie and herself. He had a number of games which she specially loved. "One was called 'Taglioni,' a sort of opera dancing on his knees." The Taglionis were a celebrated family of ballet dancers and a Taglioni was a kind of dancer's dress. Charles also laid Etty on his knees and drummed on her with "a large voice for the big drums and a little voice for the little drums." Her oddest enjoyment was to lie on the floor and have "a mysterious quaking sensation produced by his gently shaking us with his foot." He would also tickle her knees and chant to a monotonous tune, "If you be a fair lady as I do hope you be, then you will not laugh at the tickling of your knee." He had a story about an old woman with nine little pigs, and there were other games of "arms that flew about, and crocodiles which were 'very naughty beasts,' and worms that crawled in and worms that crawled out." According to Francis, his father's body was very hairy; the children would put their hands inside his shirt "and he would growl like a bear at us." After Charles's rest every day, he used to have his back rubbed, and this had a pleasant or comforting effect on him. One of Francis's earliest recollections was of beating or patting his father's back all over in time to silent tunes.

When Francis was a small boy and played soldiers with his elder brother, he was a private while George was a sergeant, and "it was part of my duty to stand sentry at the far end of the kitchen-garden until released by a bugle-call from the lawn. I have a vague remembrance of presenting my fixed bayonet at my father to ward off a kiss which seemed to me inconsistent with my military duties."

Charles loved good tunes but had difficulty remembering them. Francis only ever heard him hum one, the beautiful slow Welsh hymn "All Through the Night," which he had probably first heard as a small child when his mother took him to services at the Unitarian chapel in Shrewsbury. He remembered another little tune from another world, a song that a Tahitian girl had sung to him when he landed there on HMS *Beagle* in 1835. He wrote in his diary at the time that "Numbers of children were playing on the beach, and had lighted bonfires which illuminated the placid sea and surrounding trees; others in circles were singing Tahitian verses. We seated ourselves on the sand and joined the circle. The songs were impromptu and I believe relating to our arrival. One little girl sang a line which the rest took up in parts, forming a very pretty chorus. The air was singular and their voices melodious. The whole scene made us unequivocally aware that we were seated on the shores of an island in the South Sea."

Charles used the pictures in his scientific books to entertain his children. Etty remembered an old book of animals which they called the "monkey book." "He had a particular little story which must never be hurried belonging to most of the pictures." Francis recalled the colour plates in Andrew Smith's *Illustrations of the Zoology of South Africa*. Smith was an Army doctor and explorer with whom Charles had struck up a friendship when he met him in South Africa in 1836, on the last leg of the *Beagle* voyage. Charles made an exciting game of looking through the coloured plates "by the supposition that the birds belonged alternately to himself and the child who was his playfellow." Francis remembered vividly, "after a series of dull thrush-like birds had been calmly shared between my father and myself, the agony of seeing a magnificent green and purple one fall to his lot. I am sure he tried to cheat himself, but this was not always possible."

The magnificent bird was *Lamprotornis Burchellii,* most of whose upper surfaces were, in Smith's meticulous description, "dark duck-green with a splendid metallic lustre; the sides of the head pansy-purple, many of the feathers tipped with brilliant shining purplish red passing into flame red." The two middle feathers of the long barred tail were "bronzed purple, deadened by a gloss of green." Smith wrote: "If it be essential, in order to carry out the plan of the Creator, that certain birds should be provided with longer and more weighty tails than others . . . it will also be necessary that provision should be made to ensure them against injury or inconvenience from such [an] arrangement." The wing feathers of the *Lamprotornis* were greatly developed in order to help it fly with its lengthy tail. Whatever Charles told his children about this bird would not have included "the plan of the Creator."

CHAPTER FIVE

# THE GALLOPING TUNE

*Brothers and sisters—Out of doors—Uncle Ras and Joseph Hooker—Reading—The Governess—Annie's writing case*

AS THE YOUNGER CHILDREN GREW, they joined Willy and Annie in their play, and life in the household grew more hectic. In many wealthy homes at the time, the children were not allowed to play in the rooms used by the grown-ups and were kept in the nursery for much of the day. But at Down, according to Annie's cousin Snow Wedgwood, "the only place where you might be sure of not meeting a child was the nursery." Francis remembered having their evening meal there while Charles and Emma dined in peace. "We came down after our tea, rushing along the dark passage and descending the stairs with that rhythmic series of bangs peculiar to children . . . I have also a faint recollection of black-coated uncles sitting by the fire and not unnaturally objecting to our making short-cuts across their legs. It was no doubt a pity that we were not reproved for our want of consideration for the elderly, and that, generally speaking, our manners were neglected. One of our grown-up cousins was reported to have called our midday dinner 'a violent luncheon,' and I do not doubt that she was right."

The furniture and decoration in the family's living rooms felt the wear of the young family's life. Charles wrote once to Emma: "The children are growing so quite out of all rule in the drawing room, jumping on everything and butting like young bulls at every chair and sofa, that I am going to have the dining room fire lighted tomorrow and keep them out of the drawing room. I declare a month of such wear would spoil every thing." Louisa Nash remembered talking once with Emma about bringing up children. Emma said, "When we were young, Charles and I talked over together what we should do. The house was newly and expensively furnished. Shall we make the furniture a bugbear to the children, or shall we let them use it in their plays?" They agreed that they would not worry about things getting shabby. "So chairs and other furniture used to get piled up for railways and coaches, just as the fancy took them." And then she added, "I believe we have all been much the happier in consequence."

Etty remembered how the drawing-room furniture was pushed to one side, and a troop of little children danced and leapt round the room while Emma played "the galloping tune" which she had composed herself and was "very well suited for its purpose." Etty was fond of dressing up, especially when a Wedgwood cousin was with them. "Our plan was to ask my mother for the key of her jewel box, a simple wooden box in which her jewels, pearls and all, rattled about loose, with no cotton-wool to protect them. The key, too, worked badly, and we had to shake and bang the box violently to get in. Then we locked her bedroom doors to prevent the maids coming in and laughing at us, took out of the wardrobe her long skirts and pinned them round our waists. Out of her lace drawer, we fitted up our bodies with lace fallals, put on the jewels, and then peacocked about the room trailing the silks and satins on the floor. A favourite costume was a silver-grey moire-antique. When we had done, we hung up the gowns, put back the lace, and locked up the jewels, and returned the key, but she never looked to see whether the two little girls had lost or damaged any of the jewels."

George's earliest memory of childhood was of drumming with his spoon and fork on the nursery table because dinner was late, while a barrel organ played in the lane outside. He remembered his father's study as "a sort of sacred place not to be invaded in the morning with-

*George's recollection of the house before 1858*

out some really urgent cause, such as the absolute necessity for a piece of string or a foot-rule. We always were received with the utmost kindness, and it was only in the extreme case of three or four interruptions in half an hour that we were cautioned: 'You really mustn't come again.' The cutting of fingers was one of the urgent causes for which we went there to fetch sticking plaster. I always felt this a very serious affair and Henrietta, when in that predicament, used to wait until he had gone out for his walk and then purloin the plaster. The reason why this was so serious a matter was twofold; first that his sympathy was so strong with us when we hurt ourselves, and next that he had a morbid horror of the sight or even the word of blood."

Charles's microscope stool was mounted on castors. George remembered how he would often take it from the study, sitting on it with Etty and punting them round the drawing room with a walking stick. He wrote: "However hard my father was at work, we certainly never restrained ourselves in our romps about the house, and I should certainly have thought that the howls and screams must have been a great annoyance; but we were never stopped. There was one fearfully noisy game which invaded the whole house, called 'roundabouts,' and we generally played at this when there was a house full of cousins. It was a modified hide and seek and necessitated yells from all the players to tell where the demon of the game was."

When George's daughter Gwen came to the house in the 1890s, the nursery was still a place of quiet. She wrote in her memoir of her childhood, *Period Piece,* that the room "had a white painted floor; it had green venetian blinds, too, and a great old mulberry tree grew right up against the windows outside. The shadows of the leaves used to shift about on the white floor, and you could hear the plop of the ripe mulberries as they fell to the ground, and the blackbirds sang there in the early mornings. They lived permanently in the tree in the fruit season. I used to get out of bed to listen to them before anyone else was awake."

Annie spent as much time as she could playing in rough clothes with her brothers and sisters out of doors. They had the run of the sixteen acres. Around the house were the flower garden and lawn, the two yew trees with a child's swing hung between them, the mulberry tree, the deep well for the household's drinking water, and the orchard with apple trees, pears trained on the walls, quinces and plums. The kitchen garden was lined with gooseberry bushes, and there were cows, pigs, horses, chickens and geese in the cow yard, stables and enclosures for livestock. Farmyard cats kept down the mice. Etty "cared for all the animals about the place, seeing the cows milked, rushing past the heels of one hornless yellow cow who it was supposed would kick us if she could, grooming the donkey with worn out brushes and combs, and taming the chickens so that they would eat out of my hand." She took possession of a small garden shed just beyond the mulberry tree and made it her "little home with flower pots for seats and broken bits of crockery for china. Here my cats sometimes kept their kittens and I would sit for long hours watching them."

Charles once remarked that "Children have an uncommon pleasure in hiding themselves and skulking about in shrubbery when other people are about." Thinking of our animal ancestry, he commented: "This is analogous to young pigs hiding themselves, and [is the] hereditary remains of savages' state."

William Brooks, the cowman and gardener, lived in a cottage close to the cow yard with his wife Keziah, who was the best smocker in the village. Their youngest daughter Emily was Annie's age and they were playmates. Joseph Comfort, the second gardener, helped with the cows and pigs and drove the family's two carriages. Another under-gardener, Henry Lettington, was a great friend to Francis. "It was he who taught me to make whistles in the spring and helped me with my tame rabbits. He also showed me how to make brick traps for small birds, and a more elaborate trap made of hazel twigs."

Gwen remembered being taken to gather nosegays for the house, "down the long pebbled walk between the tall syringa and lilac bushes all wet with dew, to the kitchen garden, where the roses were imprisoned behind high box borders, near the empty greenhouses, where my grandfather had once worked. We took the wooden trug full of flowers, which smelt sweeter than any other flowers in the world, back to the house, and arranged them in water on a green iron table, in the Old Study." Long after she had forgotten all her human loves, she wrote, "I shall still remember the smell of a gooseberry leaf, or the feel of the wet grass on my bare feet, or the pebbles in the path."

Willy and Annie played with the children of Sir John Lubbock, the Whig banker, mathematician and astronomer who was squire of the parish and lived at High Elms, the grand house on the other side of the village. Sir John supported a number of "progressive" causes, and was a council member of the Society for the Diffusion of Useful Knowledge. One day when Annie was six, the Lubbocks gave a party for their neighbours' children. Thirty came, the Darwins among them. Lady Lubbock noted in her diary: "Jugglers and tumblers performed in the garden. Then they danced; Mr Taylor on the Piano Forte and a post horn was the Music."

The Lubbocks' eldest son John was seven years older than Annie, and a schoolboy at Eton. In the holidays he would ride round the neighbourhood with his gun, shooting squirrels, rabbits and wrynecks. He was also a keen collector; he netted small water creatures in the village pond and searched for fossils in quarries nearby. Charles was glad to help him with his natural history; they walked together and John came often to the house for lessons in dissecting and describing his specimens. Annie and Etty became used to seeing John's horse tethered in the stable yard; he found Charles's friendly advice and encouragement over the microscope easier to respond to than his austere father's impatience when he could not grasp abstruse mathematical ideas.

Annie was given a small garden of her own to grow flowers and vegetables in as she wanted. Girls were encouraged to take a special interest in flowers, as they would always be the ladies' responsibility when they grew up. Young ladies were supposed to leave the heavy tasks to a labourer, but a child could borrow small tools and dig and heave as she wanted. John Lubbock grew lettuces in his plot at High Elms. His mother may have mentioned to Annie a verse she copied into her commonplace book.

*It cometh forth in April showers,*
*Lies snug when storms prevail;*
*It feeds on fruits; it sleeps on flowers;*
*I would I were a snail!*

When Annie was nine, John, who was later to become a leading statesman and populariser of science, gave the first of a lifetime of public lectures on natural history, speaking in Downe village hall about the habits of the wireworm. This was "useful knowledge" for the Darwin children and all gardeners in the neighbourhood, as the small yellow grubs were, with slugs, the most voracious of all pests, devouring carnations, pinks, irises, lobelias, dahlias and almost all vegetables. The creatures were also a small problem for natural theology. A reading book for parish schools commented: "No doubt the wireworm fulfils some important and useful part in the economy of nature, but we have not been able to find any aspect in which it can be said to be otherwise than injurious to the farmer." The only solution was to destroy it by burning.

Beyond the Darwins' garden were the hay-meadow and the Sand-walk copse. "Of all places at Down," Gwen wrote in *Period Piece,* "the Sand-walk seemed most to belong to my grandfather. It was a path running round a little wood which he had planted himself; and it always seemed to be a very long way from the house. You went right to the furthest end of the kitchen garden, and then through a wooden door in the high hedge, which quite cut you off from human society. Here a fenced path ran along between two great lonely meadows, till you came to the wood. The path ran straight down the outside of the wood—the Light Side—till it came to a summer house at the far end; it was very lonely there; to this day you cannot see a single building anywhere, only woods and valleys." Gwen remembered that faint chalk drawings of dragoons could be made out on the wooden walls of the summer house. They had been drawn by her father, George, and Francis when they were small children.

For Gwen, the Light Side of the Sand-walk copse was "ominous and solitary enough, but at the summer house the path turned back and made a loop down the Dark Side, a mossy path, all among the trees; and that was truly terrifying." There were two or three great old trees beside the path which were "alright if some grown-up person were there, but much too impressive if one were alone. The Hollow Ash was mysterious enough; but the enormous beech, which we called the Elephant Tree, was quite awful. It had something like the head of a monstrous beast growing out of the trunk, where a branch had been cut off. I tried to think it merely grotesque and rather funny, in the daytime; but if I were alone near it, or sometimes in bed at night, the face grew and grew until it became the mask of a brutish ogre, huge, evil and prehistoric; a face which chased me down long dark passages and never quite caught me."

All Charles's children remembered their father on the Sand-walk. He had laid it out as his "thinking path" and he used to walk round it five times every day at noon. The undergrowth and old clay-pits were also the children's favourite playground, and they would often spend the morning there. As their father paced slowly along the path, he was usually deep in thought, but he also liked to see what they were doing and "sympathised in any fun that was going on." Francis remembered that "he walked with a swinging walk using a walking stick heavily shod with iron which he struck loudly against the ground." The rhythmic click "became a familiar sound that spoke of his presence near us," and it haunted the Sand-walk for them long after his death.

Emma was courageous, even rash, in what she let her children do on their own outside the garden, and Charles did not interfere. They "wandered about the lonely woods and lanes in a way that was not very safe." "We used also to run down the steep ploughed fields, our feet grown with adhering clay to huge balls swinging like pendulums and scattering showers of mud on all sides. Then we would come cheerfully home, entering by the back door and taking off our boots as we sat on the kitchen stairs in semi-darkness and surrounded by pleasant culinary smells . . . When we used to take long winter tramps along our flinty winding lanes, this unbooting on the back stairs was a prelude to eating oranges in the dining room, a feast that took the place of five o'clock tea—not then invented."

Charles would sometimes take Willy or Annie or one of the younger children on his walks out into the countryside in the dusk of the early morning. Francis remembered "a vague sense of the red of the winter sunrise, and . . . the pleasant companionship, and a certain honour and glory in it." George remembered his father saying that sometimes in the woods in the early dawn, "he would walk very slowly, just quietly putting down his foot and then waiting before the next step—a habit he said which he had practised in the tropical forests of Brazil. In this way, he used to see many interesting things in animal life; once he watched a vixen playing with her cubs at only a few feet distance, for some time."

Later in the day his children would often walk with him down the hill to the Big Woods, and do a little collecting as they went. "He seemed to know nearly all the beetles and was immensely interested when any of the rarer sort were found." As he thought about evolution and adaptation in nature, he found points to study in the wild orchids of the neighbourhood and their intricate arrangements for pollination by the insects that fed on their nectar. Etty remembered that her father had "a kind of sacred feeling" about the orchids they found, and taught all his children "to have this peculiar feeling." A special place above the quiet Cudham Valley where fly and musk orchids grew among the junipers, and birds-nest orchids flowered under the beech boughs, was known in the family as "Orchis Bank."

In a book about country pursuits read by the children, the author suggested that "While there are boys and birds-nests, there will always be birds-nesting . . . It is an instinct, a second nature, a part and parcel of the very constitution of a lad." Charles had the instinct from childhood. Francis remembered that he "always found birds' nests even up to the last years of his life, and we, as children, considered that he had a special genius in this direction. In his quiet prowls he came across the less common birds, but I fancy he used to conceal it from me as a little boy, because he observed the agony of mind which I endured at not having seen the siskin or goldfinch or whatever it may have been."

For the children, the main pattern of life was set by the seasons. Francis, as a child, found "something impressive and almost sacred" in the changes. The onset of winter was heralded for him by "the appearance of puddles frozen to a shining white; mysterious because the frost had drunk them dry in roofing them with ice, and especially delightful in the sharp crackling sound they gave when trodden on." A "wicked groaning crack . . . ran round the solitary pond on which we skated, as it unwillingly settled down to bear us on its surface. It had a threat in it, and reminded us how helpless we were, that the pond-spirit was our master and had our lives in its grip. Another winter sound was the hooting of invisible owls, boldly calling to each other from one moonlit tree to another." At Christmas, ivy, holly, yew and other evergreens were brought inside to decorate the hearths.

In the spring, Francis remembered "the querulous sound of the lambs, staggering half-fledged in the cold fields among the half-eaten turnips beside their dirty yellow mothers." On May Day the village children carried cherry boughs tied with bunches of flowers as they went from house to house singing and collecting pennies for the May Pole. After mid-summer the Darwin children would wake early one morning and learn from the sound of scythes being whetted that the mowers had arrived to harvest the hay. "The field had been a great sea of tall grasses, pink with sorrel and white with dog-daisies, a sacred sea into which we might not enter. But now we could at least follow the mowers, and watch the growth of the tracks made by their shifting feet, and listen to the swish of the scythes as the swathes of fallen grass and flowers also grew in length. There was something military in their rhythm, and something relentless and machine-like in their persistence." Francis rode in the hay-cart and watched as the sea-green stack grew mysteriously in the corner of the field. "The inside of the hay-cart was enchantingly polished, and also full of hay-seed."

After the hay harvest, high summer meant for Etty "the rattle of the fly-wheel of the well, drawing water for the garden; the lawn burnt brown, the garden a blaze of colour, the six oblong beds in front of the drawing room windows, with phloxes, lilies and larkspurs in the middle, and portulacas, gazanias and other low-growing plants in front; the row of lime-trees humming with bees, my father lying on the grass under them; the children playing about, with probably a kitten and a dog, and my mother dressed in lilac muslin, wondering why the black-caps did not here sing the same song as they did at Maer."

With few friends in the village, the children loved visitors. When Charles's older brother Erasmus came, they attached themselves to him all day and he drew demons and imps for them. "Uncle Ras" was very tall and slight, and his movements had a languid grace. Etty remembered his "long thin hands which were wonderfully clever and neat in all practical handiwork." His face "lit up when he spoke, from a habitually patient and sad expression." He was a lonely and unhappy man who endured chronic ill-health with opium and self-deprecating wit. Francis held him in great affection. Jane Carlyle enjoyed Erasmus's sparse, sardonic comments, and her husband Thomas called him "a most diverse kind of mortal . . . He has something of [the] original and sarcastically ingenious in him; one of the sincerest, naturally truest, and most modest of men."

Erasmus was close to Hensleigh and Fanny Wedgwood and their children. They kept a number of the drawings he made for them. One for Alfred, who was a year younger than Annie, showed monkeys, parrots and a giraffe at the zoo, with a snake, a lion looking threateningly at a grinning child in a plumed hat, and a strange toothed bird flying past.

Erasmus once wrote to Alfred: "I have nobody to play with, so I hope very soon to see you again when you have done travelling about the country. What a great many places you have been to . . . you will be able to tell me very long stories indeed—one of those nice stories without any end to them." Snow Wedgwood, Alfred's elder sister, wrote many years later of their uncle's "quaint, delicate humour, the superficial intolerance, the deep springs of pity, the peculiar mixture of something pathetic with a sort of gay scorn, entirely remote from contempt."

Annie's favourite was Joseph Hooker the young botanist. Once in the early years, he sent a cake from London for the family. When Charles wrote to thank him, he mentioned that Annie wanted to know whether it was "the gentleman what played with us so." Hooker would come to stay for a week at a stretch, bringing his own work, so that Charles could put questions to him whenever he wanted. Hooker told William many years later that Emma did everything to make him feel at home. "Often I worked in the dining room . . . through which your mother often passed on her way to the store closet in the end, when she would take a pear, or some good thing, and lay it by my side with a charming smile as she passed out. Then in the evening she always played to me, and sometimes asked me to whistle to her accompaniment of some simple air!" He added, "There were long walks, romps with the children on hands and knees, music that haunts me still." Francis remembered the children eating gooseberries with him in the kitchen garden. "The love of gooseberries was a bond between us which had no existence in the case of our uncles, who either ate no gooseberries or preferred to do so in solitude."

For everyone in the family, reading and being read to was an essential part of everyday life, and there were books, periodicals and papers of different kinds in almost every room. Long letters came from sisters,

*Drawing by Erasmus Alvey Darwin for Alfred Wedgwood*

brothers, cousins and friends; they were read aloud and passed round. Charles and Emma read the *Edinburgh Review,* the *Quarterly Review* and the *Athenaeum*. Charles was often unhappy with *The Times* but could not manage without it. He once commented that it was "getting more detestable (but that is too weak a word) than ever. My good wife wishes to give it up, but I tell her that is a pitch of heroism to which only a woman is equal."

Charles and Emma listed books to look at and noted what they thought of them. They bought books regularly from their old bookseller in London; they borrowed popular novels and books of travels from Mudie's Circulating Library in New Oxford Street; they also took books out from the London Library in St. James's Square, and they bought cheap "yellowbacks," the paperbacks of the Victorian age, at the bookstall at London Bridge Station. They read Dickens's novels and shared them with the children. Etty was nicknamed "Trotty Veck" after the old ticket-porter in *The Chimes,* who "trotted everywhere" and "loved to earn his money."

Much of Annie's reading was moral. The children had Maria Edgeworth's *Popular Tales* with their rational, liberal and gently put lessons. The best-known story was "The Purple Jar" about a thoughtless girl who pressed her mother to give her the beautiful jar which she had seen in an apothecary's window, rather than a new pair of shoes. She found that the purple colour was in the water, not the glass, and her shoes wore out. "Oh mamma," she said at the end, "how I wish that I had chosen the shoes—they would have been of so much more use to me than that jar; however, I am sure—no, not quite sure—but, I hope, I shall be wiser another time."

Annie read *Gulliver's Travels,* the *Arabian Nights, The Last of the Mohicans* and *The Children of the New Forest.* Emma had a small book of Madame d'Aulnoy's *Fairy Tales* from her childhood, and the family had a taste for magic and enchantment which would have surprised many for whom Wedgwood was a by-word for seriousness, sobriety and reserve. Emma's brother Harry, who had trained as a barrister but lived on his inheritance, wrote stories which were passed round the family. One tale of Eastern magic he called *The Bird Talisman.* Emma read it to her children and grandchildren, and our families have known and loved it ever since.

"There once was an old hermit, who lived in a hut near the source of the Ganges. He was very kind to all birds and beasts; and they were so accustomed to him that the very wild beasts were neither afraid of him nor would hurt him. One day, as he sat by the stream watching two daws that were flying about and playing together in the air, one of the birds happened to fall into the water, which was very rapid, and was swept away by the stream, and would have been drowned if the old hermit had not run to its help, and, stepping into the water, pulled out the daw with his hooked staff. He laid the bird in the sun, and as soon as it was dry the two daws both flew away to a high rock, just above where the Ganges rises. The hermit saw them fly into a little cave, halfway up the rock, and presently come out again, and fly back towards him; they alighted close to him, and one of them laid a ring down at his feet. He picked it up and put it on his finger, and he was immediately astonished to hear the daw speak to him and say . . ."

While Uncle Harry told his tales of oriental magic, Charles was exploring another world, real but also vividly imagined, with Annie, Willy and Etty. During the 1840s, Charles lived in fear that his wealth might be destroyed in a financial crisis. Many people at the time thought of emigration, and parents would read books like Captain Marryat's *The Settlers in Canada* with their children as a way of thinking through the idea with them. In 1850, Charles wrote to his former servant Syms Covington who was then settled in Australia: "You have an immense, incalculable advantage in living in a country in which your children are sure to get on if industrious. I assure you that, though I am a rich man, when I think of the future I very often ardently wish I was settled in one of our colonies, for I have now four sons (seven children in all, and more coming), and what on earth to bring them up to I don't know. A young man may here slave for years in any profession and not make a penny. Many people think that Californian gold will half ruin all those who live on the interest of accumulated gold or capital, and if that does happen I will certainly emigrate." He wrote to his cousin Fox around the same time that he fancied most "the middle states of North America."

Charles read in *The Emigrant's Manual* about "those vast districts of prairie and woodland watered by the Mississippi, Missouri, and Ohio, and their tributaries." "The attraction of the prairie consists in its extent, its carpet of verdure and flowers, its undulating surface, its groves, and the fringe of timber by which it is surrounded . . . Although Ohio has already become so populous, it is surprising to the traveller to observe what an amount of forest is yet unsubdued."

Charles bought a book for children, *Our Cousins in Ohio,* which gave a vivid impression of the life the family might be able to enjoy there. The boy and girl in the book, Willie and Florence, lived with their parents in a homestead on the banks of the Ohio River. The author, Mary Howitt, a London Quaker, based their story on letters from her sister who had emigrated to America and settled in Athens, Ohio. The homestead was a large house with grounds just like Down House, but the surroundings were described as an earthly paradise on the edge of a wilderness. Beyond the family's pasture and a neighbour's wood was an "unbroken portion of the primeval forest . . . left uncleared . . . for fire-wood." "Here grew hickory, maple, birch, and walnut trees—the splendid American linden—the red-bud or Judas tree, by the budding of which the Indian, in olden time, regulated the sowing of his corn. The wild clematis, sweet-briar, and American hawthorn, were among its abundant undergrowth; and here also were thickets of blackberries, which produced such splendid fruit as English children can form no idea of." In a wood near their home the children found "the toothwort, a beautiful waxen-like flower with extremely elegant leaves; the snake-wort and the poke-weed; and here rabbits, chippy-monks or ground-squirrels, and snakes, abounded." Annie knew the strange albino toothwort growing in the undergrowth of the Sand-walk copse, but some of the other plants she could only dream about. "In passing one deep-wooded hollow, the fragrant pine-apple odour of the pawpaw came so strong and rich that they stopped the carriage and Willie and his father went down to gather them." Other plants, though familiar to Annie in the garden at Down, sounded magical in the wilds of Ohio. "Here and there, the trunks and arms of gigantic dead trees would be wreathed with Virginian creepers of the most intense scarlet, often starting forth from the very thickest of the forest, like a fantastic scarlet tower."

Mary Howitt gave a full account of the good life for the settlers in Ohio, and the harder lot of the "movers"—farmers from Indiana with loads of wheat to sell in the city, and others moving back after failure. She also wrote sympathetically about the many former slaves who had bought their freedom or escaped along the Ohio from the slave states to the south, and had settled in shanty towns around the towns and homesteads.

When Charles noted that he had read the book, he referred to it as "life in Ohio." It was another life to talk about with the children.

Charles and Emma had also to think of formal education for their sons and daughters. Both Darwins and Wedgwoods had a long-standing interest in advanced approaches. Charles's grandfather Erasmus had been interested in Jean-Jacques Rousseau's revolutionary ideas in *Émile* and had sought him out when David Hume brought him to England in 1766 to escape from his enemies. Rousseau argued that children were naturally good, against the orthodox Christian view that they were corrupted by sin from birth. He suggested that we should "love childhood; promote its games, its pleasures, its amiable instinct." Children should be taught through their own experience rather than by instruction. "Nature wants children to be children before becoming men." Education in what he called the "age of nature," up to twelve, should be "purely negative," not teaching virtue or truth but allowing the child to run free in the country and securing his natural goodness and understanding from vice and error. Rousseau had many practical suggestions: one was that if a child was afraid of the dark you should not try to reason with him, but should help him to conquer his fear with games. Build a labyrinth of tables, chairs and screens; hide little prizes in it, and "let your child be laughing as he enters the dark; let laughter overtake him again before he leaves it."

Erasmus Darwin read *Émile* with many of his scientific friends, and when his daughters asked him to sketch out a plan for a girls' boarding school they set up in 1794, the scheme he offered reflected a number of Rousseau's themes. In one passage which Charles quoted in an essay he wrote about his grandfather, Erasmus suggested that "a sympathy with the pains and pleasures of others is the foundation of all our social virtues," and it could best be encouraged by example and "expressing our own sympathy."

The Wedgwoods were more serious and analytical. Charles and Emma's uncle Tom Wedgwood had met the radical philosopher William Godwin in 1797 and explained to him his proposal for a "nursery for genius" in which infants would be sheltered from the "chaos of perceptions" in normal life, so that the first sense-experiences could be simplified and rendered intense. Tom drew his ideas from the associationist psychology of David Hartley. The children would be kept in rooms with plain grey walls and offered "one or two vivid objects" from time to time to see and touch. Mutual friends had told Tom that two young men, William Wordsworth and Samuel Taylor Coleridge, who were living in Devon at the time, were "disengaged" and might be suitable as superintendents of the nursery. Tom got to know them both well but they were working on their *Lyrical Ballads,* and were not prepared to help in the experiment.

The next year, Tom's brother Josiah II, Emma's father, drew up a plan for his governess to follow with his children. He had no time for Tom's theories, but based his scheme firmly on Rousseau's sensibility and his practical approach. Emma, her brothers and her sisters were "allowed to act in an unrestrained manner without rules and precepts," as their father felt that "every act of interference does harm" to a child's nature.

"The children may be taught to exercise their faculties by inducing them to answer their own questions, either experimentally, or by having the subject so presented to them that the inference shall be sufficiently clear without its being drawn for them." One comment of Josiah II's was characteristic in its relaxed tone and quiet challenge. "Toads, newts, lizards, frogs, beetles, worms &c are innocent, and it is adding very unnecessarily to the evils of life to teach a child to consider them as disgusting objects." Josiah II noticed that Elizabeth had a fear of wolves in the deep evening shadows of a long room upstairs, and wrote cheerfully: "Try Rousseau's plan of sports in the dark." A few years later, when Emma was ten, a visitor observed the effects of "the Wedgwood education." Emma and her sister Fanny were "happy, gay, amiable, sensible, and . . . not particularly energetic in learning."

At the time, Charles was being brought up by his elder sisters Caroline and Susan. They admired Rousseau's Swiss follower, Johann Heinrich Pestalozzi, who had pioneered teaching children through participation and discovery. "The child must be led to see for himself that which he is to learn, and not to take it upon the mere authority of the teacher." When Caroline and Susan set up an infant school for the children of the poor neighbourhood near their house in Shrewsbury, they equipped it with the toys and learning apparatus that Pestalozzi recommended.

When Charles and Emma had their own children to bring up, they read *Levana,* a well-known book by another follower of Rousseau, the German Romantic novelist Jean Paul Friedrich Richter. He wrote about childhood with a sensitivity and appreciation equal to Rousseau's. In a chapter on "the joyousness of children," he asked: "Should they have anything else?" He suggested that "Play is the first poetry of the human being . . . Never forget that the games of children with inanimate playthings are so important because for them there are only living things." A parent who has a piano should call his children together "and let them every day for an hour hop and turn by his playing, in pairs, in rows, in circles . . . and always in any way they like. In the child happiness dances; in the man, at most, it only smiles or weeps." On quarrels and argument, "Never let the contest of parental and childish obstinacy take place." And on discipline, "A serious punishment of a child is scarcely so important as the quarter of an hour immediately succeeding, and the transition to forgiveness."

Charles and Emma's approach with their children was undemanding and liberal; they saw little value in discipline and learning by rote, but wanted to encourage their children to think for themselves. Etty remembered "rejoicing in this sense of freedom." "Our father and mother would not even wish to know what we were doing or thinking unless we wished to tell." But if one of the children did want to tell, Charles would make them feel that their opinions and thoughts were valuable to him. "He cared for all our pursuits and interests, and lived our lives with us in a way that very few fathers do . . . He always put his whole mind into answering any of our questions."

Charles was eager that his children should share his interests but felt strongly that he should not force them to. George wrote that "he never tried to make us take an interest in science . . . When however we freely exhibited any wish to learn, there was no amount of trouble which he would not take, and the result was of course far more powerful than if it had been at his urging." Charles once commented to a friend that giving specimens to children in order to give them a taste for natural history "would tend to destroy such taste. Youngsters must themselves be collectors to acquire a taste; and if I had a collection of English *Lepidoptera,* I would be systematically most miserly and not give my boys half-a-dozen butterflies in the year." But Francis remembered "the pleasure of turning out my bottle of dead beetles for my father to name, and the excitement, in which he fully shared, when any of them proved to be uncommon ones." Charles wrote to Fox about

Francis's collecting: "My blood boiled with old ardour when he caught a *Licinus*—a prize unknown to me." Another time he wrote: "I feel like an old war-horse at the sound of the trumpet, when I read about the capturing of rare beetles. Is this not a magnanimous simile for a decayed entomologist?"

Charles explained to Etty and George how the steam engine worked, and at one time when George was eight or nine, Charles read Jane Marcet's *Conversations on Optics* with him every day. Mrs. Marcet was a well-known writer for children; the Wedgwoods and Darwins had known her in London in the 1820s. The *Conversations on Optics* was part of her *Conversations on Natural Philosophy,* which were widely read and went into many editions alongside her *Conversations on Chemistry* and *Conversations on Political Economy.* One striking point about Mrs. Marcet's books is that she wrote her often highly technical dialogues equally for girls and boys. In her *Conversations on Optics,* the teacher, Mrs. Bryan, helped the twelve-year-old Emily instruct her young sister Caroline in the natural sciences, showing that they were not too difficult "when familiarly explained." She covered physics and optics with careful explanations of the principles involved, explaining, for example, how colour was a property of the light reflected from objects, not of the objects themselves. Caroline commented: "What a melancholy reflection it is, that all nature, which appears so beautifully diversified with colours, should be one uniform mass of blackness!" Mrs. Bryan replied: "Is nature less pleasing by deriving colour as well as illumination from the rays of light; and are colours less beautiful, for being accidental, rather than essential properties of bodies?"

Annie liked to name the precise colours of things she found, and often imitated her father by matching objects with the colour samples in a small book he had used for describing specimens during his travels on HMS *Beagle. Werner's Nomenclature of Colours* edited by Patrick Syme, a flower painter in Edinburgh, gave a set of colour samples with names, and for each, offered a list of examples in different parts of the natural world, animal, vegetable and mineral. Charles would have been able to show Annie the minerals from his collection of specimens. Many of the other things she knew in the garden and countryside around the house. For wine yellow, Syme offered the body of a silk moth, white currants and saxon topaz. For tile red he gave the breast of the cock bullfinch, the shrubby pimpernel and porcelain jasper; for ash grey he offered the breast of the long-tailed hen titmouse, fresh wood ashes and flint, and for bluish black, the crowberry, black cobalt ochre and a large black slug.

Charles was anxious that his boys should receive the formal instruction they would need to enter a profession, but he was unhappy with the emphasis in the school syllabus on Latin and Greek. When Willy went to Rugby School, Charles believed he could see the "contracting effects on his mind of his very steady attention to classics: formerly I think he had more extended interests, and cared more for the causes and reasons of things."

In the 1840s, most daughters of wealthy households were educated at home because boarding schools for girls, though they had been popular in the 1820s, had come to be seen to be dangerously unhealthy. A leading physician wrote that "A delicate girl submitted to such a discipline cannot escape disease. While school-boys have the advantage of a play-ground, or enjoy their recreation at pleasure in the open fields, the unfortunate inmates of a female boarding-school are only permitted to walk along the foot-paths in pairs, in stiff and monotonous formality, resembling . . . a funeral procession."

Annie and Etty's education was Emma's responsibility, but she took as little trouble about their instruction as her parents had taken about hers. Etty commented that "Our education as far as book learning was concerned was not what would now be considered to be of an advanced type; my mother was somewhat easy-going about what we learnt, and to get the best possible teaching was not a great object with her."

When Willy was six, he was "getting on a little with his reading," and Emma enjoyed teaching him with Annie. She felt she should engage a governess, but was anxious about the idea. Her Aunt Jessie recognised "the melancholy, the discomfort, and discontent of keeping a governess," and suggested there was no hurry. "We English lay much too great stress on bringing children forward in learning, by which we give them longer lessons than their little heads can take in, and only serve to weary the poor teacher . . . The learning that profits our understanding is of our own acquiring, therefore later."

Emma waited another year and a half after Aunt Jessie's advice before engaging a governess, but when she was pregnant with Francis in the summer of 1848, she decided that the time had come. A nineteen-year-old girl, Catherine Thorley, came from Tarporley, a village in Cheshire, where her father had worked as a solicitor for the Tollet family of Betley Hall, close friends of the Wedgwoods and Darwins. He had died when he was twenty-eight and Catherine was five, and the family had moved to London. Catherine was the eldest child, with three sisters and a brother. The position of the governess in Victorian households was a difficult one. Her main task was to teach the "accomplishments" of a gentlewoman—needlework, etiquette, music, French and dancing, in preparation for entering the adult world. To be able to do so with assurance, she herself had to be genteel, but by accepting payment for her work, she put herself in a clearly inferior position. Many young women became governesses because their fathers had died leaving the family unprovided for. In 1848, Lady Eastlake wrote in the *Quarterly Review* that a governess remained "a needy *lady* whose services are of far too precious a kind to have any stated market value, and it is therefore left to the mercy, or what they call the *means,* of the family that engages her." Charles paid Miss Thorley £50 a year, an average amount, neither generous nor miserly.

Annie was close to Miss Thorley; she used to go with her when she went to see her family in London, and they may have had a special understanding because each was the eldest girl in her family with brothers and sisters to look after. Miss Thorley's next sister, Sarah, remained with her mother while the others also became governesses. According to Etty, Sarah was "mad," but nothing more is known about her.

Etty considered Miss Thorley a "dull but worthy girl." She "had no gift for teaching nor for making me care for her. I well remember how I despised her for never saying in answer to a question 'I don't know' but always trying unsuccessfully, as far as I was concerned, to hide her ignorance. I used to consider, before trying to get something cleared up, whether it was likely she would know or not. As soon as the blessed hour of twelve struck and I had hardly banged the door before rushing off, she always began one of the best known and most whiny 'songs without words'—always, always the same, so that it still rings in my ears." Etty herself was a headstrong and difficult child, and she was fiercely direct in conversation. On one occasion "mild Miss Thorley" gave notice because her authority was "not sufficiently upheld." It must have been Etty who challenged her. Miss Thorley was eventually persuaded to stay, but she had been prepared to make a stand.

Miss Thorley joined in when Emma played with the children. Once, when Willy had just returned to boarding school, Emma wrote to him: "Everyone (but the stair carpet which you nearly wear out in a month) is very sorry to lose you . . . Mr Lewis has made a sliding board for the children and they enjoy it very much. They put it on the stairs and I have taken a slide or two, and so has Miss Thorley." The staircase slide became another family fixture. Francis's son Bernard wrote: "This toy, to use an unworthy and inadequate name, has had a place in every Darwin household, but I have never seen one anywhere else. It consists simply in a long strip of polished wood with protecting edges and a small flange at one end by which it can be hooked on to any step of the staircase that the slider, in his timidity or bravery, desires. It is possible to invent various feats of fancy sliding—sitting, standing, or head-first, but the ultimate test of skill and courage was always 'eight steps standing.'"

While many governesses were kept firmly on the edge of their master and mistress's life, Miss Thorley shared Charles and Emma's reading; she helped Charles in his work and he appreciated her company. He once wrote to Hooker: "Miss Thorley and I are doing a little botanical work for our amusement, and it does amuse me very much, viz. making a collection of all the plants, which grow in a field, which has been allowed to run waste for fifteen years, but which before was cultivated from time immemorial; and we are also collecting all the plants in an adjoining and similar but cultivated field, just for the fun of seeing what plants have arrived or died out." Ten days later, he added: "If ever you catch quite a beginner, and want to give him a taste for botany, tell him to make a perfect list of some little field or wood. Both Miss Thorley and I agree that it gives a really uncommon interest to the work, having a nice little definite world to work on, instead of the awful abyss and immensity of all British Plants."

Charles described this study in the first version of the work that was eventually to become *The Origin of Species.* His point was to show "the degree of diversity in our British plants on a small plot." "142 phanerogamic plants were here collected by a friend during the course of a year; these belonged to 108 genera, and to 32 orders out of the 86 orders into which the plants of Britain have been classed." It was remarkable for a man of science to acknowledge his children's governess as "a friend" in this way. Her collection was valuable to Charles as it gave strong support for his insight that "the greatest amount of life can be supported by great diversification of structure." Linked with his notion of divergence between species, the idea became one of the keys to our present understanding of the importance of biodiversity in all thinking about the protection of natural habitats.

. . .

Annie started learning to sew long before Miss Thorley came, but Miss Thorley helped her with her needlework. Sewing was practical; embroidery was an accomplishment, and needlework was one of the few demanding occupations to which a lady could devote time and attention in daily life. Every lady would have a piece of "company work" in hand, making something for the household, a gift, or a piece of some kind for the "deserving poor." Annie had a "lady's companion," a little fitted case containing a thimble, needles and cotton, and a pair of scissors. She had a workbox for other things, with a needlebook made out of pink silk with small leaves cut from a cotton printed fabric appliquéd on the cover. When she was nine she made another silk-lined booklet with paper leaves for her five-year-old cousin Hope. Inside she drew trees, houses, plants in pots, a dress and a girl. Hope later remembered the intense pleasure it gave her. Annie also did beadwork with small glass beads bought in folds of paper from a milliner. In her writing case there is an unfinished piece, the pale yellow ribbon embroidered with blue and gold beads in a zigzag pattern, possibly for a bookmark, purse or pincushion.

When Annie was only one, Charles had noted in his "natural history of babies" how neatly she took hold of pens, pencils and keys "in the proper way." When she was six, Emma taught her to write, and she gave her a diary, *The Regal Pocket Book,* for her seventh birthday. Annie made one entry in confident joined-up writing—"Erney and Charles Parker came"—but after then kept the book unmarked, with its chromolithographs of the new Houses of Parliament, two Beefeaters in front of Windsor Castle, the Queen of Portugal and the Virgin and Infant Saviour after "Permegiano." Later that year, when Charles was staying in Shrewsbury, Annie sent him a note which he asked Emma to tell her was "very nicely written."

It was generally felt that a writing-master was required to teach the finer points of a good writing hand. Charles engaged John Mumford, the village schoolteacher, to coach Willy and Annie. He was "young and active and efficient," and probably used Mulhauser's method, which was approved by the government for use in elementary schools. The manual for the method started the pupil off with the task of writing out "Peace, joy and happiness are only to be gained by the way of duty" and carried on in the same tone. The instructions echoed infantry drill in their precision, with "1st Position," "2nd Position—take distances," "3rd Position—right arm extended," "4th Position—take pens," "5th Position—write"; then "Stop writing," and "Arms at ease."

Annie's formal handwriting was distinctive. Emma and Etty both had well-rounded and "good free-running hands" of the kind "usually adopted by ladies" according to a handbook of the time. Annie varied her tilt from word to word and added little curls to some letters. Her spelling was wayward and she sometimes missed out letters and words, but she enjoyed writing without being tidy and correct. She doodled and jotted freely on scraps of paper in pencil and crayon, but letters to cousins and friends had to be written in ink on special notepaper, enveloped neatly and closed with a wax or wafer seal. Letter-writing was an essential accomplishment because it gave you a way of communicating with people outside the household; the person who received your letter might pass it round to others, and they would all judge you from it. For the Darwin children in their lonely life at Down, letters were particularly important as they were their only window into the world of friendships beyond the household and the village.

Annie's writing case was not a child's toy, but a small version of a grown-up's possession, one of the first signs that the little girl would soon be growing into a young lady. An American book for children

Thank you, pretty cow, that made
Pleasant milk to soak my bread,
Every day and every night,
Warm and fresh and sweet and white
Do not chew the hemlock rank
Growing on the weedy bank
But the yellow cowslip eat.
That will make it very sweet.
Where the purple violet grows
Where the bubbling water flows
Where the grass is fresh and fine
Pretty cow, go there and dine

*"The Cow" copied by Annie*

which first appeared in 1851 gives a vivid picture of what the case and the things Annie kept in it would have meant to her. Elizabeth Wetherell's *The Wide, Wide World* was one of the most popular novels of the time, an evangelical story of a little girl, Ellen, in New York whose mother died of tuberculosis and who eventually found faith and happiness with kind people in the country. Emma had a copy which she read to her grandchildren in the 1890s, and the story must have rung true to her.

In the first chapters, Ellen's mother had to leave her to travel to

Europe to try to regain her health. She sold her jewelled ring to buy her daughter three things to treasure—a beautiful Bible, a workbox for embroidery and a writing case. For Ellen, filling the writing case, with all the choosing of small things involved, was a part of the pleasure of the gift. Ellen's mother took her to a large fancy store. It was the first time Ellen had "ever seen the inside of such a store; and the articles displayed on every side completely bewitched her. From one thing to another she went, admiring and wondering; in her wildest dreams she had never imagined such beautiful things." Her mother first chose a case, "perfect in its internal arrangements" but empty. The shopman provided a choice of letter-paper, large and small. "Ellen looked on in great satisfaction. 'That will do nicely; that large paper will be beautiful whenever I am writing to you, mamma, you know; and the other will do for other times when I haven't so much to say.' "

Matching envelopes followed, and Ellen then suggested: "Let us have the pens. And some quills too, mamma." "Do you know how to make a pen, Ellen?" "No, mamma, not yet; but I want to learn very much." Ellen's mother bought some feather quills and steel nibs. Ellen then chose a plain ivory pen-holder. "I think it is prettier than those that are all cut and fussed, or those other gay ones either." Next came a penknife to cut the quills, and the sealing wax. Ellen's mother pointed to a tray of different colours, and told Ellen to choose her own. She made up an assortment of the oddest colours she could find. " 'I won't have any red, mamma, it is so common.' 'I think it is the prettiest of all,' said her mother. 'Do you, mamma? then I will have a stick of red on purpose to seal to you with . . . You must not mind, mamma, if you get green and blue and yellow seals once in a while.' " Her mother then laid in a good supply of wafers of all sorts; and completed the case with an ivory leaf-cutter, a paper-folder, a pounce-box, a ruler and a neat little silver pencil; also some sheets of drawing paper, drawing pencils and an india-rubber.

Annie had many of these things in her writing case, and may well have been allowed to make her own choices in a shop selling fancy stationery on a visit to London with her mother or Miss Thorley. The case itself is a small container covered in morocco leather, with a lock and key to fasten the hinged lid. On the lid is stamped "Writing Case" in gold lettering inside a red, blue, white and gold cartouche. There are four small compartments and one large one, all lined with dark blue paper decorated with gold stars. The two inner compartments have a lid for "Matches" and "Vignettes." Annie had a number of sheets of letter-paper with red and blue edges and matching envelopes. She also had little envelopes with embossed and coloured flowers on the flaps, and pre-paid envelopes for the Penny Post with an embossed design. Other sheets of fancy stationery, with embossed patterns and cut to look like lace, said "Listen to me my love," "I could love thee for ever and ever," "Your health and happiness dearest," "That is to be with you my sweet" and "Taking wine in the wood."

Annie kept a number of patent steel pen-nibs in the box, together with a wooden pen-nib holder. There are two goose-quill pens and a small penknife with a mother-of-pearl handle. The quills were probably cut by Annie with the penknife from feathers plucked from geese in the backyard. They have remains of ink on their tips. The steel nibs were harder and scratchier to write with, but they would have saved Annie the trouble of having to trim her nib again and again as she wrote.

When an envelope had to be closed, sealing with wax was still the proper way; fastening the flap with a wafer was second best, and adhesive envelopes were a practical necessity for everyday mail. To seal her letters, Annie had two sticks of red sealing wax and a stick of green. She kept her wafers in a small circular cardboard box with gold-embossed decoration on the lid; the ones there now say in minute lettering "Write or Die," "Am I Welcome," "United," "Dieu vous garde" and "Adieu."

Four of Annie's letters survive. One was written on a piece of her fancy stationery; it was to Miss Thorley's sister, and Annie wrote carefully, perhaps under her governess's eye. "Dear Sarah, The other day Dick killed a rabbit in the orchard. Yesterday George and I went to Aunt Sarah to tea, and before that went a mushroom picking but only found one. It is very fine today. We are going to have a kitten. I have got a pencil with my name upon it. Etty sends her love to you. Goodbye. I remain your affectionate Annie Darwin." Annie put the letter in a matching envelope with an embossed pattern of convolvulus, and sealed it with a little emerald wafer with a design of two love birds.

Annie also made her own confident entries in two notebooks kept by the grown-ups in the household. Emma had a book of recipes, in which Annie entered hers for "Italian Vegetable Soop." "The heart of 6 lettuces & cucumbers pared & cut in quarters. 1 Pint green pease, a little onion peper & salt to your taste. Put all together in a stew pan over a very slow fire for 2 hours. Then boil a pint of older peas in good broth gravy with a lump of sugar. Pass them threw a sive into the broth, then warm. Add a litle cram to the yolks of 2 eggs boild in, mix with the stewed vegetable and heat up all together."

A book of medical preparations and dosages was kept in the medicine cupboard. Annie must sometimes have had the difficult task of purging her younger sisters and brothers, and she seems to have managed, like her mother, with a carrot rather than a stick. She wrote: "If you boil Castor Oil with an equal quantity of milk, sweeten with a little sugar, and stir it well and give it when cold, children will never suspect it to be medecine, but will like it almost as well as custard."

CHAPTER SIX

# FAITH, CRICKET, AND BARNACLES

*Villagers—Animals—Church and Dissent—Charles's doubts about Christianity—Barnacles*

WALKING IN THE VILLAGE, Annie met the girls and boys she had played with since early childhood, but as she grew, she became the young Miss Darwin, daughter of Mr. and Mrs. Darwin, gentlefolk of independent means. Charles had no place with Sir John Lubbock in the "squirearchy" of landlord and tenant, but the Darwins were respected customers of the village shopkeepers, the blacksmith, miller, brewer and carpenter. Charles went to meetings of the parish vestry which set the rates for poor relief. In one year his gifts to charities and poor people he was supporting amounted to £62; he spent £34 on beer for the household, and £15 on "science." All three sums were small in relation to his total income; of the £3,250 he received that year, he spent only £1,900 and reinvested the rest.

According to Etty, her father liked the "friendly recognition he met with" in the neighbourhood, and he remarked with pleasure on the many people he did not know who used to nod to him "in a friendly way." "There was always a great deal of curtseying and grinning of the children in the village; I think he sometimes used to give them pennies, but it was more his friendly greeting." Charles talked with the young labourers who came to work at Down House, and when Annie was nine, he encouraged a group of them to form a Friendly Club which for a monthly contribution provided payments in sickness and on death. The club met every month in a pub called the George and Dragon, opposite the church, and Charles acted as their treasurer. They had to calculate their benefits carefully, and the Society for the Diffusion of Useful Knowledge provided tables of sickness and death rates on which to base their sums. According to the society's manual, "Frugality and providence give to a man a moral independence, and a happiness of which a mere pauper can scarcely form an idea." It was appropriate that Charles, who had rejected the view that God had a special providence for each person, should help his neighbours in this way to make one for themselves.

The members of the Friendly Club were keen cricketers, and Charles provided a pitch in his meadow for them to play on. In the Kent countryside in the 1840s, cricket was a game for the villagers, and all the children of the parish watched with admiration and excitement. James Pycroft, author of *The Principles of Scientific Batting,* wrote: "The game is free and common as the light and air in which it is played." Social precedence was suspended as "the cottager stumps out his landlord."

Emma visited the "deserving poor" in the parish, and provided bread for the hungry and medicines for the sick. She took Annie and Etty with her to show them how needy villagers lived and what kinds of help charitable gentlefolk should give. One of her concoctions for poor women "weak after lying, or with pains in the back" was a gin cordial with peppermint, laudanum, sugar, bitters and wine. She was generous but businesslike. When I gave a talk to Downe Women's Institute in 1999, I was introduced to an elderly lady who told me that her mother-in-law came from a poor family and Mrs. Darwin had paid for her schooling in the 1890s, on the understanding that she would work for her as a housemaid during the summer months.

The treatment of animals was a concern for the Darwins in village life. There were working horses and donkeys, dogs and cats; pigs being reared for slaughter; sheep, goats and cattle, geese and chickens; farmyard vermin, and the rabbits, squirrels and hedgehogs in the fields around. Some villagers had songbirds in cages, and there was talk of cock-fighting and badger-baiting, both recently outlawed but carried on in private gatherings. Many people treated their animals with indifference to their pain, or wilful cruelty, but a strong current of care had been flowing in evangelical and liberal circles for some years. Children's books like the Wedgwood cousins' *Cobwebs to Catch Flies* reflected that feeling, and the Society for the Prevention of Cruelty to Animals led a campaign for humane treatment which was supported by many prominent figures. The society's inspectors carried out many prosecutions before local magistrates. Their efforts were widely welcomed among the "respectability," and in 1840 Queen Victoria made the society "Royal."

For some who supported the cause, kindness to animals was purely an "extension of humanity to the brute creation," showing the quality of your compassion by stretching it beyond mankind. Charles and Emma, on the other hand, had no doubt that animals felt pain just like humans. Emma's certainty was a feeling or intuition like her Christian faith. Charles had developed the idea as a central theme in his view of man's common nature with animals. On this subject, they could speak to their children with one voice.

When he saw cruelty, Charles could not restrain his anger. Francis remembered his father returning one day from his walk pale and faint from the agitation of arguing violently with a man he had seen ill-treating his horse. Charles was also prepared to take neighbours to law. One, a gentleman farmer, was said by some villagers to have allowed some of his sheep to die of starvation. Legal actions against gentry were rare, but when Charles heard about the matter, he went round the whole parish, collected all the evidence himself, had the case brought before the magistrates, and secured a conviction. Willy always remembered his father's action, and how as a child he had been "immensely impressed."

Annie's world was firmly Christian. Etty wrote that in their childhood, their mother "was not only sincerely religious . . . but definite in her beliefs. She went regularly to Church and took the Sacrament. She read the Bible with us and taught us a simple Unitarian Creed, though we were baptised and confirmed in the Church of England."

A prayer book used by Unitarians at the time told parents that a daughter should be told to love and do good to all people, "because all are equally the children of God with herself, and the objects of his fatherly kindness and care: that she is not born only for herself but for others; to serve her country and mankind by promoting truth and virtue, and the public good." Joseph Priestley, the chemist and Dissenting minister who was a close friend of Charles and Emma's grandfathers, Erasmus Darwin and Josiah Wedgwood, set out a simple Unitarian Creed in his *Catechism for Children and Young Persons.* "Jesus Christ was a person whom God sent to teach men their duty, and to persuade and encourage them to practise it." "How doth God govern the world by his providence?" "He suffers nothing to come to pass, but what tends to promote his design of making mankind virtuous and happy. His providence extends to the meanest creatures that he has made, and even a sparrow falls not to the ground without his will."

Priestley dealt with the problem of suffering. "If nothing come to pass without the will of God, why doth he suffer storms and tempests, pain and sickness, which occasion such distress and misery to his creatures?" "The evils and miseries of which we complain are intended for our good, though we are not always sensible of it. They are the corrections of a wise and affectionate parent." While Charles had abandoned the idea that pain and suffering had any Divine purpose for the individual victim, Emma always believed that suffering was linked to sin, and may have given their children Priestley's mysterious and gentle message.

Emma taught the children to pray and gave them the poems of Ann and Jane Taylor which dwelt on goodness and grace, things of beauty and their loss, in simple language and rhythms that made them the most popular verses of their time. The two sisters wrote in a poem "for a very little child" that

> *. . . God looks down from heaven on high,*
> *Our actions to behold;*
> *And he is pleased when children try*
> *To do as they are told.*

In one unflinching poem "about dying," a child asked:

> *"Tell me, Mama, if I must die*
> *One day, as little baby died,*
> *And look so very pale, and lie*
> *Down in the pit-hole by its side?"*

Willy and Annie understood clearly what this meant, because every Sunday as they walked to the church door, they passed the family grave where Emma's "little baby" Mary had been buried in the family's first weeks at Down. Jane Taylor suggested that children should face up to death. "Let not young persons think this subject inapplicable to them. For, not to mention the uncertainty of life at every age, it is of the highest importance to be early impressed with just ideas of death and futurity; that it may become a subject of familiar and agreeable reflection, rather than of dread and terror."

All the days of the week were the same for Charles, but Emma saw "keeping Sunday" as a duty. She held prayers for the household and the family went to the morning and afternoon services in the church. They had a large pew lined with green baize near the clergyman's desk. The parish clerk, John Osborne the village wheelwright, gave them "the full flavour of his tremendous amens," and Francis played with the india-rubber threads of his elastic-sided boots, gently tweaking them like miniature harp-strings. There were silverfish in the prayer books and among the cushions, and Francis watched them move like minute sardines running on invisible wheels.

The church had a small choir in which Mr. Osborne sang bass and Parslow sang tenor with the Lubbocks' footman. They had a barrel organ to accompany the "chants" to which they sang the canticles. Their book of chants had a setting for the Athanasian Creed, and Emma with her Unitarian beliefs must have winced when the villagers intoned the mysteries of the doctrine of the Trinity on the appointed days. "And yet there are not / three e/ternals: but / one eternal. / As also there are not three incomprehensibles, nor three / uncre/ated: but one uncreated, and / one in/compre/hensible."

During the Sunday service, Emma insisted on one small assertion of the family's independence. When the congregation turned towards the altar to recite the Creed, the Darwins "faced the other way and sternly looked into the eyes of the other church-goers." Francis commented: "We certainly were not brought up in Low Church or anti-papistical views, and it remains a mystery why we continued to do anything so unnecessary and uncomfortable." After the service, the children would come out of church cold and hungry. People would stand about the porch, Sir John Lubbock in a fluffy beaver hat and the labourers in their green or purple smock-frocks.

We think of the village church in Victorian England as one of the sturdiest pillars of the accepted social order. But Downe, like many other villages, was riven by disagreements, doctrinal and political, and one of the places where they were shown most clearly was in and around the pulpit. During Annie's childhood, Mr. Innes, the vicar, was an assertive young High Churchman. Charles once went to a vestry meeting "to defy Mr Innes and all his works," and Emma hated him for what she saw as his bigotry, saying that it made her feel "desperately vicious against the Church."

A book of sermons preached by Mr. Innes at Downe in 1851 gives a flavour of what Annie heard each Sunday. On Innocents' Day he spoke about children and their nature. They were "wayward, and prone to evil; yet are there signs of good in them, marks which tell us they have lately come from a pure and holy God, and that an element of good is theirs, as well as a liability to sin." He also preached urgently about death, judgement and the afterlife. "And what a fearful question it is! Heaven or Hell! We do not know indeed what are the joys of the one or the pains of the other . . . Neither can we form an idea of the lot of the condemned. Body and soul in torment, the worm that never dies, the fire that is never quenched, the bitter remorse of a too late awakened conscience." A time would come "in the eternal death of every condemned being, when he will have suffered as much torment as has been endured by every sufferer and every criminal in this world." But then, "the punishment is no nearer to its end than it was at the beginning, for all the time that is past is nothing in the ages of eternity."

Mr. Innes was losing his congregation to Dissent and indifference. A book for children that he gave as a prize to his Sunday school pupils described Dissenters as "those unhappy persons who . . . have become guilty of the dreadful sin of schism." In 1851, a National Ecclesiastical Census was held, and while nonconformist ministers in the neighbourhood willingly declared their attendance on the census day, no figures were given for the parish church. At the end of the year, Mr. Innes looked out over his congregation and preached gloomily that "the same seats in the ale house are still filled by the same drunkards: the same oaths are heard, the same obscene jests and filthy conversation goes on." The year had been "troubled and stormy" for the church. "There is always great mischief in controversy and theological disputes." Argument might lead to good but was dangerous for the uneducated. "They are led to talk about things they do not understand, and question and dispute when they require to be taught; and to consider religion to consist of subtle questions and fine drawn distinctions . . . instead of a true faith, showing itself in practical holiness."

The main Dissenters in the village were a group of around thirty who met in a small chapel on the lane between the village and Down House. Their minister, James Carter, was a painter and glazier from Wiltshire who had married in the 1820s and settled down with his wife in Tooting, "in a religion of their own making." This phrase reflected the freethinking of the times. It could have been used of William Blake as he lived quietly in London with his personal prophecies, or of Charles in Downe for that matter, with his own private thinking about the sources of our moral sense. Mr. Carter had become an itinerant preacher and had been invited to Downe in 1836 to form a little church "upon New Testament principles, with the exception of believers' baptism." He lived in a cottage near the chapel and came to Down House whenever there was woodwork to be painted or a broken window to be repaired. According to his daughter Marianne, he was a man "of most tender conscience," and lived a troubled life. While at Downe, he had "much to try his faith; many battles to fight; many enemies to encounter; trials from within, trials from the world, trials through the duplicity of professed friends; Satan often permitted to set in upon his soul with all his fiery darts; and unbelief would so cloud his mind, that he was sometimes led to question whether he knew anything for himself after all."

In June 1851, Mr. Carter and his congregation reformed themselves on Strict Baptist principles as the Church of Christ assembling at

Downe Chapel. They set out their beliefs in a "declaration" which went far beyond Mr. Innes's fierce claims about human nature and destiny. "All mankind in their own nature are totally carnal and unclean, utterly averse to all that is good, being not only enemies, but enmity itself against the Blessed God and all goodness, and utterly incapable of performing one spiritual act, being dead in trespasses and sins." The group believed that before the world was made, "Jehovah did elect a vast but certain and definite number of the human race unto eternal salvation." By contrast with Emma, they placed no value on their own responsibility for their actions, declaring that their regeneration, conversion, and sanctification "never was, and never can be, the act of man's free will and humane ability, but they are in all cases the effect of the mighty efficacious and invincible Grace of God." The Darwins' gardener, Henry Lettington, who helped Charles with many of his plant experiments in later years, was a deacon of the chapel.

The other main challenge to Mr. Innes's influence among his parishioners was the village school for boys which Sir John Lubbock provided and maintained, and the Darwins supported by paying the fees of a few needy children. At the time, parish schools were a battleground in the struggle between "church" and "secular" approaches to education and improvement for poor people. The clergy managed the schools wherever they could, and often insisted on church attendance in order to draw pupils and their parents back from Dissent. Progressive Whigs like the Lubbocks and Darwins wanted the schools to help poor people to acquire "useful knowledge" and improve themselves. They saw religion as a hindrance rather than a help because it led to doctrinal arguments and sectarian feeling. Sir John had set up his school on the "undenominational model" of the British and Foreign School Society which forbade religion in the classroom. An inspector of schools for the Archbishop of Canterbury who visited the school in 1851 found that Mr. Mumford the schoolmaster had succeeded in making his ls work "vigorously and cheerfully." The discipline was good, but the general bearing of the boys was "far from pleasant." "The general knowledge is superficial, though extended. More quickness than real intelligence is shewn in their answers." Noting that "no religious instruction whatever is permitted," the inspector commented with alarm that "the ignorance of scripture in the second and third classes is heathenish."

While Emma brought up the children in her faith, Charles thought carefully about his own beliefs, and his doubts grew. There was a history in Britain of questioning the grounds of faith and the witness of the Gospels. In the previous century, David Hume had argued in his *Enquiry Concerning Human Understanding* that there were logical difficulties in accepting the New Testament as evidence of Christ's miracles, and Edward Gibbon had treated early Christian history with corrosive scepticism in his *Decline and Fall of the Roman Empire.* The German "Higher Criticism" raised more questions and, by the 1840s, anyone disposed to doubt could choose between a number of approaches.

During Annie's childhood, Charles looked at a number of books which dealt with the so-called "evidences" of the Christian Revelation, especially Christ's miracles. In 1845, he read the Unitarian James Martineau's *Rationale of Religious Enquiry,* which explored the ways in which reason could be used to support faith in Christ. Charles came to feel that "the clearest evidence would be requisite to make any sane man believe in the miracles by which Christianity is supported." He was, though, as he said later in his *Autobiography,* "very unwilling to give up my belief." He had "daydreams of old letters between distinguished Romans and manuscripts being discovered at Pompeii or elsewhere which confirmed in the most striking manner all that was written in the Gospels." His dreams may have been prompted by Sir

Edward Bulwer-Lytton's novel *The Last Days of Pompeii,* which he read in 1847. Lytton had introduced a secret Christian group into his story of the catastrophe which overwhelmed the Roman city in a.d. 79; many of his readers would have known that the greater part of the city still lay buried beneath the debris of the eruption, and the possibility of finding documents that somehow confirmed the Gospel narratives must have seemed real.

In spring 1848, Charles and Emma read *The Evidences of the Genuineness of the Gospels* by Andrews Norton, an influential Unitarian theologian at Harvard University. Norton recognised that the Gospel texts were probably corrupt in places, but aimed to prove that they had been composed by the Apostles, and that the surviving texts preserved their essential content. Emma made notes in her interleaved Bible about a number of passages that Norton suggested were not original. His comments may have helped her over some difficulties, and they may have brought her to the same point of view as her Aunt Jessie, who wrote that she found in the Bible "all my heart wants, without believing that every word is inspired . . . What puzzles me too much, or appears contradictory, I lay to the faults of the many hands through which it reaches me, and still clasp it to my heart as a divine book, however it may have been perverted by the perverse."

Charles judged Norton's book "good" but, as he dwelt on the question of proof of the Gospel narratives, he found it more and more difficult, even giving free rein to his imagination, "to invent evidence which would suffice to convince me." It is important to recognise how fundamental his doubts had become. He was not concerned about what particular evidence there happened to be of Christ's miracles, but whether one could envisage any historical evidence that could ever prove that a truly miraculous occurrence had taken place. This was David Hume's notorious philosophical doubt about the grounds for belief in any supernatural happening.

Charles's brother Erasmus, together with Hensleigh and Fanny Wedgwood, knew many people in London whose Christian faith was evolving into forms of theism and humanism that shocked orthodox believers. Books were appearing about faith and doubt, and the Darwins and Wedgwoods often discussed them with their friends. In the summer of 1848, Charles read a memoir of John Sterling, who had been an undergraduate at Cambridge two years before him. Energetic and impetuous, Sterling had wanted to become a priest but could not because of ill-health. He had doubts about the Old Testament; he discovered the German Higher Criticism and came to question his whole Anglican faith. Facing a wasting death from tuberculosis, he rejected the complacency of Paley's natural theology. "I do not pretend to believe that in a system of things full of corruption and curses, all is for the best . . . Paley's saying 'It is a happy world after all,' which some might attribute to the goodness of his heart, seems to me one of the most cruelly heartless of all human utterances." Sterling eventually found the faith he yearned for in his own feelings, rather than in the authority of Scripture.

In November 1848, Charles's father died an "unbeliever," and Charles and Erasmus faced the question whether he would be eternally punished for his rejection of Christ's message of salvation, as Scripture suggested. Erasmus was confident that God had been kind to their father in death, writing to Fanny Wedgwood that he could not "feel anything but how good God was to take him without suffering more." He does not seem to have been concerned that his father might be doomed to eternal punishment for rejecting the Christian message, and there is no indication that Charles was worried either.

Early the next year, Charles continued with his sceptical reading. Erasmus and Fanny's friend Harriet Martineau wrote about a journey to the Holy Land in *Eastern Life, Present and Past.* She had started her writing career as a Unitarian but in the 1840s grew increasingly doubtful of revealed religion and moved through a theist phase to an atheistic humanism. In *Eastern Life* she tried to show the genealogy of the Christian faith in the Egyptian and Judaic religions, and pointed towards the purely human ideals which she saw as the final aim of moral values. Harriet Martineau had offered the text to the publisher John Murray, but he had refused to take it on because "it was a work of infidel tendency, with the obvious aim of deprecating the authority and invalidating the veracity of the Bible." Charles, on the other hand, found the book "curious and interesting."

Charles made two comments towards the end of his life about when and why he had finally abandoned his Christian faith. Both point to 1849 and the years before and after as the critical time. In his *Autobiography,* Charles wrote that in the 1840s "disbelief crept over me at a very slow rate, but was at last complete. The rate was so slow that I felt no distress, and [I] have never since doubted even for a single second that my conclusion was correct." His second remark was reported by a visitor to Down a few years later. Charles told him: "I never gave up Christianity until I was forty years of age." Asked why he had given up his belief at that time, he replied simply that he had investigated the "claims of Christianity" and they were "not supported by evidence." Charles became forty in February 1849. His comment that he felt "no distress" as he came to reject Christianity was remarkable for the period, as the pain of doubt and the struggle for faith were pressing issues for many. For Charles, the issue was purely one of "the evidences."

One writer whose books Charles read with interest in 1849 and 1850 was Francis Newman, Professor of Latin at the "Godless College" and the younger brother of John Henry, later Cardinal Newman. The two brothers had been estranged in the 1830s as John became a leading figure in the Oxford Movement with its emphasis on Church, authority and religious forms, and Francis argued with equal conviction for a liberal and critical faith. He was a central figure in "advanced" Unitarian circles and a friend of Hensleigh and Fanny Wedgwood. His *History of the Hebrew Monarchy* used the Higher Criticism to argue that the Old Testament could not be the word of God as it was a patchwork of texts bearing "plentiful marks of the human mind and hand." His work on *The Soul, her Sorrows and her Aspirations* was subtitled *An Essay towards the Natural History of the Soul, as the True Basis of Theology.* Like Charles, he believed that in "organic life," we often suffer pain or loss through the operation of natural laws which take no account of the individual's moral state. He rejected all scriptural or doctrinal arguments for a future life, feeling that the belief in one could only ever be a personal hope based on intuition alone.

Annie had no inkling of her father's private thoughts about religion but watched every day as he carried on his scientific work in his study. After putting away his essay on his species theory with his note to Emma in 1844, he had gone back to his *Beagle* notes and completed his *Geological Observations on South America* in which, using his maps of the continent, he showed how Lyell's geology could be used to tell the history of the whole land mass. His account of the signs that the western rim of the continent had gradually risen through the geological eras was a *tour de force* of Lyell's new method. Charles planned to start next on the evidence for his species theory, but, before doing so, he decided to look at a few remaining specimens from HMS *Beagle.* He had always meant to describe them, but they had sat untouched in their jars of spirits for the ten years since his return.

Charles examined some arrow worms and flatworms and wrote two papers about them. He then turned to a small barnacle. When he put it under his microscope and examined it, he was intrigued and he compared it with some others. He talked to experts in London and went back to his specimens. Within a few months, he embarked on a taxonomic study of barnacles which he acknowledged to a friend "will put off my species book for a rather long period."

Charles's meticulous examination of the one minute sea creature and its cousins was a remarkable change of focus from the essay of 1844 which had ranged freely over the whole natural world, but Charles had reasons for his interest, and when he eventually completed the task he had set himself, his findings fitted into place as an essential part of the work that led to *The Origin of Species.*

Barnacles were familiar to any Victorian child who knew the seashore of the southern or western coasts of England or Wales. The encrustations of small cones marked rocks below the high-tide mark, and any Victorian seaman would know how they also fouled ships' bottoms, buoys and pier pilings. They looked like limpets, which were classed as molluscs, but in the 1830s it had been pointed out that they had free-floating larvae like crabs and other Crustacea, and when Charles started working on them in the mid-1840s, they were seen as a very curious class of animals which, like the zoophytes, which Charles had studied in Edinburgh, combined the characters of two major groups of organisms in a way that offered a challenge to taxonomy.

Charles had found his *Beagle* barnacle on one of the Chonos Islands off the southwest coast of Chile. He had been collecting molluscs and had been surprised to see that the thick shells of one species were completely drilled by a soft orange organism less than a tenth of an inch long. When he put it under his microscope back on board ship, he was fascinated to see that it was a barnacle without a shell. He then found eggs attached to it in four stages of growth. In the fourth stage, they were coffin-shaped with two "thick, clumsy legs" which reminded him of some crustacean larvae. This link, and the lack of a shell, were points to look into and write up, but when he came to do so at last in 1846, Charles found more to think about at every step.

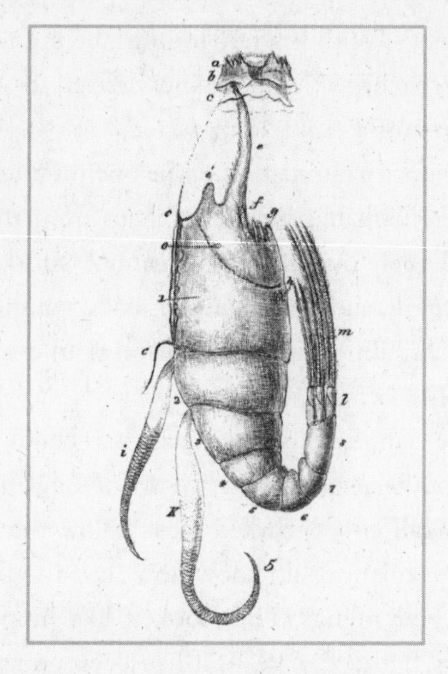

*Barnacle from the Chonos Islands*

Trying to compare the soft parts of his barnacle with other species, he found that little was known about any of the group, and the taxonomy of the whole class of Cirripedia which includes the barnacles was in a muddle. Professor Owen encouraged him to do a study; the keeper of the zoological collections at the British Museum gave him full access to the museum's specimens and lent his personal collection. Charles wrote to leading naturalists in France, Denmark, Holland, Germany and the United States, and the postman was soon bringing specimens from all over the world.

Charles had a strong reason of his own to take on the task. In 1845 he had been discussing the difficulties of distinguishing species with Hooker, and Hooker had insisted that "to be able to handle the subject at all, one must have handled hundreds of species with a view to distinguishing them, and that over a great part, or brought from a great many parts, of the globe." Charles took this comment as a suggestion tha himself lacked the experience needed, and wrote back, with revealing sensitivity, that it did not "alter one iota my long self-acknowledged presumption in accumulating facts and speculating on the subject of variation, without having worked out my due share of species." But he needed to gain experience and prove his ability as a taxonomist, and the cirripedes were a perfect opportunity.

Charles was to spend eight years on his specimens while dust gathered on his essay, and for all of Annie's life from October 1846 until 1851, her father dissected barnacles under his microscope at the window of his study. George remembered that when he was small, the children all regarded this as "the natural occupation of the head of a family." When they went one day to play with the Lubbock children at High Elms, he asked where Sir John "did his barnacles." Charles wrote to a fellow naturalist that "most of my friends laugh." After the first volume of his study was published in 1851, he was amused himself when Sir Edward Bulwer-Lytton put a character called Professor Long, author of "Researches into the Natural History of Limpets," into one of his novels. Another character commented that Professor Long's subject required deep research; it was one "on which a learned man may say much without fear of contradiction," and "the history of limpets is to a man" "what the history of man might be to a limpet."

When Annie came into her father's study, she found him using simple methods and few instruments. Francis wrote that "if any one had looked at his tools etcetera, lying on the table, he would have been struck by an air of simpleness, make-shift, and oddness." His dissecting table was a low board, let into the right-hand window of the study. He sat there on his microscope stool, which had a revolving seat so that he could turn easily from side to side. He kept a few implements on the main table and he had odds and ends in the drawers of another table to the left of his microscope. The drawers were labelled "best tools," "rough tools," "specimens," "preparations for specimens" and so on. Francis wrote: "The most marked peculiarity of the contents of these drawers was the care with which little scraps and almost useless things were preserved. He held the well-known belief, that if you threw a thing away you were sure to want it directly, and so things accumulated."

In later years Francis was to help his father with his botanical observations and experiments. He noticed then how his father's "eager desire not to lose time was seen in his quick movements when at work." In one experiment which required some care in manipulation, "fastening the little bits of card . . . was done carefully and necessarily slowly, but the intermediate movements were all quick . . . all these processes were performed with a kind of restrained eagerness." Francis recalled him, "as he recorded the result of some experiment, looking eagerly at each root &c, and then writing with equal eagerness. I remember the quick movement of his head up and down as he looked from the object to the notes . . . I can recall his appearance as he counted seeds under the simple microscope with an alertness not usually characterising such mechanical work as counting. I think he personified each seed as a small demon trying to elude him by getting into the wrong heap, or jumping away altogether; and this gave to the work the excitement of a game."

Charles worked on living and fossil barnacles, hard parts in boxes, and soft parts preserved in "spirits of wine." He used two microscopes, one simple and one compound, with wooden blocks to support his wrists while he was dissecting. The compound microscope was of the kind familiar to us with lenses mounted in a cylindrical tube, but he always preferred the simple one with a single lens. While he was working on the barnacles, he got a London manufacturer to make a simple microscope to his special design; he found it ideal for his work, and the manufacturer sold the model for many years later as "Darwin's Dissecting Microscope."

In June 1851, Charles described his technique to a dentist in Swansea who had sent him drawings of his own dissections of barnacle larvae. "I have been accustomed to preserve the results of most of my minute anatomical researches . . . in common water without any spirits, with a bit of thin glass over the object . . . and gold size all round the rough edge; objects thus prepared will sometimes keep for a long time . . . Every cirripede that I dissect I preserve the jaws &c &c in this manner." Four drawers of Charles's slides of barnacle parts are now kept in a polished oak cabinet in the basement of the Museum of Zoology in Cambridge. He placed each specimen in a drop of water on a glass slide and made a temporary or permanent mount with a thin glass cover slip and a seal as he wanted. He sealed the temporary mount with gold size or asphaltum, which discoloured with age. For a permanent mount, he carefully dehydrated the specimen with alcohol and an aromatic oil, placed it finally in Canada balsam, and sealed it under a cover slip with gold size or asphaltum. The balsam bottle with its glass rod and domed lid would have been a familiar sight on the table.

The main difficulty for Charles in his survey was how to group different kinds of barnacle in the initial classification. Plant and animal taxonomy were central concerns for natural historians. Carl Linnaeus's *System of Nature* had appeared in 1735, and Baron Cuvier's *Animal Kingdom* in 1817. Professor Owen was developing Cuvier's approach to reveal ever more interesting and complex links between the structures of different organisms, both living and fossil, and to suggest principles on which a sound classification should be based. Both Cuvier and Owen rejected the idea of common descent and believed that the likenesses between species reflected Divine design. Owen developed an idea of archetypes and functional homologies to show that animals conformed to general plans, but he found that the shared elements and "design" they displayed were often varied and obscure, and he had to admit that Nature was "not so rigid a systematist as man."

Charles had little use for Professor Owen's abstractions, and wrote to a friend in 1843 that he had long felt that the problem of taxonomic groupings lay "in our ignorance of what we are searching after in our natural classifications." Linnaeus had confessed profound ignorance and most writers since had said that the aim was "to discover the laws according to which the Creator has willed to produce organised beings." "But what empty high-sounding sentences these are. It does not mean order in time of creation, nor propinquity to any one type, as man. In fact it means just nothing." Charles went on to declare his own view, that "classification consists in grouping beings according to their actual *relationship*, i.e. their consanguinity or descent from common stocks." Many animals would show similarities that were due to other factors, and it would always be difficult to identify true relationships for a proper natural classification, but "I know what I am looking for."

When Charles tackled the barnacles, he worked out an "archetypal cirripede" with all the common features of the group, but instead of presenting it as Professor Owen would have done, as a Platonic ideal, God's blueprint, he made it a common ancestor, and looked to see whether all existing species could be placed on a branching family tree as descendants from it by gradual changes from that earlier common form. The work that Annie watched her father carry on from day to day at the window in his study was one of the first careful explorations of a part of the natural world on the understanding that organisms had evolved. As Charles dissected and examined each specimen, he was testing his theory in a precise and determined way, taking hundreds of organisms that showed marked similarities but also bewildering variations and transformations, focusing on the minutest details of their anatomy and seeing whether he could group them as first, second, third and fourth cousins in a single family tree. If he was right to assume that they had all evolved by variation from a common ancestral form, he should be able to classify them according to their family relations. If he was wrong, it should be immediately obvious that some other factor was involved.

Barnacles posed two main challenges to the taxonomist: the extreme variety between different species, and the transformations they went through from larva to adult. Charles found that he could make sense of both. He suggested, for instance, that the ancestor had a body with seventeen segments. The three segments of the genus *Alcippe* were the three front ones, and the fourteen behind had wasted away in the course of evolution, while *Proteolepas* had the last fourteen segments and had lost the front three. He wrote to Hooker in 1848 that he was becoming "rapidly a complete cirripede in my mind." He had worked out that the shell of a *Balanus,* and even the whole peduncle and shell of *Lepas,* were "the three anterior segments of the head, wonderfully modified and enlarged so as to receive the fourteen succeeding cephalic, thoracic and abdominal segments." He declared: "I know of no more surprising metamorphosis, and it is perfectly clear and evident."

When Charles looked at the antennae of larvae and adults, he found that they were used by the larva for locomotion and touch, and in the adult they became organs for attachment to the surface on which the barnacle lived. In all these ways, by looking carefully at anatomically linked structures in different species, and by considering the possibility that organs could change their form and their function over time or waste away if they lost their use, Charles found that it might be possible to explain the extraordinary range of forms in the different species as a series of transformations of the ancestor's original structure. The explanation worked.

Charles found other surprises under the microscope: strange sexual arrangements which were full of meaning for someone taking an evolutionary view, but could only seem ludicrous to someone looking for evidence of the wisdom of God in Creation. Early in his speculations about the origin of species, he had guessed that separate sexes had evolved from hermaphroditic forms. He recognised that different patterns of reproduction might be important for the inheritance of variations, and was always watching out for evidence of change and its effects. Most barnacles were hermaphroditic, but a few had separate males and females, and Charles was intrigued to find that his first little orange barnacle had two penises for no reason that he could see. In 1848, he found some unidentifiable parasites in a species of the genus *Scalpellum*. When he found that hermaphroditic species of the genus *Ibla* had extra microscopic males to complement their male organs, he looked again at the *Scalpellum* parasites and discovered that they also were minute males. He wrote to Hooker about his discovery, saying that he never would have made it had not his species theory convinced him that a hermaphroditic species must pass into a species with separate sexes "by insensibly small stages, and here we have it, for the male organs in the hermaphrodite are beginning to fail, and independent males ready formed." He went on: "You will perhaps wish my barnacles and species theory al Diabolo [to the Devil] together, but I don't care what you say; my species theory is all gospel."

In September 1849, Charles wrote to Lyell about another barnacle he had found with separate sexes in which the female had "two little pockets, in each of which she kept a little husband." He mentioned the hermaphrodites with complemental males, noting that one had "no less than seven of these complemental males attached to it. Truly the schemes and wonders of nature are illimitable." In his eventual book, he summarised what he had found out about these strange forms and concluded: "In the series of facts now given, we have one curious illustration more to the many already known, how gradually nature changes from one condition to the other—in this case from bisexuality to unisexuality." He felt he had seen evidence of evolution at work, that he could trace part of the route along which one of the fundamental features of natural life had come into being.

The leading populariser of microscopy at the time was Philip Gosse, the father of Edmund Gosse, who described his extraordinary upbringing in his memoir *Father and Son*. Philip was a naturalist who made a living in the 1850s writing about marine life for a popular readership. He combined a deep and absorbing interest in the natural history of the English and Welsh coastline with a Christian faith that bewildered his son in its strength and simplicity. He greatly admired Charles's work on barnacles and included a chapter on the remarkable metamorphoses of their larvae in *Tenby, a Sea-side Holiday.* In his later *Evenings at the Microscope,* which was published just before *The Origin of Species,* he described how the free-floating larvae attached themselves to a rock and changed to their adult form. "And this is a most wonderful process; so wonderful, that it would be utterly incredible, but that the researches of Mr Darwin have proved it incontestably to be the means by which the wisdom of God has ordained that the little Water-flea should be transformed into a stony Acorn Barnacle."

Charles would not have suggested to Annie that the fixing of the barnacle had been ordained by the wisdom of God, but he would have been delighted by Gosse's final comments. "Marvellous indeed are these facts. If such changes as these, or anything approaching to them, took place in the history of some familiar domestic animal; if the horse, for instance, were invariably born under the form of a fish, passed through several modifications of this form, imitating the shape of the perch, then the pike, then the eel, by successive castings of its skin; then by another shift appeared as a bird; and then, glueing itself by its forehead to some stone, with its feet in the air, threw off its covering once more, and became a foal, which then gradually grew into a horse . . . should we not think them very wonderful?"

CHAPTER SEVEN

# WORLDS AWAY FROM HOME

*Leith Hill Place—Chester Terrace—Etruria—Malvern*

Away from Down, Annie knew other family worlds through visits to her Wedgwood cousins. Uncle Joe lived the life of a country gentleman in Surrey; Uncle Hensleigh was a man of letters in London, and Uncle Frank managed the Wedgwood works in the Staffordshire Potteries.

Their father, Josiah Wedgwood II, had wanted to hand the works over to Joe, the eldest son, but Joe had no wish to take on the burden and sold his share in the partnership to Frank, his youngest brother. With the proceeds Joe bought Leith Hill Place in Surrey, a Georgian mansion with a four-hundred-acre estate on the south slope of Leith Hill, the highest point in southern England. When Emma first saw the house, she felt it was "really grand-looking, but I think rather too high for comfort when it grew very cold, and the house is somehow just like the open air."

Uncle Joe was elderly, silent and autocratic. His wife, Aunt Caroline, was Charles's elder sister and had brought him up after their mother died. Like Uncle Joe, she was a figure of awe for the children. Sophy, the eldest child, had been Annie's closest cousin since they were baptised together at Maer in 1841. Margaret, known as Greata, was two years younger, and Lucy was born three years after her. Sophy and Greata were tall and large-eyed, musical and closely attached to each other, but anxious and painfully shy with others. Emma wrote after one visit during Annie's childhood: "Sophy has still some odd morbid feelings, chiefly about anything relating to Heaven or God, but Caroline has persuaded her with great screwing up of her courage to stay in the room while she repeats a very short little prayer. It is too much to expect her to venture on saying a prayer for herself yet."

For Annie, the excitements of Leith Hill Place were all out of doors. There was a swing in the garden, and it was said that thirteen counties could be seen from the tower at the top of the hill. Next to the home farm there was a four-acre kitchen garden enclosed by a high brick wall with glasshouses and potting sheds. In the woods to the west, Uncle Joe and Aunt Caroline planted rhododendrons and azaleas from the Himalayas, and on the common to the north the children could sometimes see kangaroos leaping through the bracken. The animals were kept there by the Evelyn family, lords of the manor and enthusiastic naturalists.

Sophy shared all four grandparents with Annie. They were two shakes of the same genetic dice, but they could not have been more different. When Sophy was in her thirties, her Uncle Frank said: "I never met anyone silenter." She lived a frugal, miserly and increasingly eccentric life, making do on oranges and biscuits and insisting that others should too. According to one story told of her, she once found a bottle of the bitter medicine ipecacuanha when clearing out a cupboard; there was one dose left and she swallowed it, saying "A pity to waste a good emetic." After visiting Leith Hill Place in 1880, Emma wrote to George: "Poor Sophy strikes one anew every time one sees her as utterly dead, and quite as much dead to mother and sisters as to the outsiders. I felt the house with that long dark passage and no carpet so depressing, and wondered how they would ever get through the winter." But Sophy's music was important to her, and when her nephew

Ralph Vaughan Williams was a small child in the household, she gave him his first lessons at the piano.

Hensleigh and Fanny Wedgwood had moved with their six children from Gower Street to a house in Chester Terrace, the grandest of the imposing stucco buildings in Regent's Park. They led an active social life in a circle of friends that included Elizabeth Gaskell, Harriet Martineau, Leigh Hunt, Thomas and Jane Carlyle, Robert Browning, John Ruskin, Florence Nightingale and Richard Monckton Milnes. Their daughter Snow, precocious and reflective, was encouraged in her writing by Harriet Martineau and befriended by Elizabeth Gaskell and Florence Nightingale. For Annie, eight years younger, the house was a place from where she could go out with Snow's younger brothers and sisters to enjoy the excitements of the great city: the Zoological Gardens, the exhibition halls, the shops and the pantomime.

In the first days of January 1849, while Charles and Emma stayed with the younger children at Down, Willy, Annie and Etty came to London. The three were taken to Claudet's photographic studio in the Adelaide Gallery to sit for their portraits. At a guinea each in their red morocco leather and velvet cases, the three daguerreotypes cost the equivalent of £150 today. Willy was wearing a tight jacket with check trousers and waistcoat. The cut of his jacket was slightly old-fashioned but the check pattern of his trousers and waistcoat was up to date. Annie and Etty took their turns in the studio armchair, wearing matching loose dresses with crochet collars. When sisters were dressed to match as they were, it was common for the two fabrics to be of different colours. Etty's hair was cut simply and brushed firmly down like Willy's. She looked quite directly and unselfconsciously at the photographer and his apparatus. Annie was quite different, with her rich brown hair elaborately plaited and looped for the occasion. She held herself upright and still, tensely aware that her picture was being taken.

Uncle Frank lived with his wife and seven children at the Upper House in Barlaston, seven miles from the Etruria works in Staffordshire. The neighbourhood was one of the first landscapes of the Industrial Revolution, dominated by coal mines, iron foundries, pottery kilns and canals. Uncle Frank had tried to sell the works by auction in 1844, but trade was bad and when the property failed to reach its reserve, he took on the business and ran it conscientiously for the next thirty years. He was reckoned by his workmen to be a fair and good employer, partly perhaps because he left them to manage things largely as they wanted. He was cast in the same Wedgwood mould as his eldest brother, but had a warmth and sense of duty that Uncle Joe lacked. While others in the family joined the Church of England for social respectability, he remained true to the Unitarian faith of his grandfather, Josiah I, and fellow liberals in manufacturing and trade. He lived very simply, with plain wooden furniture and no carpet in his bedroom. The children were shy like Sophy and Greata. Among them, Cecily, though three years older than Annie, was her special friend.

The Wedgwood factory at Etruria, next to the Trent and Mersey Canal, had a place in the history of manufacturing. In the late 1700s, Josiah Wedgwood had pioneered mass production there, producing long runs of "useful" and "ornamental" ware of consistently high quality for ever-growing markets in Britain and overseas. But his works were not "dark Satanic mills." When Josiah II gave evidence to a parliamentary committee on children working in factories, he explained that his buildings were "very different from those of cotton works, and other manufactories in which machinery supplies the power. They are very irregular, and very much scattered, covering a great space of ground, and, in general, only of two storeys in height." They were a rabbit warren of yards and passages, kilns, workshops and drying rooms for the different stages of manufacture. One of Annie's London cousins called them "that nasty old pot shop."

Josiah Wedgwood IV, a radical Liberal MP in the 1900s, described how when he was young, the boys of the family would be taken around the works. "We turned and threw and blunged and painted and 'mussed ourselves up,' coming home in clothes patched with white, and clutching in our hands the largest lumps of cold, moist clay." His niece, the historian Veronica Wedgwood, explained in a radio talk in the 1950s how it had felt for her as a child of the family who was living away from the Potteries to be taken round for the first time. She had been there in the 1910s, but spoke for her cousins back through the generations. "In the labyrinthine stairs and corridors of the old works, it was fun to walk hand in hand with some purposeful striding elder, some godlike uncle or cousin, who had brought you in for a treat and was introducing you to everything and everyone. Pottery-making has an instant appeal. It was fascinating to see people going to and fro with huge plank-like trays of bowls and cups and coffee pots and sugar basins, looking like enlarged ghosts of the ones we had at home—ghosts because of the grey-white colour of the unbaked clay, and larger than life because they shrink in the firing. Then there was the breathless joy of watching the potter at his wheel. His assistant—how I longed to be that assistant!—slaps together a shapeless handful of wet clay, exactly the right amount, and puts it before him. The skilful hands close around it, and in a matter of seconds, as the wheel turns, it springs up like a live thing, lengthens, takes on a graceful, familiar contour, spins into a bowl or a vase. Then there was the light, open room where girls sat at work, deftly, symmetrically, painting the patterns on plates; in another room craftsmen were pressing clay into tiny delicate moulds of classical wreaths and nymphs and goddesses; and they miraculously released these fragile dancing figures by patting them out with a spatula."

When Annie went to Etruria in 1847 and 1848, the works were at a low ebb. There were two sections, "Earthenware" for everyday crockery and "OW" standing for Old Wedgwood, the Jasper and Black Basalt urns, vases, ewers and figures that had sold well in the past. The Wedgwood display at the Great Exhibition in 1851 consisted only of the established lines; it attracted little attention and Uncle Frank was not concerned, so long as the business ran smoothly. On his office door in the main courtyard of the works was a notice: "Please do not knock but come in." He sat inside at the table his grandfather had used, and a dusty showcase of old Wedgwood ware stood in the corner.

When Annie was taken round the factory, she may have met the eyes of children working there: a few girls, perhaps, of little more than her own age, familiar figures to her uncle during the working day, but living a life harshly different from his children's at home. Josiah II had told the parliamentary committee that he employed three hundred and eighty-seven people, of whom a hundred and three were between ten and seventeen and thirteen were under ten. The children worked the same twelve- to thirteen-hour day as the men "because generally the young children are employed in attending on the men, and assisting them in carrying their moulds and other little services that they can perform for them, the men working by the piece and paying those children themselves."

Josiah II tried to shrug off responsibility for the way the children were used. "I believe that the employment of children under ten years of age is never desired by the masters; that the employment of children under ten years of age is an accommodation to the workmen themselves, and perhaps in most instances they are employed under the eyes of their own parents." He recognised that one part of their work, "dipping" in the glaze before firing, was "unwholesome." The glaze contained white lead and the children, "if careless in their method of living, and dirty, are very subject to disease." They were employed on that work for one or two years only and if they showed "any symptom of suffering from the nature of their employment," their parents removed them. Some girls worked "with a camel hair pencil in painting patterns upon the ware, sitting at a table." The committee asked: "Is the state of health of the children in the painting rooms affected by the work? Are not consumptions found to exist very frequently?" Josiah II said he could not say because he had no information, and tried to suggest that the problem was not peculiar to the pottery industry. The children who painted had "that sort of delicacy which is universal in sedentary employments."

In the spring of 1849, Charles and Emma announced to the children that they would all be going to stay for a few months at Malvern, the spa village in Worcestershire, while Charles tried Dr. James Gully's water treatment for his chronic ill-health. The news was so remarkable for the children that Etty could remember sixty years later "the exact place in the road, coming up from the village, by the pond and the tall Lombardy poplars, where I was told."

After his illness at Maer in 1840, Charles had continued to suffer repeated bouts of sickness. William remembered that "It threw a certain air of sadness over the life at Down." In 1845 Charles wrote: "Many of my friends, I believe, think me a hypochondriac," but he was often in acute pain. Joseph Parslow the butler told a neighbour: "Many's a time when I was helping to nurse him, I've thought he would die in my arms." Charles and Emma understood that his health was often directly affected by his state of mind. When he was worried or upset about something, he could not control his thoughts, and he and Emma saw the physical effects of his mental turmoil. Underlying all his other concerns were his anxieties about the theory of evolution, the strain of living with the secret and his anticipation of the attacks when he announced it and people saw the implications. When he came at last to work on the final text of *The Origin of Species,* he felt it was the cause of "the main part of the ills which my flesh is heir to."

Charles had been particularly ill during the winter of 1848–49. He wrote to Hooker: "All this winter I have been bad enough, with dreadful vomiting every week, and my nervous system began to be affected, so that my hands trembled and head was often swimming. I was not able to do anything one day out of three, and was altogether too dispirited to write to you or to do anything but what I was compelled. I thought I was rapidly going the way of all flesh." He was suffering from insomnia, and whatever he did during the day "haunted" him at night "with vivid and most wearing repetition." In 1848, an old shipmate from HMS *Beagle,* Bartholomew Sulivan, had told him about some friends with stomach ailments who had been treated by Dr. James Gully at Malvern. When Charles then heard from his cousin Fox about two other people who had benefited from the treatment, he decided to find out about it. He read Dr. Gully's book, *The Water Cure in Chronic Disease,* and wrote to him. None of their letters to each other survives, but Dr. Gully's first reply appears to have been cautious and non-committal. Emma commented that he wrote "like a sensible man" and did not "speak too confidently." Charles wrote to Fox that he had resolved to see "whether there is any truth in Gully and the water cure. Regular doctors cannot check my incessant vomiting at all. It will cause a sad delay in my barnacle work, but if once half-well, I could do more in six months than I now do in two years."

In his treatment at Malvern, Dr. Gully used water in ways that were conventional in medicine at the time, but he challenged accepted medical practice in rejecting all drugs. Most active preparations in use at the time were emetics, laxatives or purgatives. Some contained heavy metal poisons like arsenic, lead and mercury, and others were toxic substances from plants: opium, strychnine and quinine. Dr. Gully had become interested in the water cure after the death of his two-year-old daughter from croup in 1840. She had been treated with "emetics to prolong nausea, mercury to increase salivation, endless blistering and drastic purgatives," and she died in extreme pain. Shortly afterwards, Dr. Gully met Dr. James Wilson, who argued that many illnesses were simply "drug diseases." "Many of the most desperate cases . . . I have seen, owed their forlorn state to little else than mercury, quinine, arsenic, and purgatives."

Dr. Wilson linked his concerns about drugs with the evils of alcohol and inebriation. "How often have I observed the undertaker's house placed between a gin palace and a druggist's shop, and heard at the same time the curse and drunken hiccup, the undertaker's hammer, and the pestle and mortar of the druggist, blending into strange unison." One evening in London, Dr. Wilson observed "a quiet, retired-looking undertaker's, a single candle in the front shop dimly lighting a solitary figure dressed in rusty black. In his hand he held a small hammer, with which he produced a monotonous and unceasing rat-tat . . . on a coffin lid. Did you ever hear while alone at night, the *tic-tic-tic* of what is called the spider death-watch, or observe it while half hidden at the mouth of its net, watching the passing flies? So seemed to me the position of this solitary figure in black, as the crowd of anxious-looking men and women were hurrying by; numbers of whom suddenly arresting their steps, entered on one side into a gin palace, on the other into a druggist's shop, both brilliantly illuminated, of festive appearance and inviting aspect. They looked to me, indeed, like gaudy baits, held out by the solitary figure of the hammer and nails, to entice the human flies into the deadly trap—where 'FUNERALS ARE PERFORMED.'"

Searching everywhere for a method of treatment without drugs, Dr. Wilson had visited Vincent Priessnitz, who offered a simple water cure at Graefenberg in the mountains of Silesia. On Wilson's return to England, he persuaded Dr. Gully to join him in setting up a clinic for hydrotherapy. They took a hotel in Malvern for their enterprise, because the village was already a fashionable watering place with a reputation for a healthy climate in spring and summer. Some years before, a physician had discovered the exceptional purity of the water which flowed from the springs below the steep Malvern Hills to the west. As a local wit put it:

*The Malvern water, says Dr John Wall,*
*Is famous for containing just nothing at all.*

The air was invigorating, the hills gave shelter, and there were unrivalled views to the east over the Vale of Severn, and west to the hills of Herefordshire. Princess Victoria had spent some time there in 1830, and Queen Adelaide came a number of times in the following years.

This royal patronage had given Malvern a cachet, but Dr. Gully and Dr. Wilson's patients gave it a style and atmosphere quite different from other watering places. The spa had two springs; there were also a bathhouse, public buildings and a number of villas, all recently built in a heavy classical style. The streets were thronged with invalids old and young, "all resolutely bent on the business of getting health, 'building themselves up,' as the phrase of the place is."

Each of the two doctors had his own theory of the workings of the human body, and their water treatment consisted of a strict diet and a range of special ways of washing and soaking different parts of the body to achieve effects on the nervous system and circulation. Lying wrapped in wet sheets for hours on end was the best known. Sweating with a lamp under a blanket was another, and a third was the "douche," a special shower with a sudden fall of icy water from a great height. Among Dr. Gully and Dr. Wilson's patients in the 1840s and 1850s were many prominent writers and thinkers, politicians and churchmen. Sir Edward Bulwer-Lytton, Thomas and Jane Carlyle, Henry Hallam, Alfred Tennyson, Florence Nightingale, Charles Dickens and his wife, and Bishop Samuel Wilberforce all came for the water treatment. Joseph Leech, a journalist in Bristol, described the cure in *Three Weeks in Wet Sheets, Being the Diary and Doings of a Moist Visitor to Malvern.* He said of those who had lost faith in orthodox medicine: "It is, I think, in medicine as in religion. Let a man once forsake his old faith, and he is sure to make a great many more changes—soon run through the whole cycle of systems. A man may go on for years satisfied with his old family physician—swallow the potions given him with implicit reverence, but let him become a sceptic for a moment, and refuse to believe in drugs, and if he tries one 'opathy,' he'll try them all." Thomas Carlyle caught an element of the appeal of the water cure to people among his and the Wedgwoods' freethinking friends when he commented that many of his fellow patients enjoyed the "strange quasi-monastic—godless and yet *devotional*—way of life which human creatures have here." It was "useful to them beyond doubt. I foresee this 'water cure,' under better forms, will become the Ramadan of the overworked, unbelieving English in time coming, an institution they were dreadfully in want of, this long while!"

Dr. Gully and Dr. Wilson abandoned their partnership after a few years, and a third doctor, James Marsden, came to compete with them. "The three hydropathic doctors . . . have each had their portraits lithographed, and from the window of the bookshop and the bazaar, and even from the walls of the inns, they seem to bid for the possession and management of the visitor's body on his arrival. With whiskers silkily curled, sitting by a table, and serenely musing, the 'Great Original,' Wilson, seems intrepidly to assure the invalid and hypochondriac of a cure. Standing up with arms a-kimbo, pert and pragmatical, Dr Gully appears to push himself forward, and say 'I am your man—try me': while Marsden, who unites homoe-opathy with hydropathy, may be said to have a mezzotint manner between both, and looks from his frame upon you as intently as if he were listening to your case."

Charles and Emma took the family, Miss Thorley and the servants to Malvern in the early summer. They travelled from London to Birmingham on the London & North Western Railway, a three-hour journey by the fastest service. Railway travel was an excitement for the children. Mrs. Marcet wrote about a small boy's first rail journey in her book *Willy's Travels on the Railroad*. When he looked out the window he saw "the houses and trees and fields, looking as if they were moving. 'I know they do not, but in the railroad, I think everything seems to be moving. And do, Papa, look, how little the cows are in that field. And are those sheep? they seem to be no bigger than lambs; and I declare those houses,' said he, pointing to them, 'look almost like baby houses at the toy shop.' 'Those houses are really small,' replied his father, 'but not so very small as you suppose, for they are large enough for people to live in; everything seen from the train when it is moving fast, appears smaller than it really is.' "

At Birmingham the family changed to the Bristol & Birmingham Railway for the short journey to Worcester. When Dickens passed through the city in 1851, he wrote of its quiet "with the lights and shadows of its cathedral architecture, cut sharp by the strong sunlight. Even the central streets are quiet, in comparison with Birmingham; much more so the clean, old-fashioned, red-brick houses within the precincts, where the very pavement seems to be never soiled by the tread of less dainty feet than those of clergy and ladies."

The last leg of the journey was an eight-mile ride to Malvern in a four-horse coach, the Hereford Queen, the Hereford Mail or the Mazeppa. The road passed first through apple orchards, then past hop-grounds and pear orchards until the Malvern Hills loomed to the west. As Dickens described the approach to the village, he saw first the blue mass of the Malvern Hills, "growing browner and greener with every mile; then the black surface of rich woods, rising from the skirts; then the long, straight row of dwellings, with their white walls shining in the sun."

The Darwins arrived at Archer's Royal Kent & Foley Arms Hotel on the Worcester Road, where families were boarded in private apartments. The hotel was royal because Queen Adelaide had stayed there when she came to Malvern. She had allowed the hotel to place her coat of arms on the cast-iron balcony above the entrance, and it is there now with a lion and unicorn supporting the Teck family crest.

Charles took a villa for the family and they were settled there in a few days. The Lodge, as it was then called, was one of a number of large white stucco houses built along the Worcester Road in the 1830s and 1840s for well-to-do visitors to the spa. The house was set in its own grounds on the wooded slope above the road, a three-storey building almost identical to Down House except for the imposing classical ornament and the darkness of the cavernous hall. Charles wrote to Fox: "We have got a very comfortable house, with a little field and wood opening on to the mountain, capital for the children to play in." George remembered the slope of the garden down to the road and a little fountain halfway between the house and the gate on the carriage drive. The fountain is still there under a rough brick arch set into the steeply rising ground beneath a large rhododendron. The water trickles from a marble boss into a stone basin and falls from the lip into a small pool fringed by Solomon's seal.

Charles wrote to Fox: "I much like and think highly of Dr. Gully. He has been very cautious in his treatment and has even had the charity to stint me only to six pinches of snuff daily. Cold scrubbing in morning, two cold feet baths and compress on stomach is as yet the only treatment, besides change of diet &c. I am, however, tomorrow to commence a sweating process . . . I expect fully that the system will greatly benefit me, and certainly the regular Doctors could do nothing." To his sister he wrote: "I like Dr Gully much—he is certainly an able man. I have been struck with how many remarks he has made similar to those of my Father. He is very kind and attentive; but seems puzzled with my case—thinks my head or top of spinal cord cause of mischief."

Annie followed the treatment, and watched with a daughter's close concern how her father reacted to it. Emma wrote to Fox: "Annie was telling Miss Thorley all her Papa had to do about the water cure and how he liked it. 'And it makes Papa so angry.' Miss T must have thought it a very odd effect. He said it did make him feel cross."

Annie enjoyed the stay. She had a Leghorn hat and a black "polka" knitted jacket. The family explored the neighbourhood, and Etty was struck by a stream widening out into a pool overshadowed with trees, as streams were unknown to the children in the dry chalk country around Downe. George went one day with his father and mother to a toy-bazaar. "They bought me a little jingling organ which made a twanging with wires stretched inside. I remember that with a passion for realism I had it mounted on a stick and tied round me with a piece of string to imitate an organ-grinder, and that I broke it open to examine the inside." The shop was probably Henry Lamb's Royal Library and Bazaar, which sold an "immense variety of the newest and most elegant and useful goods, and almost every article connected with the fancy, bookseller's and stationer's business." It had a music saloon with pianos for hire, and a reading room with London and provincial newspapers.

One of the springs, St. Anne's Well, lay in a hollow in the hillside reached by a path climbing steeply from the centre of the village. According to Joseph Leech, "The water itself, which dribbles away into a carved stone basin at the rate of about a glass a minute, through a kind of penny whistle placed in the mouth of a pleasant dolphin, is quaffed by crowds in a little house which is half a pedlar's shop and half a pump room, attached to a cottage where knives and forks are hired out to tourists, and kidney surreptitiously grilled between meals for hungry patients under water treatment."

Dickens overheard talk at the spring. In the cool chamber a woman sat, "plying her needle, while enjoying pious conversation with a lady who has some tracts in her hand. They are saying 'how very 'andsome' the clergyman was that preached last Sunday." A German band, brought over by Dr. Gully, also played there every morning, and invalids could then "slowly imbibe the pure element to an andante of Haydn's, or toss off tumblers from the 'sacred rill' to a Pot-pourie of Donizetti, or the measured time of the Presburgh Polka."

A guide for visitors explained that the paths up behind the well "wind in every direction among the hills, constantly traversed during the season by parties of pedestrians, or invalids on donkeys, enjoying the many splendid views with which the country abounds, and courting the invigorating breezes, there far removed from those sources of contamination too frequent in the vales below." Leech wrote that, "the moment you set your foot on the summit, commerce cries and clatters around you with its importunate clamour, in the shape of basket-girls pestering you to become the purchaser of ginger-beer, biscuits, and walnuts, or the owners of return donkeys tempting you with a cheap ride down again." The donkeys could be hired by the day or the hour from the Donkey Exchange near Lamb's Bazaar. They stood patiently, as many as forty or fifty, with white cotton cloths covering the side saddles for ladies.

During the family's stay, Willy went for lessons with a clergyman, Mr. Fancourt, who took in pupils at the Ankardine House Academy. There were dancing lessons on Fridays for all the children. These were a treat for Annie, as there was no dancing instruction in Downe. The Misses Clinnick at Pomona House had "secured the assistance of the most eminent Masters in the county for Music, Singing, Dancing, and Drawing." Annie enjoyed music; watching her face as she listened to others playing, Charles felt sure that she had a strong taste for it. He wrote to a friend, "I believe . . . that she is a second Mozart," and then, thinking of his own weakness with tunes, he added, "She is more than a Mozart, considering her Darwin blood."

Dancing was an essential accomplishment for young ladies at the time, and it taught a child important points about behaviour, expression and grace. John Locke had written a hundred and fifty years before in his *Treatise on Education:* "Nothing appears to me to give children so much confidence and behaviour, and so to raise them to the conversation of those above their years, as dancing." Children were first taught the positions, which "constitute the alphabet of dancing." *The Young Lady's Book,* a manual of feminine accomplishments, said: "The Battements &c in the positions form a series of very graceful domestic morning exercises, and we strongly recommend their frequent practice." Annie danced quadrilles; according to *The Young Lady's Book,* "the lady who joins in a quadrille aspires . . . to glide through the figure with easy and unobtrusive grace."

Another entertainment for the family was Dr. Gully's belief in clairvoyance. George remembered that "he bothered my father for some time to have a consultation with a clairvoyante, who was staying at Malvern, and was reputed to be able to see the insides of people and discover the real nature of their ailments. At last he assented to pacify Dr Gully, but on condition that he should be allowed to test the clairvoyante's powers for himself. Accordingly, on going to the interview he put a bank note in a sealed envelope. After being introduced to the lady, he said 'I have heard a great deal of your powers of reading concealed writings and I should like to have evidence myself; now in this envelope there is a bank note and if you will read the number I shall be happy to present it to you.' The clairvoyante answered scornfully 'I have a maid-servant at home who can do that.' But she had her revenge, for on proceeding to the diagnosis of my father's illness, she gave a most appalling picture of the horrors which she saw in his inside."

After three months of the treatment, Charles wrote to John Herschel that it had had "an astonishingly renovating action" on his health. "Before coming here, I was almost quite broken down, head swimming, hands tremulous and never a week without violent vomiting. All this is gone, and I can now walk between two and three miles. Physiologically, it is most curious how the violent excitement of the skin, produced by simple water, has acted on all my internal organs. I mention all this out of gratitude to a process which I thought quackery a year since, but which now I most deeply lament I had not heard of some few years ago."

As his health slowly recovered, Charles went out for long walks on his own. He told Fox that he had looked for beetles for old times' sake, but could not find one. When he walked in the early morning on the great hill above the house he would have seen, as Dickens did, "the sheep come running up the shaded side to meet the sun, instead of crouching into dark nooks. Then the lark springs up from some grassy crevice, and the swallows are innumerable . . . On the east the mists still shroud the landscape; but on the Herefordshire side all is clear and bright, both within the shadow of the hills and beyond it. What a vast shadow it is! and how cool lie the farmsteads and orchards and dark pools within it!"

Some features of the landscape reminded Charles of the ancient history of the earth and his species theory. When he had accompanied

Professor Sedgwick on his geological expedition to Wales in 1831, Sedgwick was working out the sequence of the oldest strata and the fossils they contained, because he realised that life forms appeared for the first time at a certain level, and the forms of progression they showed, such as that from invertebrates to vertebrates, must have some significance for the history of the natural world.

A young clergyman, the Reverend William Symonds, knew Charles from the Geological Society and gave lectures in Malvern about the rocks of the neighbourhood. In his book *Old Stones,* he explained the origins of the Malvern Hills as Charles would have understood them. They were a ridge formed from some of the oldest rocks known, at the time called "Plutonic" from the ancient god of hell, which had been thrust up through the overlying strata in the remote geological past. To give a sense of the forces involved, Symonds quoted "the words of Mr Darwin in *The Voyage of the Beagle* on the effect of earthquakes, and the gradual uprising and sinking of land in South America . . . 'Daily it is forced home on the mind of the geologist, that nothing, not even the wind that blows, is so unstable as the level of the crust of this earth.' " Symonds wrote that "forces rest beneath, which, if called forth, might rend a world; but which are yet so beautifully and so perfectly under *His* control, that slowly and imperceptibly whole continents are uplifted and depressed." Symonds pointed to evidence all around the neighbourhood "of the vast, the unmeasured lapse of time since first our planet was called into being." He felt with Sedgwick that "geology impresses the mind, perhaps more than any other subject of natural history, with the truth of a beginning, and the far-reaching and *eternal* agency of the First Great Cause . . . Surely, then, we learn here an important fact in geologic history, viz. that everything was *created,* and that the lowliest animal appeared not upon the wide world's surface without the fiat of the Creator."

Another local naturalist, Edwin Lees, had similar visions of the distant past in the plant-covered rubble of a limestone quarry on the south side of the hills. If Charles found a quarry on a walk, he would always look round for any rock exposures of interest. Lees wrote: "Before the eye are exposed the supporting ribs of the earth's framework, moulded within the depths of the primeval ocean, and now lifted up today by the force of the . . . outbreak that reared that lofty citadel . . . Broken debris of ancient marine life strew the ground all about . . . But strange is the mixture of the extinct life of the past, that formed the foundations of the hills, with the evanescent encroachments of the present. The ancient heraldic forms of nature, the Trilobite and the Encrinite, with numerous shells and corals, lie in upturned confusion on the sides of the quarry, covered with the rank weeds and thistles that strive again to entomb them, and with monstrous coltsfoot leaves, that spread thickly around, though only the growth of a single season—the hasty scum of yesterday. But the deposit of past life is thrown into the kiln, the smoke of ancient forests trails in the air, and lime goes to form new combinations of life in this wonder-working scene."

When Annie walked with her father over the hills, he could tell her about their history and the succession of past worlds of which they had been part. He kept a young person's wonder in his understanding of nature, and he was glad to share it with his children.

CHAPTER EIGHT

# THE FRETFULNESS OF A CHILD

*Annie's last summer—Illness—Ramsgate—*
*Winter months—Annie's birthday*

WHEN THE FAMILY CAME HOME from Malvern in 1849, Charles carried on with the water treatment on his own, and John Lewis the builder erected a wooden hut for his daily shower near the well in the garden. Lewis's fifteen-year-old son came to work in the household as Charles's page. Every morning he would go out to the hut and pump gallons of water up into a little steeple attached to the roof. Charles undressed in the hut; he pulled a string, and the water fell on his back with great force. Etty remembered that she and Annie used to stand outside to "listen to his groans, and I have an image of his coming out half running and half frozen to take his usual morning walk in the Sand-walk, where we meant to accompany him."

Charles and Emma's eighth child, Leonard, was born in January 1850. Chloroform had been introduced for relieving pain two years before, and Charles had arranged for the doctor to give it to Emma for the birth. But, as he wrote to his cousin Fox, "Her pains came on so rapidly and severe, that I could not withstand her entreaties for chloroform and administered it myself, which was nervous work, not knowing from eye-sight anything about it or of midwifery. The Doctor got here only ten minutes before the birth. I thought at the time I was only soothing the pains, but it seems she remembers nothing from the first pain till she heard that the child was born. Is this not grand?" He wrote later to Hooker that when he gave the treatment to Emma, "I was perfectly convinced that the chloroform was very composing to oneself as well as to the patient."

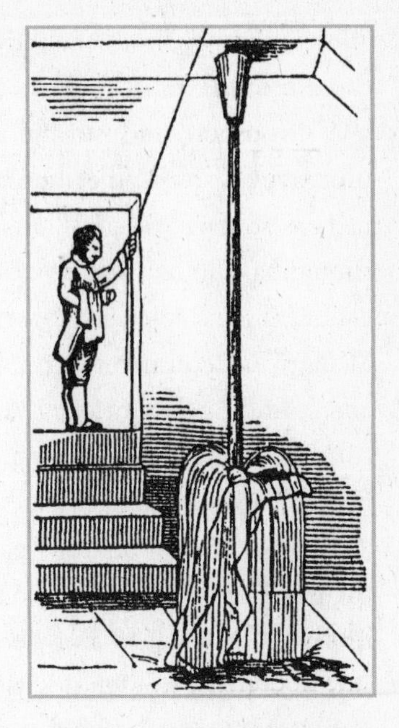

*Douche bath at Malvern*

Charles had been worrying for some time about his boys' future. He was well off now that he had inherited part of his father's fortune, and he hoped to pass on his wealth to his children, but with seven now to provide for and the ever-present fear that one of his investments might fail, he wanted each boy to have a profession. He could not bear to think of them wasting "seven or eight years in making miserable Latin verses," but Willy had to master Latin to be admitted to a boarding school, and Charles found a tutor "who teaches nothing on earth but the Latin Grammar." At the end of January, when Willy was just ten, he went to lodge with the Reverend Mr. Wharton, Vicar of Mitcham in Surrey. Mr. Wharton had six other pupils between eleven and fourteen. All Willy remembered later of his time with them was the stag beetles which were common in the neighbourhood and kept by the boys to play with.

When Willy left for Mr. Wharton's, Annie was the eldest child at home. Etty was six, George four, Betty two and Francis one. The younger children were all close in age; the gap of two and a half years between Annie and Etty was a reminder of Mary who lay in the family grave near the door of the parish church.

Annie now led the others out to play; she suggested games and kept the peace. The child whom everyone remembered later with such affection must have been the confident and forthcoming Annie of those months. Charles wrote later that "her cordiality, openness, buoyant joyousness and strong affection made her most lovable." His unmarried sister Catherine said she often used to think of living nearby in her "solitary old age," and "what a bright affectionate little niece I should have in her, and how she would not despise me, but be always so candid and kind-hearted which was so entirely her nature." Fanny Wedgwood commented on her "bright, engaging, qualities, so open and confiding and lovable." And Emma's brother-in-law, Charles Langton, commented on her "responsive and confiding nature," adding that he had "always found her a child whose heart it was easy to reach."

Charles was an anxious parent, feeling that "Nothing comes up to the misery of having illness amongst one's children." But up to Annie's ninth year, they had not been seriously affected by the diseases that struck so many households. Some illnesses came "more or less to all" as one doctor wrote, and when each of the Darwin children was born, an entry was made in the family Bible with headings for smallpox vaccination, chickenpox, measles, scarlet fever and whooping cough. Willy, Annie, Etty and George had chickenpox in 1845, and Annie, Etty and Betty had scarlet fever in 1849. Charles and Emma were acutely worried when the scarlet fever came because it was often fatal, but their children recovered. When Charles wrote to Syms Covington in Australia later in the year, he mentioned that his children were "all, thank God, well and strong."

In May 1850, Miss Thorley took Annie on a trip to London. The "great icon of the day" was Obaysch, the first hippopotamus to be seen in England, or at least the first for half a million years, as some palaeontologists pointed out. The British Consul General in Cairo had obtained it from Abbas Pasha, Viceroy of Egypt, for the collection at the Zoological Gardens, and Charles had tickets as a Fellow of the Zoological Society. Everyone wanted them and Charles passed on to his brother Erasmus those that the family did not use. When Professor Owen saw the animal, "it now and then uttered a soft complacent grunt, and lazily opening its thick smooth eyelids, leered at its keeper with a singular protruding movement of the eyeball." The historian Thomas Macaulay wrote: "I have seen the Hippo both asleep and awake, and I can assure you that, asleep or awake, he is the ugliest of the works of God." But Queen Victoria, who eventually saw Obaysch five times, found that his eyes were "very intelligent."

For Annie, the first three weeks of June were quiet; aunts and uncles came and went while she looked after the younger children. At the end of the month, Willy came back from Mr. Wharton's for the summer holiday, and five Wedgwood cousins came to stay: Ernie and Effie from London, and Cecily, Amy and Clement from Barlaston. Willy had escaped from his Latin lessons and, for the cousins, the easy and open way of life at Down was also a release from the confines of their households in London and the Potteries. There were six older children together, but Annie had Cecily and Effie as her special friends.

The cousins' arrival was a sudden change of pace for Annie; the weather was very hot, and five days later, as the dry spell broke with thunder and lightning, Emma saw that she was unwell. Emma wrote later in her diary: "Annie first failed about this time," and in her recollections of Annie's last months, she noted that she "never was well many days together afterwards, finding her lessons a great effort and frequently crying especially after she went to bed." Emma remembered that during this time, "from her clinging affectionate nature, if she felt uncomfortable she was never easy without being with one of us."

In August Emma took Annie and Miss Thorley on a nine-mile ride in the family's phaeton to Knole. The great house was open to visitors, and the Darwins admired the thousand-acre park. A guide published in 1839 described how from a furze-bank in the grounds the visitor could "catch the chimneys and long roofs of the house, prominent above the many broad masses of oak and sycamore and the feathery tops of beech which rise from the intervening valley; and it will well repay an evening walk to view its towers and gables standing forth in bold relief against the setting sun, or the long lines of the grey twilight clouds."

A few days later, Charles and Emma took Willy, Annie, Etty and the five-month-old Leonard to stay with Uncle Joe and Aunt Caroline at Leith Hill Place. Sophy was now nine, Greata seven and Lucy four. Emma remembered that Annie was very happy. The children clambered over the heath above the house looking for the season's bilberries in the low-growing shrubs. The berries were small and purplish black, and their sweet juice stained the children's hands and mouths; but Annie was "overfatigued" by the effort.

Back at Down, Emma found that she was pregnant with her ninth child. In September, Charles noted that Willy was "showing the hereditary principle by a passion for collecting Lepidoptera," and bought him an entomological box. He also gave him a pony, and started teaching him to ride. "We began without stirrups, and in consequence Willy got two severe falls, one almost serious. So we are thinking of giving him stirrups, more especially as I am assured that a boy who rides well without stirrups has almost to begin again when he takes to stirrups."

On the last Sunday in the month, the villagers turned out to watch when the Archbishop of Canterbury came to confirm the elder Lubbock children in the parish church. He gave an address "both extempore, and very plain and easy, even for the most uneducated to understand," as Lady Lubbock noted in her diary. He spoke to the children on the verse:

*Teach me to live that I may dread*
*The grave as little as my bed.*

At the beginning of October, Charles bought a canary for Annie and Etty. The way to teach a bird to sing was to place its cage in a room by itself and drape it with a cloth. A short tune should then be whistled or played on a flute or bird-organ, five or six times a day. The canary would take a few months to learn. While the children played with their pet, Charles watched with his own interests. He had been reading books about caged birds and talking to bird fanciers ever since he recognised the importance of the Galapagos finches for thinking about the origin of species. The marked differences of appearance and character between the different breeds of "canary finch" were a point for his argument that such variations were the raw material for natural selection. He was interested to know the results of crossing canaries with other species, "mule-breeding" in the phrase of the fanciers who made a hobby of it. He was intrigued by how some birds learnt songs by imitation, and suggested that the "singing of birds, not being instinctive, is hereditary knowledge like that of man." He was also intrigued by reports of canaries singing to their reflection in a mirror. Putting a mirror in your bird's cage to encourage it to sing was a trick many cottagers in Downe would know. For Charles there were parallels with his experiments with Jenny the orang, Willy and Annie ten years before. What did the bird make of its reflected image? Why did it sing to it? Out of rivalry or love? How to guess?

*Picture in the Darwin children's scrap book*

. . .

By early autumn, Charles and Emma had recognised that Annie had an illness of some kind which she was not managing to shake off, but she had no clear symptoms pointing to any particular cause for her distress. One recognised treatment for a delicate child was sea-bathing. Charles had once written to a friend whose children were ill, that "the sea will do all good." A medical encyclopaedia of the time said that bathing was an effective treatment for "languor and weakness of circulation," "many of those symptoms usually called nervous," and "a listless and indolent state of the mind." "A sudden plunge into the ocean causes the blood to circulate briskly, and promotes the heat of the body." The treatment also worked through "the free exposure of the body to the bracing sea-breeze."

In the first week of October, Miss Thorley took Annie and Etty to spend a few weeks at Ramsgate on the Kent coast. The town was served by the new London & South Eastern Railway and was, in a vaunting description of the time, "one of the most elegant resorts for sea-bathing in the kingdom." The stone east pier reaching out into the English Channel was "perhaps the finest marine parade in the universe: and here, in fine weather, on the departure, but more surely on the arrival of the steam-vessels, it is daily graced with an assemblage of fashion and beauty perfectly startling to those who see it for the first time. The sands . . . are considered as fine as any in existence, if not finer." In the summer a tent on the beach was "supplied with newspapers, magazines, and books, on easy terms. Chairs on the beach are allowed to subscribers, and at a small gratuity to non-subscribers. Donkeys and light vehicles are in constant attendance, caparisoned gaily, if not gorgeously: and the whole scene bears an exceedingly animated and joyous aspect."

The rules of modesty made sea-bathing a delicate and laborious procedure. Bathers gathered in the bathing room on the promenade. A guide to Ramsgate explained: "As most of the company prefer the morning for bathing, a slate is put up in the lobby to receive the names of the bathers, who take their turns in entering their machines, which are carried by horses a convenient depth into the sea, under the conduct of careful guides. At the back of the machine there is a door through which the bathers descend a few steps into the water, and an umbrella of canvas being contrived to fall over them, it conceals them from public view. The pleasure and advantage of this salubrious exercise may be enjoyed in a manner consistent with the most refined delicacy." A lady or child taking a machine with a guide would be charged 1s 3d. The price for two or more young children with a guide was 9d each. A writer described a conversation in the bathing room. "Most of the company had talked over their own case, which *invalids* are particularly fond of doing, and all had given a judgement on the sea; but in general so contradictory, that had I formed my opinion on others, it would have amounted nearly to this— that it thinned and it thickened the blood—it strengthened it—it weakened it—made people fat—it made them lean—it braced—it relaxed—it was good for everything—and good for nothing. *'It will wash you all clean however,'* says a gentleman in the gallery, 'if it does nothing else.' "

On the beach, there was "a strange hum of human voices wherever you wend your steps; while shouts arise from the boys who urge their horses seaward and landward." The bathing women would help unwilling children to enter the water by singing to them. "Here is a blue-dressed dipper doing what she can to assuage the fears of some delicate girl, whose plump and robust mamma has resolved that she should bathe." Hawkers went "hither and thither with cheap and gaily-coloured books for all ages; wooden spades for the children's favourite manufacture of 'sand-pies' and hills; sand shoes for delicate feet of both sexes; and shells of divers hues, and forms, and gay iridescence, with sea-weeds and corallines in rich variety and abundance."

A fortnight after the children had arrived with Miss Thorley, Charles and Emma came to join them. Emma remembered afterwards Annie's "bright face on meeting us at the station." They had a windy walk on the pier and bathed twice. Etty made doll's shoes out of seaweed on the beach, and walked with her father, who "entered into daily life with a youthfulness of enjoyment which made us feel we saw more of him in a week of holiday than a month at home." The children also collected shells. A booklet of the time with pictures of each kind that could be found on the beach showed knotty cockles, white and beaked piddocks, truncated gapers, Sandwich beauties, keeled worm shells, pig cowries and potted trumpets. Poor women made a living by collecting and selling the scarce kinds, and boys on the fishing boats hawked shells which came up in their nets. Annie and Etty found piddocks, limpets, tellins and scallops, and Charles would have been able to help name them, remembering all the shells he had collected and listed with Syms Covington during the *Beagle* voyage.

For Charles, the days in Ramsgate were a release from his work at his dissecting microscope, but there were crustings of barnacle shells on the stonework of the pier to remind him of it. *Pickings on the Sea-shore,* a guide to the natural history of the beaches of Kent and Sussex, quoted a poem by Jane Taylor about a naturalist walking on a seaside promenade.

*Now 'tis high water, and with hundreds more,*
*He goes to catch a breeze along the shore;*
*Or pace the crowded terrace, where one sees*
*Fashion and folly, beauty and disease.*
*Then one, perchance, who differs from the rest,*
*As much as—O, too much to be express'd;*
*He, Nature's genuine lover, casts his eye,*
*Lit up with intellect, on sea and sky,*
*Drinks in the scene, and feels his bosom swell*
*With what he could not, what he would not tell.*

Two days after Charles and Emma arrived at the lodgings, Annie developed a fever and headache. A mattress was fetched so that she could sleep in her parents' room. The next morning there was a fierce autumn storm which battered the house all day. Charles, Etty and Miss Thorley left for home, and Emma followed them with Annie a few days later when Annie was fit enough to travel.

Annie now was seriously unwell; Charles and Emma's concern for her was filling their minds and they needed the best advice they could find. Although Charles was willing to follow Dr. Gully's instructions for his own treatment, he had reservations about his judgement on others' illnesses because of the water doctor's interest in clairvoyance, homoeopathy and other unorthodox treatments. Charles wrote that homoeopathy was "a subject which makes me more wrath, even than does clairvoyance: clairvoyance so transcends belief, that one's ordinary faculties are put out of question, but in homoeopathy, common sense and common observation come into play, and both these must go to the dogs, if the infinitesimal doses have any effect whatever." Charles may have been afraid that Dr. Gully would suggest both for Annie. "It is a sad flaw, I cannot but think, in my beloved Dr Gully, that he believes in everything. When his daughter was very ill, he had a clairvoyant girl to report on internal changes, a mesmerist to put her to sleep, an homoeopathist . . . and himself as hydropathist!"

So Charles and Emma looked first to orthodox medicine. At the beginning of November, just a week after they came home from Ramsgate, Emma took Annie to London to see Dr. Henry Holland, who had attended at Annie's birth. Emma noted later that Annie's "nights became worse about this time."

Annie was not kept in her bedroom or the nursery, but carried on her life around the house and garden as best she could. One day in late

November, she and Etty went to see their Great Aunt Sarah who had taken a house called Petleys in the village, to spend her last years close to Emma. She was the last surviving child of the first Josiah Wedgwood, "tall, upright and thin" according to Etty. She was an austere figure who had impressed Samuel Taylor Coleridge when they met in the 1790s, but found it difficult to show any warmth of feeling for anyone, and devoted her life to religion and good works. She once wrote: "It is my misfortune to be not of an affectionate disposition, though affection is almost the only thing in the world I value." She kept several pairs of gloves by her, "loose black ones for putting on coals and shaking hands with little boys and girls, and others for reading books and cleaner occupations." Her three servants, Mrs. Morrey, Martha and Henry Hemmings, were Etty and Annie's good friends, and Etty remembered that "whenever life was a little flat at home, we could troop off, crossing the three fields that separated our house from Petleys, sure of a warm welcome from them." Mrs. Morrey's gingerbread was like no other Etty ever tasted, and Martha would sing them songs "which only gained by repetition."

Aunt Sarah paid no attention to her garden, but the flowers that grew there seemed to Annie and Etty to have "a mysterious charm." Every autumn, the children picked small wild plums called bullaces in the hedge of her little field. That November afternoon, Annie found some to eat, but she could not remember their name, as she wrote to Effie the next day.

During the following week Brodie gave her a present, a little pocketbook which she had made herself, embroidered with flowers and leaves in chenille and silver thread, and tied with a red silk ribbon. Annie started a diary on the first page, and wrote:

23. Cicely wrote to me. Went Aunt S. Rainy morn. but fine afternoon.

24. Very rainy all day.
25. Cold but rather fine morn.
26. Thremomiter 46.
27. Greata wrote to me.

Was Cecily writing from Barlaston and Greata from Leith Hill Place because Annie was ill?

Annie replied to Greata on a piece of fancy notepaper from her writing case. "We have got a new pony. It is rather a little one. I think your donkey sounds a very nice one. I should like to see your little white guinea-fowl. Are all the little turkeys sold? On Sunday it was raining dreadfully, and the pit in the sand walk was full of water. Is your swing taken down? Ours has been taken down a long while. Have the calfs grown much bigger since I was at Leith Hill? I suppose they have."

Annie and Etty arranged their shells from Ramsgate, and Charles gave them a handful more from the collection he had brought back from the *Beagle* voyage. It had been in store ever since Syms Covington copied a list of the specimens at their lodgings in London over ten years before. Charles had tried to find an expert conchologist to describe them, but no one had been interested. If no scientist wanted to examine the shells, the children could now have some of them to look over with their sharp young eyes.

When any of the children were unwell, Charles and Emma were "unwearied in their efforts to soothe and amuse" them. Etty remembered that when she was ill, "my father played backgammon with me regularly every day, and my mother would read aloud to me . . . I remember the haven of peace and comfort it seemed to me . . . to be tucked up on the study sofa, idly considering the old geological map hung on the wall." She remembered her father sitting in the horsehair armchair by the corner of the fire. When a child was ill, while the others had their tea in the schoolroom, the sick one was allowed to have hers with Charles and Emma in the study, with their old blue teacups on the mahogany Pembroke table.

During November, when Annie came into the study, Charles was working on a collection of barnacles that Covington had sent him from Twofold Bay on the coast near where he had settled in New South Wales. When Charles wrote to thank him for them, he explained: "I have received a vast number of collections from different places, but never one so rich from one locality. One of the kinds is most curious. It is a new species of a genus of which only one specimen is known to exist in the world, and it is in the British Museum." Charles named the species *Catophragmus polymerus,* and wrote later in his monograph on barnacles that the genus was remarkable among barnacles of its kind, "from the eight normal compartments of the shell being surrounded by several whorls of supplemental compartments or scales: these are arranged symmetrically, and decrease in size but increase in number towards the circumference and basal margin. A well-preserved specimen has a very elegant appearance, like certain compound flowers, which when half open are surrounded by imbricated and graduated scales." He may have shown Annie the beauty of the flower-like form.

Emma took Annie to London again in early December for a second visit to Dr. Holland. It is not known what he said, but Charles and Emma were left with their worries. A few days later, Emma noted in her diary: "Annie began bark." The cough may have prompted Charles to make a note in the family book of medicines. "Annie's gargle—Alum 2 drachma to pint of water." Alum was a sulphate of aluminium and potassium; it had a sweetish but astringent taste.

When Dr. Holland, one of the leading physicians in London, could not help, Charles set aside his reservations about Dr. Gully and wrote for his advice. Annie's illness was now chronic and Dr. Gully claimed that his water treatment was especially suitable for chronic conditions. Neither Charles's letter nor Dr. Gully's reply survives, but it appears from Charles's notes during the following months that Dr. Gully suggested a course of water treatment for Charles to give to Annie, and they agreed that she should come to Malvern in the spring for Dr. Gully to see and treat her himself.

Charles was at the time giving himself daily treatment under Dr. Gully's guidance, and he noted its effects carefully on foolscap sheets so that he could report them to Dr. Gully. In late January, he took a fresh sheet; he headed it "Anne," and set columns across the page for the day and date, the treatment given, Annie's condition during the day, and how she was during the night.

Dr. Gully's instructions for Annie were elaborate, and Charles used six of his special methods in a regular sequence. Dr. Gully explained the methods in his books with a show of medical science, portraying the patient's body as a physical mechanism, finely tuned but only precariously balanced by the influence of nervous forces, and needing to be corrected by external adjustment of the influences. The first method was the "dripping sheet," a wet sheet, slightly wrung out, wrapped around the body and then rubbed vigorously for five minutes. The aim was "to stimulate the nervous and circulatory systems of the body." Dr. Gully wrote that "to very delicate persons I often apply, in the first instance, only friction of the trunk and arms with a wet towel; dry and dress those parts, and then have the legs rubbed in like manner."

The "spinal wash" was a process invented by Dr. Gully which was particularly suitable for delicate persons. A wet towel was rubbed "up and down the length of the spine, not waiting until it is warmed by the patient's body, but constantly changing the water in the towel, so as to renew the shock of the cold as long as possible." Dr. Gully explained that the treatment "clears the head when it is confused, pained, or lethargic; it gives alacrity to the limbs, and spreads over the skin a sense of comfort which is due to the stimulus propagated along its nerves."

"Packing" in damp towels and sheets was supposed to "reduce excess of blood in one organ in order to send sufficient to another." The patient was wrapped up for an hour to an hour and a half. Dr. Gully wrote that the treatment was "one of the most agreeable, because one of the most soothing, of all the water remedies. By it, the nerves proceeding from the brain and spinal cord to the skin, and which are morbidly sensitive in all chronic diseases, are relieved, for the moment, from the irritation of the air."

The "shallow bath" and the "footbath" included rubbing and washing. "The feet and hands, the soles and palms especially, contain an accumulation of animal nerves and of blood vessels . . . in order to bind them by the closest sympathies with the great centres of thought and volition, so that their applications and movements may be accurately directed by the mind."

The last method, "sweating by the lamp," was in Dr. Gully's view "rightly esteemed one of the most effective means of arousing torpid and obstructed viscera into activity, by throwing an immense amount of irritation on the exterior surface." Joseph Leech wrote in *Three Weeks in Wet Sheets* how he sat on a chair and was draped with a tent of sheets for which he served as the pole. A lamp containing spirits of wine was placed under the chair and lit. "For two or three minutes I felt as though I were more likely to roast than melt." He began to think of "a martyrdom in singed blankets . . . when suddenly, as though it could bear no more, the skin opened its pores . . . and I ran like a shoulder of mutton before the fire."

Brodie and Bessy Harding were nursing Annie and helped wash her every morning. Bessy would have been up first; according to a guide to servants' duties, "The youngest nurse or nursery maid usually rises about six o'clock to light the fire, and do the household work of the nursery before the children are up, perhaps about seven o'clock, at which time the head nurse is dressed and ready to bathe and wash them all over with a sponge and warm water; after which they are rubbed quite dry and dressed." Each morning from 21 January, Annie was given the dripping sheet and spinal wash; every three or four days she was packed in a damp towel; once a week she was sweated by the lamp, and in a change of treatment from mid-February, perhaps after another letter from Dr. Gully, she also had a shallow bath and footbath every morning.

In his daily notes on the treatment and how Annie responded, Charles chose from a set of phrases to say how she was feeling, ranging from "well very" through "well almost very," "well," "well not quite," "good," "pretty good" and "poorly a little" to "poorly." From day to day he noted many "cries," tiredness, coughs and the strength of the pulse. Annie's nights were often "good," but from time to time she was "good not quite," "little wakeful," "wakeful" or "wakeful and uncomfortable." Observation and note-taking were second nature to Charles, but in watching Annie as he did, in varying the treatment and in noting the results, he was using his practice as a naturalist and experimenter in a desperate effort to work out what was affecting her, to help Dr. Gully find a pattern of treatment and cure. She still had no clear illness, but she could not shake off her persistent and distressing malaise.

On a visit to London in early January, Charles called at the London Library in St. James's Square and borrowed some books to read to Annie. *Geneviève* by Alphonse de Lamartine, the French Romantic poet and critic, was a simple story about the life of a poor woman in a village in the French Alps. Echoing Wordsworth, Lamartine believed that such tales "should be at once true and interesting"; they should describe "the lives, sufferings and joys of ordinary people," and should be "written almost in their language, a species of unframed mirror of their own existence, in which they might see themselves in all their simplicity and in all their candour." *The Book of the Seasons* by William Howitt, husband of Mary Howitt who had written *Our Cousins in Ohio,* offered a "calendar of nature" with notes on the plants, insects and animal life that could be seen each month through the year. In *The Boy's Country Book* William Howitt set out "all the amusements, pleasures and pursuits of children in the country." In a chapter on "Employments of the children of the poor," he wrote how the life of all village children was far preferable to that of town children. "No, I have no pity for country lads in general. They have, it is true, to blow their fingers over turnip-pulling on a sharp frosty day . . . but, bless me! What are these things to a cotton mill!—to a bump on the bare head with a billy-roller, or the wheels of a spinning-jenny pulling an arm off!"

Charles returned to his own reading about religion. His brother Erasmus, Fanny and Hensleigh Wedgwood and their friends were all talking about another book by Francis Newman which had just appeared. *Phases of Faith* was a close and clinical account of the slow decay and dismantling of his Christian beliefs, as he had questioned element after element and found each wanting. Charles had been following the same path as Newman since 1838, and found his book "excellent." One of Newman's points must have called Annie to mind and cut Charles to the quick. Newman argued that the Christian doctrine of eternal punishment had no clear Scriptural authority, and suggested that it posed intolerable moral difficulties. One was that every sin was "infinite in ill-desert and in result, because it is committed against an infinite Being." "Thus," he said, "the fretfulness of a child is an infinite evil!" On 24 January, Charles had noted that Annie had had "two little cries"; on 27 January, "late evening tired and cry" and the next day "early morning cry." Each time Annie had recovered and was "well" shortly afterwards. There was no evil in Annie's distress or her efforts to overcome it. The evil lay hidden in whatever it was that was afflicting her.

During those weeks, Annie opened her mother's diary, and put in three jottings. On 5 February, "I was unwell." On 11 February, "Beautiful day. Children rode pony. Got up." And on the page for 2 March, she wrote "Annie's birthday." She would be ten years old.

January and February had been mild but the first day of March was cold, and snow fell that night. On Annie's birthday, the family woke to brilliant sunshine. Annie was poorly at first but soon felt better. Emma had a book for her, and put it by her plate at breakfast. Willy had come home from Mr. Wharton's the day before, and Annie played outside with him, "romping with him through the hedge in the Sand-walk." Parslow gave Annie her first ride on Willy's pony. She wanted to play in the open air, and her parents were glad for her to.

But her cough returned during the following week and two days later she had 'flu. The weather was exceptionally stormy and there were heavy falls of rain. Charles stopped the water treatment, writing that Annie was "Poorly with cough and influenza." He continued his notes on her illness with a string of dittoes for eight days, and stopped on 21 March. Annie must have made some recovery, because three days later, Charles took her and Etty with Brodie on the two-day journey to Malvern. Emma, now over seven months pregnant, had to remain behind with the others. She remembered Annie "lying on the bed with me" when she had 'flu. And on the day Annie left, "Sitting crying on the sofa."

CHAPTER NINE

# THE LAST WEEKS IN MALVERN

*Malvern—Sickness and fever—Charles's arrival—*
*Rally and decline—Death*

CHARLES, ANNIE, ETTY AND BRODIE travelled to London and stayed the night with Erasmus at his home in Park Street, close to Hyde Park. Etty remembered how, when she visited Uncle Ras as a child, "we came into that simply furnished, somewhat ascetic London drawing room, looking out on the bare street, knowing he was weary and ill, and had been alone, and would be alone again, and yet went away with a glow . . . There was no possibility of forgetting the respect due to an elder, but he met us so entirely on our own level, that in our intercourse with him we felt as free as if he were our own age."

The next morning they called on Fanny and Hensleigh Wedgwood, Effie and the other children, in their grand terraced house in Regent's Park. There was talk of Dr. Gully and the water cure, and Annie gathered that a family of friends had gone to Malvern because the father, Alexander John Scott, wanted to try Dr. Gully's water treatment. Mr. Scott was a charismatic preacher who had worked with Edward Irving, founder of the Catholic Apostolic Church, in the early 1830s. His wife Ann was sensible and liked by the Wedgwoods. Their daughter Susy was near Annie's age, and Annie looked forward to playing with her.

Charles always used to complain of the horrid sinking feeling he had at the beginning of a journey, but Francis remembered that his discomfort was "chiefly in the anticipation," and once they were under way,

Charles would enjoy the journey "to a curious extent" and "in an almost boyish way." Charles, Annie, Etty and Brodie would have taken a "Growler" to Euston Station in the morning. This was a low-slung, four-wheeled one-horse cab which could take four people with luggage—"coarse, noisy, odoriferous and jumpy as regards the springs," according to a writer of the time.

The train brought them to Birmingham by lunchtime and they arrived in Worcester by late afternoon. The city streets were littered with handbills announcing the arrival of Wombwell's Royal Menagerie "accompanied by its splendid brass band" on Thursday morning. The circus had "the largest elephant and rhinoceros in England, and the only full-grown Caffrarian lions" in a collection of "five hundred beasts and birds contained in fourteen large vehicles built expressly to convey them from town to town." The vehicles were drawn by forty horses led by forty men. They would "parade through the principal streets, and afterwards be stationed on the Upper Quay, where no doubt a goodly number of the admirers of nature will avail themselves of the opportunity of witnessing some of her noblest productions."

Charles, Brodie and the children passed through the city and took a stage coach to Malvern from one of the main hotels. When they arrived in the village, Charles found lodgings for them at Montreal House, a stucco villa on the Worcester Road. The landlady, Eliza Partington, knew them from their visit in 1849. On Thursday, Charles wrote to Fox: "I have brought my eldest girl here and intend to leave her for a month under Dr. Gully; she inherits, I fear with grief, my wretched digestion." Annie wrote to her mother to say that they had arrived and were lodging with Miss Partington, but she had not yet seen Dr. Gully. The postal service between Malvern and Downe was quick and reliable. Letters handed in at the post office before half past six in the evening would be delivered to Downe by the local postman around noon the next day. Letters given to him then would reach Malvern the next morning.

When the postman brought Annie's letter to Downe, Miss Thorley had left to join the children in Malvern, and Emma wrote back to Annie the next day: "I was very glad to have your nice letter and to learn that you are at Miss Partington's. It is most unlucky Dr Gully being gone out . . . When Miss Thorley comes, you had better ask her to take you to see Susy Scott. It is very nice for you Mrs Scott being there and I dare say Susy will often walk with you." Thinking back to the family's walks around Malvern in 1849, Emma ended her letter: "I should so like to see St Anne's Well and the hills. It makes me quite thirsty to think of it."

Montreal House stands on the other side of the road from the Lodge where the family had stayed before. At the Lodge, the children had played in the garden climbing up the steep slope to the Malvern Hills; at Montreal House the garden fell away below, and they looked out over the distant patchwork of the Severn Vale. In 1999, I met an elderly lady who had lived and worked in the house as a young maid in the 1920s when it was kept as a boarding house by her aunts. She remembers the large light kitchen in the basement with a red-tiled floor which she scrubbed, and a massive black range which she black-leaded every day. There were twelve rooms for the guests on the three floors above. On the ground floor there were two private rooms, a drawing room and a dining room; on the first and second floors, two rooms faced the hill rising to the west, and two had the wide view to the east out over the river plain. The stairs were at the back of the house and there was a small lift to deliver food from the kitchen to the upper floors. The lower part of the garden was planted with neat rows of vegetables which could be seen from the bedroom windows.

After settling the children and Brodie in their lodgings, Charles returned to London on Friday. He stayed with Erasmus in Park Street, and on Sunday they called again on Fanny and Hensleigh. Fanny's half-sister Mary Rich and the Wedgwoods' Aunt Fanny Allen were staying with them, and both mentioned the brothers' visit in letters to others.

Aunt Fanny wrote that Charles was looking well, and "there is something uncommonly fresh and pleasant in him." Thomas and Jane Carlyle also came to the house, and Ruskin's *Stones of Venice* was discussed. Mary wrote: "Charles Darwin dined here yesterday looking so well, and in excellent spirits, and agreeable as he always is . . . Poor Annie, their eldest girl, has been very far from well for some time, and they have just sent her with the governess to Dr Gully to try the water cure, which was of so much use to her father."

The Great Exhibition was due to open in London at the beginning of May, and interest was growing. Newspapers and periodicals advertised special maps and guides to the metropolis, describing and celebrating it as a proper setting for the grand display of the "Works of Industry of All Nations." But the journalist Henry Mayhew struck a note of harsh discord in his *London Labour and the London Poor,* which appeared in instalments in the early months of the year. His aim was to reveal the true nature of life for the many hundreds of thousands of people who lived in poverty in the parts of London that the guides glossed over. He found people prepared to tell him about themselves, and printed their words, often almost as they spoke. Charles read the book when he arrived back at Down. The life of one small girl whom Mayhew had talked to was strikingly different from Annie's. The child was an eight-year-old watercress girl who, as Mayhew wrote, "had entirely lost all childish ways, and was, indeed, in thoughts and manner, a woman. There was something cruelly pathetic in hearing this infant, so young that her features had scarcely formed themselves, talking of the bitterest struggles of life with the calm earnestness of one who had endured them all." After hearing about her daily work, bargaining fiercely with people as needy as herself to get her watercress at a price on which she could hope to make a profit, and then taking it through the streets to sell to passers-by for whatever she could get, Mayhew talked to her about the parks, and asked whether she ever went to them.

" 'The parks!' she replied in wonder, 'where are they?' I explained to her, telling her that they were large open places with green grass and tall trees, where beautiful carriages drove about, and people walked for pleasure, and children played. Her eyes brightened up a little as I spoke; and she asked, half doubtingly, 'Would they let such as me go there—just to look?' " Mayhew commented that "All her knowledge seemed to begin and end with watercresses, and what they fetched."

Erasmus lent Charles a book by his friend Harriet Martineau which had just appeared. Her *Letters on the Laws of Man's Nature and Development* was an enthusiastic exchange of ideas with Henry Atkinson, a wealthy man who believed that phrenology and mesmerism offered the key to understanding the human mind. He argued for a materialist view of mental activity, and insisted that "philosophy finds no God in Nature." Miss Martineau welcomed his freethinking, and the book caused a great stir among people interested in issues of faith and doubt. Charlotte Brontë wrote to a friend: "It is the first exposition of avowed atheism and materialism I have ever read; the first unequivocal declaration of disbelief in the existence of a God or a future life I have ever seen . . . Sincerely, for my own part, do I wish to find and know the Truth; but if this be Truth, well may she guard herself with mysteries, and cover herself with a veil." Atkinson had written: "Man has his place in natural history . . . his nature does not essentially differ from that of the lower animals . . . he is but a fuller development and varied condition of the same fundamental nature or cause." Miss Martineau praised his courage in looking into human nature with no prejudices about the dignity of man, and no worries about how his ideas might undermine it. "The true ground of awe is in finding ourselves what we are."

The last day of March was enumeration day for the ten-year National Census, so just as everyone had been listed where they were living or staying a few days before Annie's christening in June 1841, they were now to be recorded again ten years on. On the Thursday when Charles was in Malvern, the local newspaper had reported that every householder had received a schedule to fill in for the enumerator. When the day came and the forms were collected throughout the country, they showed Charles back in London staying with Erasmus; Emma at Down House with the servants and her four youngest children, and Willy with his fellow pupils at the vicarage in Mitcham. At Montreal House in Malvern, Eliza Partington was listed with her cook, the housemaid, four lady guests and their ladies' maid. Annie, Etty and Brodie were not named, probably because the form had been filled in before they arrived.

As Annie and Etty settled in and went out to the places they remembered in the village and on the Malvern Hills, they found the spring season well under way. According to the local newspaper, farmers at their meetings were remarking that the season was "particularly abundant and lucky for lambs."

On the following Thursday, the local newspaper reprinted a report from *The Gardener's Chronicle* headlined "Safety of Dr Hooker." "We are happy to announce by the last Indian mail, Dr Hooker and his friend Dr Thomson arrived safely from Chittagong. We have had frequent occasion to speak of the important and dangerous travels of the former of these gentlemen, from which the most valuable contributions to physical geography and natural science are to be expected. We understand that Dr Hooker's collections, which are on their way home by the Cape of Good Hope, are equally remarkable for their interest and great extent." Annie and Etty had not seen Hooker for three years, but they knew all about his adventures from his long letters to their father. He had been away since November 1847 collecting plants and surveying on the northern edge of British India, in the high Himalayan valleys of Sikkim, Nepal and Tibet. He had mapped the mountains of Sikkim and found extraordinary numbers of plants, including twenty-five new species of rhododendron.

Botanists and geologists in England had been reading his reports, and plant specimens that had been shipped back were being studied eagerly at Kew. On the last leg of his journey he had explored the orchid-rich tropical jungles and bare uplands of the Khasi Hills in Assam; head-hunters were roaming in the higher hills, and had taken thirty heads from one village a week before he arrived. He had now reached safety in Calcutta and was at last on his way back to England. In his last letter to Charles he had sent his "best remembrances" to the children. Annie and Etty could look forward to his stories when he came to Down, and would then be able to press him to tell them about the elephants, tigers and Naga warriors, bridges of creepers over mountain chasms, orchids, butterflies and the great white peaks to the north.

That day in Malvern, Annie and Etty were both happy. Etty wrote to their mother on a piece of fancy notepaper embossed with floral patterns, "My dear Mama—we are going to buy the combs this morning. Yesterday I fell down twice. We bought som orangs this morning. Yesterday we bought some canvas, and I and Annie are making a pattern out of our own heds. We saw the Marsdens playing in a garden. There are a great many ladies in this house. Will you send my mits. I remain your aff Etty." Brodie had made the little pocketbook she gave to Annie by embroidering a pattern on a piece of wire mesh, and she probably helped the two children stitch their own patterns on their pieces of canvas. The Marsdens were the children of James Marsden, one of the three water doctors whose portraits were displayed around the village. His daughter Lucy was twelve, Emily was eleven, Marian ten, Rose seven and Alice six. The "great many ladies" who so impressed Etty must have been the four lady guests on the census return.

Miss Thorley arrived by the next day, and Etty wrote to George: "My dear Georgy—we went a donkey ride. When Miss Thorley and Annie rode on before, my donkey would not go on, and I was obliged to get off. Brodie and I had to drag it along. I have got some little ladybirds, and I keep them in a little box, and I feed them milk, some sugar. There is a shop at St Anne's Well. Brodie sends her love to Frankey and Lizzie. I remain your affec. Etty. Brodie sends her love to you."

There was a service in the village church on Sunday morning. The vicar, John Rashdall, was a young high churchman, recently arrived from London where people had admired his sermons in a fashionable Knightsbridge church. That day in Malvern, he preached to a large congregation on a text from Hebrews: "But Christ being come an high priest of good things to come, by a greater and more perfect tabernacle, not made with hands, that is to say, not of this building; Neither by the blood of goats and calves, but by his own blood he entered in once into the holy place, having obtained eternal redemption for us."

The next day Annie had a sudden and sharp attack of vomiting. She had been receiving the water treatment from Dr. Gully and had been "going on very well" as Charles wrote later to Fox. The attack "was first thought of the smallest importance, but it rapidly assumed the form of a low and dreadful fever," and by the following Sunday, Annie was very weak. Dr. Gully was seriously concerned about her condition, and Miss Thorley reported everything she could gather from him to Emma. Miss Thorley was desperate to talk to another woman about Annie, and hastened through the village to see Mrs. Scott for reassurance.

On Monday, Miss Thorley wrote to Emma: "Dr Gully's opinion is that Annie is very slowly progressing; this has much relieved my mind. He came early today . . . and I again asked if he thought there was danger. 'No,' he said, 'it is a smart bilious gastric fever, but she has turned the corner.' These are the words he said. He has many similar cases on hand now, and he said it is quite an epidemic; hers has not been brought on by her treatment with him, but it is always more or less general at this season." Now that the fever was passing, they saw Annie's "extreme weakness." Brodie had been ill with distress the previous day and Miss Thorley had thought of sending for Charles, but she was now in much better spirits and had taken Etty out for a walk. "Dear Annie sends her love; how I long for the time to be able to say she is strong enough to get up. Adieu, ever yours, my dear Mrs D."

By suggesting that Annie was suffering from a "smart bilious gastric fever" which was affecting others in the neighbourhood, Dr. Gully hoped that the illness was a normal ailment which would be cured by the healing power of nature, the *vis medicatrix naturae* as the textbooks called it. It was believed that if the patient survived fourteen days after the onset of a fever, he or she would throw it off. The epidemic that Dr. Gully referred to was probably not serious; it was not mentioned in the local newspapers, nor was there any pattern of deaths in the parish records.

When Dr. Gully came to see Annie on Tuesday, he was alarmed by her condition. He felt her life was in danger and wrote at once to Charles suggesting that he should come immediately. He told Miss Thorley about his fears and she "gave way sadly." Charles received the letter at Down at midday on Wednesday and left at once for Malvern. As he was on his way, Miss Thorley wrote again to Emma. Annie was "a shade better" but Dr. Gully said she was not out of danger. There had been crises in the early morning and the afternoon, but she had got through both well. Dr. Gully had come to see her three times, and saw many good signs. "We are giving her now a dessert spoon of white wine every hour, and a medicine he prescribed last night . . . I have just enjoyed feeding her with some orange juice which the dear child thoroughly relished." Annie had been dozing and her mind wandered at times, but that may have been the effect of the wine. "All fever is removed which is a great thing." Her "excitable temperament" was Dr. Gully's only reason for fear. Miss Thorley longed for Charles to arrive. She gave Emma a glimpse of Etty who was "amusing herself with her new doll, beads &c. She is very anxious about the dear child Annie."

Charles reached Montreal House the next afternoon. Etty later remembered her shock and bewilderment as a small child at "his coming in, and after Miss Thorley saying something, his flinging himself down on the sofa on his face, and Miss Thorley sending me out of the room in a frightened way." By four o'clock, he had composed himself, seen Annie in her bedroom and talked to Miss Thorley. He wrote to Emma: "I am assured that Annie is several degrees better. I have in vain tried to see Dr Gully as yet. She looks very ill: her face lighted up and she certainly knew me." They had stopped giving her wine but had given "several spoon-fulls of broth, and ordinary physic of camphor and ammonia. Dr Gully is most confident there is strong hope. Thank God she does not suffer at all—half dozes all day long." Camphor and ammonia were used as stimulants and to suppress vomiting.

When Dr. Gully came in the evening, Annie's pulse was irregular and for a time he feared she was dying. He stayed in the house through the night to give what help he could. The next day was Good Friday. Charles wrote to Emma that Dr. Gully had been "most kind." Annie had a bad vomiting attack at six o'clock in the morning which showed nevertheless that she had more "vital force" than before. She was very quiet for the rest of the morning but her pulse was firmer. Charles and Brodie gave her spoonfuls of gruel with brandy every half hour. "She does not suffer, thank God," but "It is much bitterer, and harder to bear than I had expected."

A letter had come from Emma, and Charles replied: "Your note made me cry much, but I must not give way, and can avoid doing so by not thinking about her. It is now from hour to hour a struggle between life and death. God only knows the issue." With the treatments then available, there was nothing Dr. Gully could do to cure Annie's illness, but as an experienced physician he was accustomed to reading the signs of deterioration and improvement. Charles looked to him for that intuitive sense, and hung on every word. "Sometimes Dr G. exclaims she will get through the struggle; then, I see, he doubts. Oh my own, it is very bitter indeed."

During the afternoon, Annie vomited green fluid, bile from her liver. Thinking back to his long sickness in the summer and autumn of 1840 before Annie was born, Charles wrote to Emma: "Her case seems to me an exaggerated one of my Maer illness. We must hope against hope. My own poor dear unhappy wife." In the evening, Charles scrawled a note to Emma. "Dr Gully not come. She appears dreadfully exhausted, and I thought for some time she was sinking, but she has now rallied a little. The two symptoms Dr G. dreads most have not come on—restlessness and coldness. If her three awful fits of vomiting were not of the nature of a crisis, I look at the case as hopeless. I cannot realise our position, God help us." Dr. Gully came some minutes later; he examined Annie and told Charles that she was weakening, but he understood Charles's desperate need for hope and gave him the few straws he could offer. Charles continued his note, picking up more and more ink on the quill and writing more urgently. "Dr Gully has been and thank God he says though the appearances are so bad, *positively* no one important symptom is worse, and that he yet has hopes—*positively he has Hopes.* Oh my dear be thankful."

After that, Annie fell asleep and had a restful night. When Dr. Gully came and saw her sleeping, he told Charles: "She is turning the corner." When the morning came, Charles could not wait for the post to let Emma know. He sent a message to Worcester for an "electric telegraph" to be sent along the railway line to London. It was delivered to Erasmus.

From: Gentleman Montreal House Malvern
To: E. Darwin Esquire 7 Park St Grosvenor Sq
Please send man to Sydenham station thence in fly to Down to say that Annie has rallied—has passed good night—danger much less imminent.

Erasmus sent his manservant, John Griffiths, to take the message to Down, hiring a fly, a light carriage, for the last stage of the journey.

Annie was calm but emaciated by the dehydration from her vomiting. Charles wrote to Emma: "This morning she is a shade too hot; but the Doctor . . . thinks her going on very well. You must not suppose her out of great danger. She keeps the same; just this minute she opened her mouth quite distinctly for gruel, and said 'that is enough.' You would not in the least recognise her with her poor hard, sharp, pinched features; I could only bear to look at her by forgetting our former dear Annie. There is nothing in common between the two . . . Poor Annie has just said 'Papa' quite distinctly." He told Emma what Dr. Gully had said in the late evening. "I cannot express how it felt to have hopes last night . . . I then dared to picture to myself my own former Annie with her dear affectionate radiant face . . . My dear dear mammy, let us hope and be patient over this dreadful illness."

Emma had asked Fanny Wedgwood to go from London to Malvern to be with Charles. She was desperately afraid that Charles's anxiety might injure his health, and also believed that Fanny, as a mother, had her own "eye for illness," an eye that might see things Charles and the physician would miss. Fanny arrived in Malvern with her lady's maid, who then returned to London, taking Etty back with her.

Emma had also arranged for her Aunt Fanny Allen to come to Down to be with her, as she was now due to give birth in four weeks. When Fanny Allen arrived, she found her niece looking well "as to health" but "very much overcome at times." "Poor Emma is very low, but her health is not injured . . . Pray Heaven their child may be preserved to them!"

The room in which Annie lay was probably one of those looking east out over the Severn Vale. In her *Notes on Nursing,* Florence Nightingale wrote about caring for seriously ill patients in a sickroom at home. "It is a curious thing to observe how almost all patients lie with their faces turned to the light, exactly as plants always make their way towards the light; a patient will even complain that it gives him pain 'lying on that side.' 'Then why do you lie on that side?' He does not know—but we do. It is because it is the side towards the window." Florence Nightingale also wrote about sounds in the sickroom, noises that Annie would have heard as Brodie, Miss Thorley and Fanny Wedgwood were all wearing the very full skirts of the time. "Compelled by her dress, every woman now either shuffles or waddles . . . The fidget of silk and crinoline, the rattling of keys, the creaking of stays and shoes, will do a patient more harm than all the medicines in the world will do him good . . . The noiseless step of woman, the noiseless drapery of woman, are mere figures of speech in this day. Her skirts (and well if they do not throw down some piece of furniture) will at least brush against every article in the room as she moves."

Throughout the rest of Saturday, Charles wrote to Emma hour by hour. At two o'clock, "We expect Dr Gully every minute; but he is fearfully overworked . . . Annie has kept just in same tranquil, too tranquil state: she takes a table-spoonful of gruel every hour and no physic. All trace of fever is now gone and yet she is not chilly. She begins to drink a little more this afternoon and I think that is good." At three o'clock, "The Doctor has been. He says she makes no progress, but no bad symptoms have appeared. But I am disappointed." At four o'clock, "She has taken two spoonfuls of tea with evident relish. And no sickness, thank God. I find Fanny an infinite comfort."

Fanny wrote to Emma in the evening with her "eye for illness." "My dearest Emma—Charles tells you everything of your darling child, but you will like to hear any other impressions . . . She has been sick again since four o'clock when Charles closed his letter, but is looking more comfortable and has seemed quite to like being turned on the other side . . . I do not think her so emaciated as I expected . . . She has looked about more today and her face has to my eyes a more natural expression—extreme languor and prostration but no oppression about the head or eyes. She has just asked Miss Thorley quite loud something about her watch, but much of what she says we cannot make out from the roughness of her poor mouth—but I do generally make out the 'thank you' almost always . . . Dearest Emma, how thankful I am to be able to be of the least use to Charles. He looks really not ill, though sometimes of course most sadly overcome and shaken. He has been two little walks today. I do not try to prevent him doing a good deal about dear Annie. It seems as if it was some relief to be doing something, though occasionally it may be too much . . . May God grant you both the life of your child."

Charles's two letters on Friday had reached Emma on Saturday, some time before the telegraph message came. She replied to them at once. "Now dear Fanny is with you, you must let her experienced eye do some of the watching, though I know what an effort it must be to leave her for a moment, but you will be quite exhausted. Aunt F. helps me through the long hours of suspense, and I feel quite unnatural sometimes in being able to talk of other things. Poor little sweet child. I often think of the precious look she gave you, the only one I suppose. No wonder she would brighten up at your sight. You were always the tenderest of human beings to her and comforted her so on all occasions."

Emma then went out of the house thinking of Annie, and was in the garden when John Griffiths came along the road from the village with Charles's message. She read his words and wrote a note for John Griffiths to take back with him. "The message is just arrived. What happiness! How I do thank God! But I will not feel too hopeful. I was in the garden looking at my poor darling's little garden to find a flower of hers when John Griffiths drove up . . . We shall hear nothing more now till Monday, but I shall wait very well now . . . I hardly dare think of such happiness. I hope you will sleep tonight, my own."

The next day was Easter Sunday. Emma wrote to Fanny Wedgwood: "I cannot express the happiness of yesterday's message though I know how much there is still to fear. I feel very anxious about Charles for fear he should quite break down, but your being with him is such a comfort. I don't know what I should have done without Aunt Fanny. By oneself one's thoughts lose all control . . . The waiting for post time is the worst, but one gets used to every thing in a degree and now I have so much more hope, I feel greedy to hear your impression also of my poor dear's looks. God grant that dreadful sickness may keep off."

Charles wrote to Emma: "I do not know, but think it is best for you to know how every hour passes. It is a relief to me to tell you, for whilst writing to you, I can cry tranquilly." He carried on with his account. They had called the surgeon, Mr. Coates, to draw Annie's water off. "This was done well and did not hurt her, but she struggled with surprising strength against being uncovered &c. Soon it evidently relieved her." During the night, she "slept tranquilly except for about ten minutes when she wandered in slightly excited manner." She was "fearfully prostrated." "Yet when Brodie sponged her face, she asked to have her hands done and then thanked Brodie and put her arms round her neck, my poor child, and kissed her." Dr. Gully said what he could to encourage Charles, but spoke with extreme caution. "You must not trust me, for I can give no reason for my intuition, but yet I think she will recover."

Fanny sat with Annie until two o'clock in the morning. Charles commented to Emma that "poor dear devoted Miss Thorley" was able to have a full night's rest. When Charles took his turn by Annie's bedside, he did not sit calmly but was constantly up and down. "I *cannot* sit still."

Annie had vomited early in the morning. Each bout must have racked her body but Charles wrote: "It is certain she suffers very little—dozing nearly all the time. Occasionally she says she is very weak." At ten o'clock, "I grieve to say she has vomited rather much again; but Mr Coates has been and drawn off again much water, and this he says is a very good symptom. Last night he seemed astonished at her 'fearful illness' and he made me very low; so this morning I asked nothing and he then felt her pulse of his own accord and at once said 'I declare I almost think she will recover.' Oh my dear, was not this joyous to hear!"

Annie was drifting in and out of delirium. Charles remembered clearly from his own childhood "the wretched feeling of being delirious" when he had had scarlet fever at nine. Annie's senses were clear, and Charles felt that was good "as showing head not affected." "She called 'Papa,' when I was out of room unfortunately, and then added 'Is he out?'" Thinking of the hope of recovery if she survived for two weeks after the first attack, he now wished fervently for the critical fortnight to be over. "But I must not hope too much. These alternatives of no hope and hope sicken one's soul: I cannot help getting so sanguine every now and then to be disappointed."

At midday, Annie was sick again and said she felt tired. "She is very sensible; I was moving her, when she said 'Don't do that please,' and when I stopped, 'Thank you.'" "We have put mustard poultice on stomach, and that has smarted her a good deal, which shows more sensibility than I expected." The mustard poultice was a paste of mustard and vinegar applied as a stimulant in low fevers. In the afternoon, Charles wrote: "She is a little chilly and we have given her a little Brandy and hope she is asleep and I trust will warm. I never saw anything so pathetic as her patience and thankfulness; when I gave her some water, she said 'I quite thank you.' Poor dear darling child." An hour later, "The chilliness has pretty well gone off and no more sickness, refreshing sleep."

Dr. Gully came early in the evening, and while he was in the room Fanny wrote to Emma: "Dr G. is now here. There is a decided improvement in the tongue—a most important point. Pulse quick and sharp still, but perhaps from a little brandy. Dr G. thinks the vomiting not so unfavourable—it is better under the circumstances than the matters being absorbed." After Dr. Gully left, Charles and the others bathed Annie with vinegar and water, and he wrote to Emma that "it was delicious to see how it soothed her."

On Monday morning, Charles wrote: "To go on with the sick life . . ." During the previous evening Annie had "rambled for two hours and became considerably excited." Mr. Coates had failed to draw her waters, but her bladder and bowels acted of themselves during the night. When Dr. Gully came in the morning, he was discouraging. Charles told Emma all, "for it will prevent the too strong and ultimately wretched alternations of spirits. An hour ago I was foolish with delight and pictured her to myself making custards (whirling round), as I think she called them. I told her I thought she would be better, and she so meekly said 'Thank you.' Her gentleness is inexpressibly touching." Annie had called her whirling round "making custards" because when Daydy the cook or she prepared a custard in the kitchen at home, they had to stir the mixture well until it thickened.

Charles was frank about her condition. "Poor Annie is in a fearful mess, but we keep her sweet with Chloride of Lime; the Doctor said we might change the under-sheet if we could, but I dare not attempt it yet. We have again this morning sponged her with vinegar, again with excellent effect. She asked for orange this morning, the first time she has asked for anything except water. Our poor child has been fearfully ill; as ill as a human being could be . . . If diarrhea will but not come on, I trust in God we are nearly safe." He thought back to Emma's last letter. "My own dear, how it did make me cry to read of your going to Annie's garden for a flower. I wish you could see her now, the perfection of gentleness, patience and gratitude—thankful till it is truly painful to hear her. Poor dear little soul."

At midday, Annie appeared "rather more prostrated with knees and feet chilly and breathing laboured, but with some trouble we have got these right, and she is now asleep and breathing well. She certainly relishes her gruel flavoured with orange juice, and has taken table-spoon every hour . . . She wanders—and talks—more today a good deal." At three o'clock, Charles added: "She is going on very nicely and sleeping capitally with breathing quite slow. We have changed the lower sheet and cut off the tail of her Chemy [chemise] and she looks quite nice and got her bed flat and a little pillow between her two bony knees." In the evening, Fanny gave her a spoonful of tea and asked her whether it was good. "She cried out quite audibly 'It is beautifully good.' She asked, so says Brodie, 'Where is poor Etty?' "

Emma received Charles and Fanny's letters of Saturday and Sunday on Monday morning, and wrote to Charles while the postman waited. "Your account of every hour is most precious. Poor darling, she takes much more notice than I expected. I am confused now and hardly know what my impression is, but I have considerable hope. I suppose a dose of opening physic has never been thought of. One must trust entirely to Dr. Gully." After the next day's post, she wrote: "Your two letters of Monday are certainly better. Poor sweet little thing! I felt more wretched today than any day, but now I do think looking at the accounts of the last four days that there has been progressive improvement from that time . . . I shall write a few lines in the afternoon, but I always feel bewildered at first, but my impression is considerably better." She thanked Fanny for writing so fully and telling her about Charles. "Your impression of our poor child's looks was a comfort. Your being there is an immense comfort for Charles and I think you are quite right to let him do as much as he can, as it must be the greatest relief he has . . . I feel today very awful, being the end of the fortnight."

Monday night was quiet for Annie and she was less delirious than the night before, but her bowels were loose. On Tuesday morning Charles started the day's letter to Emma in hope, but when he came back to Annie's bedside, he saw small signs that she might be failing; his hope broke and he could not write any more. Fanny wrote to Emma at the end of the day: "I am thankful that you felt there was much to fear in your note yesterday for I grieve to write you a worse report this evening. There has been a change today and signs of sinking. I tell you everything just as it is, my dearest Emma, and thankful also for the mercy that is given us of there being not the least appearance of any suffering in your sweet patient darling . . . The effort of the fever throwing itself off from the bowels is more than her strength seems able to bear and she has lost strength every time. We are now giving brandy and ammonia every quarter of an hour, which she takes well with no difficulty." Fanny had persuaded Charles to lie down because he had "gone through much fatigue."

In the late afternoon, Dr. Gully came and found that Annie had not gained ground. Fanny wrote to Emma: "He thinks her in imminent danger." If there was any change for the better "you shall have a message—But I have told you the worst. Oh that I should have to send you such sad, sad news . . ."

On Wednesday morning, Emma wrote to Charles: "The oftener I read over your letter of Monday, the more hopeful it made me. Your minute accounts are such a comfort and I enjoyed the sponging our dear one with vinegar as much as you did." Emma had been thinking about meals that might suit Annie when she could take a little food, "but it is more for the pleasure of fancying I have something to do for her or think of for her." Rice gruel might be flavoured with cinnamon or currant jelly. "Whey from milk is another harmless drink and very digestible, I believe . . . Aunt F. says it is slightly opening, so that I am doubtful about it."

Emma left the letter unsigned until the post came at midday. When she read Fanny's letter to her, she added at the bottom: "After post. Alas, my own, how shall we bear it. It is very bitter but I shall not be ill. Thank dear F." Aunt Fanny Allen wrote to Fanny in Malvern: "Your letter is come, and poor dear Emma bears the destruction of her hope, which was stronger, I fear, than was quite reasonable, with great sweetness, crying much, but gently. I hope and think it will not shake her frame so as to cause her confinement before her natural time. Poor

Charles must now think only of his own weight of suffering. Emma suffers, but is not ill . . . I fear, after your letter today, there is but one account to expect tomorrow."

During Tuesday night, Annie had been delirious. Fanny sat with her and wrote later to her daughter Effie: "I heard her twice trying to sing, so I think her wandering could not have been distressing to her. She talked a great deal but we could seldom make out anything."

On Wednesday morning, there was thunder in the air, but the end came quietly. Charles wrote: "My dear dearest Emma, I pray God Fanny's note may have prepared you. She went to her final sleep most tranquilly, most sweetly at twelve o'clock today. Our poor dear dear child has had a very short life but I trust happy, and God only knows what miseries might have been in store for her. She expired without a sigh. How desolate it makes one to think of her frank cordial manners. I am so thankful for the daguerreotype. I cannot remember ever seeing the dear child naughty. God bless her. We must be more and more to each other, my dear wife."

Fanny wrote to Effie that it was "just at twelve o'clock that we heard her breathe for the last time, while the peals of thunder were sounding. Poor Miss Thorley was very ill after, and Brodie too. I never saw anyone suffer as she did, but they are both going tomorrow, Brodie to Down and Miss Thorley to her own home. She wants rest."

In the evening, Fanny wrote to Emma that she had been sitting with Charles for half an hour. "He is able to find relief in crying much, and at first I am sure it is best for him. And next, I think, to go to you which he may possibly do tomorrow if able, and you may trust to Dr Gully who is full of care for him, that he will not set out if not . . . He will have told you, doubtless, of the soft and gentle departure of your dear child. It was only ceasing to breathe, and no change was perceptible in her face which, ill as she has looked, has never shewed a shade of pain, hardly of discomfort, since Friday night . . . Dearest Emma, I trust that I have not given you any additional pain in these few words. I can never forget the comfort of having been able to be here for you . . . most tenderly, F.E.W."

She wrote to Hensleigh that the funeral had been arranged for Friday. It was a comfort that it would be so soon, but Charles had not yet decided whether he could go and Fanny had suggested that he should go home to Emma on Thursday. "Poor Brodie and Miss Thorley are quite knocked up and useless—Miss T. had one of her attacks just after Annie died, and I had to look after and manage her and Brodie; so after my four nights you will expect to hear I am a good deal knocked up, and cannot do much writing."

The postman had no letter for Emma on Wednesday. At some time during the day, she walked out into the garden and picked a small daffodil—a jonquil, bright yellow and sweetly scented. She wrapped it in a fold of paper and wrote on the paper, "Gathered Ap. 23. 1851." It was perhaps from Annie's garden. Dried fragments of the flower remain in the fold.

On Thursday morning before Charles's letter arrived from Malvern, Emma wrote: "My dearest, I know too well what receiving no message yesterday means. Till four o'clock I sometimes had a thought of hope, but when I went to bed, I felt as if it had all happened long ago. Don't think it made any difference my being so hopeful the last day. When the blow comes, it wipes out all that preceded it and I don't think it makes it any worse to bear. I hope you have not burnt your letter. I shall like to see it sometime. My feeling of longing after our lost treasure makes me feel painfully indifferent to the other children, but I shall get right in my feelings to them before long. You must remember that you are my prime treasure (and always have been). My only hope of consolation is to have you safe home to weep together. I feel so full of fears about you. They are not reasonable fears, but any power of hoping seems gone. I hope you will let dearest Fanny . . . stay with you till the end. I can't bear to think of you by yourself."

Emma wrote to Fanny: "I do feel very grateful to God that our dear darling was apparently spared all suffering, and I hope I shall be able to attain some feeling of submission to the will of Heaven."

When the postman brought Charles's letter, Emma wrote simply on her diary page for Wednesday, "12 o'clock." She wrote to Charles: "I feel less miserable a good deal in the hopes of seeing you sooner than I expected, but do not be in a hurry to set off. I am perfectly well. You do give me the only comfort I can take in thinking of her happy innocent life. She never concealed a thought, and so affectionate, so forgiving. What a blank it is. Don't think of coming in one day. We shall be much less miserable together."

APRIL, 1851.

20 SUNDAY [Easter Sunday.]
Sick 3 or 4 times & took brandy once

21 MONDAY
much better

22 TUESDAY
Diarrhea came on
Eliz came

23 WEDNESDAY
12 o'clock

CHAPTER TEN

# LOSS AND REMEMBERING

*Funeral—Consolation—Charles and Emma's thinking—*
*Emma's keepsakes—Charles's memorial of Annie*

THE NEXT DAY EMMA BORE her grief in her own way. Her sister Elizabeth had come to be with her, and wrote to Charles: "The only comfort I can try to give you is telling you how gently and sweetly Emma takes this bitter affliction. She cries at times, but without violence, comes to our meals with the children and is as sweetly ready as ever to attend to all their little requirements. I do not fear, taking it as she does, that she will be made ill. It will be the greatest comfort to her to see you home very soon, and to have the additional anxiety of absence from you no longer. It is very happy that Willy is here now, and I felt quite glad last night to hear his voice talking to her out of his bed after she was in hers."

After a few hours of uncertainty, Charles had accepted Fanny Wedgwood's suggestion that he should return at once to Emma. He set off back to London for Down early on Thursday morning, leaving a note for Fanny to tell Miss Thorley he had taken some books, and to ask Brodie to look round his bedroom and bring some clothes he had left. He arrived home at half past six and spent the evening alone with Emma. The next day he wrote to Erasmus: "Poor Emma is well bodily and very firm, but feels bitterly and God knows, we can neither see on any side a gleam of comfort." Emma wrote to Fanny: "We have done little else but cry together and talk about our darling . . . I think everybody loved her. Hers was such a transparent character, so open to kindness and a little thing made her so happy . . . From her having filled our minds so much for the last nine months it leaves such an emptiness . . . I suppose this painful longing will diminish before long. It seems as if nothing in this life could satisfy it."

Charles played with the little ones; he talked calmly about Annie, and was even able to read for a short time, which was "a rest from bitter thoughts, even for a few minutes." Elizabeth thought he would be longer getting over it than Emma. "She escapes the last painful impressions of the utter change, though I do not know whether the bitter longing to have seen her again is not worse . . . She is able to read a little too, and goes about as usual amongst the children, with even a cheerful smile for them." Charles and she both talked "of indifferent things."

Charles's decision to leave Malvern before the funeral preyed on his mind. Fanny was to take his place at the service, and he wrote to her on Friday with an assurance that hinted at uncertainty. "I cannot resist writing one line to thank you for having so tenderly advised me to return to home. I am sure I have acted best for Emma's sake. It is some sort of consolation to weep bitterly together." He was comforted to know that Fanny would follow Annie's coffin to the grave. "I know of no other human being whom I could have asked to have undertaken so painful a task." He ended with painful diffidence, "Sometime I should wish to know on which side and part of the Church-yard, as far as you can describe it, the body of our once joyous child rests."

Annie was to be buried in the graveyard of the village church, and Cox & Co., linen drapers, silk mercers and undertakers, managed the arrangements. The graveyard with its yews, cypresses and cedars around the grey and brown stonework of the church was a place which drew many visitors. The journalist Joseph Leech forgot his cynicism about the water cure when he remembered it. He had become very fond of the "beautiful and tranquil burial-ground." "Seated on one of the wooden benches that stood by the broad gravel walk or its grassy slope, and overlooking the quiet landscape, while the noble old building—choir, nave, tower and transept—flung out their deep shadows and sheltered me from the warm sun, I passed many and many a dreamy hour in its quiet sanctuary."

It was the duty of the church sexton, William Whiting, to make and fill up the graves for the dead. Fanny chose the place for Annie. It was on sparsely covered ground under a cedar of Lebanon facing the north side of the chancel. Two paces away was another fresh grave which Mr. Whiting had dug for his own twenty-two-year-old daughter Frances a month beforehand after she had died of tuberculosis.

The details of a funeral were usually left to the undertaker. He would be instructed simply to "provide what is customary," and the price would be agreed. Charles paid £57 12s 6d to Cox & Co.; a large sum for the 1850s which bought a funeral with full pomp for a child of the gentry—a hearse, a coach for the family mourners, black horses with ostrich-feather plumes, and two "mutes," paid mourners with black gowns, kid gloves, silk hatbands and standards of crape, the black fabric of mourning. Charles found the ceremony of the Anglican burial service "very impressive." He appreciated the form that would be followed for Annie, and the familiar words that would be spoken over her grave.

The coffin was placed in the hearse in the driveway of Montreal House; Fanny, Hensleigh, Miss Thorley and Brodie sat in the carriage, and the procession made its way slowly along the cobbled road into the village past the stucco villas, the Foley Arms and Lamb's Bazaar. They came to the church at nine o'clock; the tolling bell fell silent and Mr. Rashdall conducted the service as villagers and visitors went about their business in the streets around. As the Book of Common Prayer ordained, "The priest and clerks meeting the corpse at the entrance of the churchyard, and going before it, shall say or sing 'I am the resurrection and the life, saith the Lord: he that believeth in me, though he were dead, yet shall he live; and whosoever liveth, and believeth in me, shall never die . . . We brought nothing into this world, and it is certain we can carry nothing out. The Lord gave, and the Lord taketh away; blessed be the name of the Lord.' " At the graveside, as the bearers prepared the coffin to be laid in the earth, Mr. Rashdall said: "Man that is born of a woman, hath but a short time to live, and is full of misery. He cometh up, and is cut down like a flower; he fleeth as it were a shadow, and never continueth in one stay. In the midst of life we are in death." And as the earth was cast on the coffin by the bearers, "We therefore commit her body to the ground; earth to earth, ashes to ashes, dust to dust, in sure and certain hope of the resurrection to eternal life."

When Fanny and Hensleigh returned to their lodgings, she wrote to Charles and Emma: "I think everything was rightly arranged as you would have wished. I feared for Miss Thorley and still more for poor Brodie, and she has suffered, poor thing, most sadly, and had to be lifted into the carriage. But since she has been relieved by a long fit of crying, and is lying down now. Miss Thorley was more composed than I expected, only now and then with bursts of grief—poor thing. There never could have been a child laid in the ground with truer sorrow round her than your sweet and happy Annie." Fanny had suggested to Brodie that she should go straight back to Down House the next day. "She longs to be there, and when she finds you want her, I hope she will be able to put restraint on herself." Fanny and Hensleigh had taken Miss Thorley for a drive in their carriage. She was better, and was to go home to her mother in London. "We all leave this sad place together at nine tomorrow morning. Dr. Gully has been to inquire if I had heard how you and Charles were. He is full of kindness."

Charles put a notice in the "Deaths" column of *The Times.* "On the 23d inst., at Malvern, of fever, Anne Elizabeth Darwin, aged 10 years, eldest daughter of Charles Darwin, Esq., of Down, Kent." The gravestone bore a roundel with the emblematic letters for Jesus, "IHS." While most other inscriptions in the churchyard quoted Scripture or referred to the Christian faith in other ways, Annie's read simply:

ANNE ELIZABETH
DARWIN
BORN MARCH 2. 1841.
DIED APRIL 23. 1851.
A DEAR AND GOOD CHILD

Fanny's maid had taken Etty with Fanny's children from London to Leith Hill Place. Joining Uncle Joe and Aunt Caroline's three daughters, they were seven children in all. Aunt Caroline asked Fanny to tell Charles that Etty seemed very comfortable and at ease with the others. "I will take the greatest care of all the dear little set and they shall never be long out of my sight or hearing." On Wednesday, she wrote: "They are all gone cowslip-gathering in the fields . . . Etty seems quite content and excellent friends with all the cousins." Delphine Bonehill, the Wedgwoods' seventeen-year-old half-Belgian, half-English nursery governess, was in charge, but she made herself unpopular because Aunt Caroline told her to speak French and "all the children were too full of play to bear the trouble."

Charles, Emma and Fanny wanted their children to face death and understand it, but whether the young ones would show their feelings was uncertain and the parents watched carefully. Etty came back to Down from Leith Hill Place on Friday, and Emma saw her "quiet grief." Charles was grateful "to see her show so much feeling about Annie, crying and quite sobbing when she heard she had asked for her." Fanny gathered from her maid that the twelve-year-old Effie "cried very much when she went to bed" the night the news came. She cried again when Etty left Leith Hill Place, but after a short time went out of doors with the others and seemed cheerful and happy. Later, she went to talk with Aunt Caroline, and "asked some particulars which made her look again very low."

On Saturday, Charles wrote to Miss Thorley's mother: "I must beg permission to express to you our deep obligation to your daughter and our most earnest hope that her health may not be injured by her exertions. I hope it will not appear presumptuous in me to say that her conduct struck me as throughout quite admirable. I never saw her once [yield] to her feelings as long as self restraint and exertion were of any use. Her judgement and good sense never failed; her kindness, her devotion to our poor child could hardly have been excelled by that of a mother. Such conduct will, I trust, hereafter be in some degree rewarded by the satisfaction your daughter must ever feel when she looks back at her exertions to save and comfort our poor dear dying child. I earnestly hope that her health will be pretty soon established. My wife joins in kindest remembrance to yourself."

Brodie came back to Down, but found that she could not recover her composure after Annie's death. She was sixty years old and needed to rest after her working life of care for others; Charles gave her an annuity and she returned to live in her home town of Portsoy in northeast Scotland. She kept in close touch with the Darwins and Thackerays, and came south often for long visits to both families.

Fanny returned to her life with Hensleigh among their liberal friends in London. In a few weeks, she was lobbying for the Italian republican Giuseppe Mazzini who was then seeking support in England. Aunt Jessie wrote to Elizabeth with alarm: "Underneath that refreshing quiet, that delicious calm, Fanny has a lava of living fire that has made her give battle to all the governments of Europe under the banner of Mazzini. She is of his Committee in London! How could Hensleigh permit it? It is so contrary to the modesty of her nature to associate her name with such notoriety that I am sure she will suffer." But purpose and confidence in her judgement were also part of her nature, and Hensleigh did not intervene.

For parents in the 1850s, the loss of a child was not the utter shock it may be for parents now, because death in childhood was a fact of life. But when it came, the pain was deep in other ways, and Christian faith offered challenges as well as consolations. There were different ways of facing up to loss depending on one's understanding and beliefs, and the feelings of close family and friends.

Some people whom Emma and Charles knew declared firm confidence in Christ's salvation as an assertion of human faith and will against the unknown. Mrs. Carter, the Baptist minister's wife who lived along the lane into the village, died in extreme pain a few months after Annie. A friend gave an account of her last hours in a Baptist periodical, *The Earthen Vessel*. "Being troubled with continued retchings, her strength rapidly declined, and her sufferings were very great indeed, arising from inward convulsions. So great was the pain, that her countenance was at times distorted with it; and a truly distressing scene it was to witness, more especially as she drew near the borders of the eternal world; the hosts of hell were marshalled, and permitted to have their last attack . . . But at length the powers of darkness were foiled, the hosts of hell were put to flight, and suddenly a heavenly ray overspread her pallid countenance, and with her hands and eyes upraised, she appeared to be conversing with invisible beings; after which she bid us all affectionately adieu; saying, 'I am going to Jesus.' And then turning to her husband, as if to tell him what she saw, with her hands still upwards, and her lips moving; but mortal speech had failed, the battle had been fought, and the victory won."

Lady Lubbock in the grand house on the other side of the village had voiced the same assurance in her orthodox Anglican piety when her young daughter was close to death with scarlet fever. "I consider children so entirely from Heaven that I could not or at any rate ought not to repine should my Heavenly Father see fit to recall one of his gifts, and my poor Mary is so sweet and gentle that I feel as if she was in a fit state for a world of purity."

One "advanced" Unitarian divine suggested that the grief of loss itself had value. The Reverend John James Tayler was a prominent figure in the north London circles in which Hensleigh and Fanny Wedgwood moved. He was a close friend of Francis Newman, and co-editor with James Martineau of the *Prospective Review* of "free theology." His *Christian Aspects of Faith and Duty* appeared in 1851, and Emma noted the title in her diary for 1852. In a chapter on the "Blessings of sorrow," he suggested that "Christianity in the highest sense is the Religion of Sorrow. It baptises the heart with a holy sadness, and prepares it for the descent of the Spirit of God." He explained that "a crippled and suffering child, looked at from without, seems the heaviest of domestic afflictions. Yet once confided to our care, what an object of tender interest it becomes! What gentle and holy affections hover over it!" When a child dies, "the heart then learns the deep blessing of sorrow." Those who have experienced the grief of loss and "seen the mortal breath pass from the pale lips" of a child, "know well that in such an hour, whatever faith is latent in the heart, comes forth in all its strength, and rises up to the demand of our wants, and enables us to say in the depth of heavenly trust, 'Father, thy will be done.' Never are the beloved so dear, never so inseparable from our inmost spirit, never can we so little conceive the possibility of their perishing from us for ever—as in the moment when death throws his dark veil between us and them, and faith glows into intensity under the breath of affection."

Other people's religion pointed to Divine punishment as well as salvation. The conviction that death was due to sin, either the victim's, or another person's or Adam's, was deeply ingrained and gave a fearful twist to the pain of a bereavement for any who felt that they or others might be to blame in some way. When in 1850 William Gladstone, the future prime minister, lost his four-year-old daughter Jessie to tubercular meningitis, he wrote in his journal how he believed her suffering and death were bound up with universal moral issues. "It was, I must own, a heavy trial to flesh and blood to witness her death-struggle; to see that little creature who had never 'sinned against the similitude of Adam's transgression,' paying the forfeit of our race . . . What a witness was before us to the immensity of sin and the wide range of its effects, when she was so torn by their force." In 1856, months after the desolating loss of five of his six daughters to scarlet fever, Dean Tait of Chester, who later as Archbishop of Canterbury played a part in reconciling Anglican beliefs with the science of the time, was tormented by a belief that his family's tragedy was a divine chastisement for his sin of worldliness. He feared that his sin "necessitating this judgement" had caused the deep grief for his wife.

When it came to the hope for eternal life, many Christian believers faced critical uncertainties. The Anglican Church and most Dissenters looked to Holy Scripture as the main authority and guidance on matters of faith, but the Bible had little to say about what happened after death and much of what it said was vague, figurative and inconsistent. Richard Whately wrote regretfully in his *View of the Scripture Revelations Concerning a Future State* of "the brief, dry, unpretending, uncircumstantial manner, in which a future state is everywhere spoken of by the sacred writers; a manner eminently unfitted to excite the passions, to amuse the imagination, or to gratify curiosity." The point was also made to children. Ann and Jane Taylor wrote in one of their *Hymns for Infant Minds:*

*. . . where my living soul would go,*
*I do not and I cannot know:*
*For none was e'er sent back to tell*
*The joys of heaven, or pains of hell.*

Within the Wedgwood circle, Emma's Aunt Jessie had felt deep uncertainties when her husband died in 1842. She had written to Emma that if she could have had "firm faith that he was only passed from the visible to the invisible world, and already lives and is waiting for me, oh what happiness it would be." But "Alas, my faith seems all hope only and no firmness, and in such discouragement as mine, even hope itself cannot wear her cheerful face." Fanny Wedgwood's father, Sir James Mackintosh, was troubled by similar worries after the death of a friend. Listening to an uninspiring sermon at the man's funeral, he found that the reasons for the "venerable and consolatory" belief in the immortality of the soul had previously seemed strong and sound to him, but "in the preacher's statement they shrunk into a mortifying state of meagreness," and thoughts occurred to him "which I should be almost afraid to communicate to any creature."

Others who accepted Christian teaching on the afterlife still found when a loved one died that they were overwhelmed by a sense that the death was final. Frederick William Robertson, a liberal Anglican clergyman admired by Emma, preached in 1852: "Talk as we will of immortality, there is an obstinate feeling that we cannot master, that we end in death; and that may be felt together with the firmest belief of a resurrection. Brethren, our faith tells us one thing, and our sensations tell us another." The feeling undermined faith and prayer. "Everyone who knows what Faith is, knows too what is the desolation of doubt. We pray till we begin to ask, Is there one who hears, or am I whispering to myself? We hear the consolation administered to the bereaved, and we see the coffin lowered into the grave, and the thought comes, What if all this doctrine of a life to come be but the dream of man's imaginative mind?" Men spoke of faith "as a thing so easy," but Robertson, who had at one time suffered grave doubts about his beliefs, declared firmly that "To feel faith is the grand difficulty of life." For many believers there was "cold dark watching," "struggle when victory seems a mockery to speak of," and "times when light and life seem feeble, and Christ is to us but a name, and death a reality."

By contrast with all Christians, some freethinkers among the Darwins' friends welcomed the idea that death was the end. Harriet Martineau wrote in *Letters on the Laws of Man's Nature and Development* that she agreed with Mr. Atkinson on "the fallacy of all arguments for a conscious existence after death." People took their wish for a life after death as evidence for it; "the desire itself is a factitious thing," and "many (and this I know) do not desire it at all." Her friend, Charles's brother Erasmus, was almost certainly one of those she was thinking of.

When a child died, the feelings and difficulties for Christian believers were especially strong. Elizabeth Birks, a devout evangelical, came to terms with the deaths of her father and her sister Frances in eager faith that they would find bliss in Heaven, but when her baby son died suddenly in 1854 she wrote to her sister: "I did not know the parting would be such a pang—a peep into a gulf I had not looked down before. It is something very different from our former losses; something all my own; part of my daily life." She came to accept "God's chastening hand resting on us" but three months later she found, "I dare not at times trust myself with what brings back to me my *baby*. My child in heaven I can think of, but not the *baby* I have lost."

For people without a clear Christian faith, two of William Wordsworth's so-called "Lucy" poems suggested a way in which a young girl might be seen as living in nature, and a kind of consolation might be found in seeing her death as a natural event. In the first he wrote:

*Three years she grew in sun and shower,*
*Then Nature said, "A lovelier flower*
*On earth was never sown;*
*This Child I to myself will take . . ."*

*Myself will to my darling be*
*Both law and impulse: and with me*
*The Girl, in rock and plain,*
*In earth and heaven, in glade and bower,*
*Shall feel an overseeing power*
*To kindle or restrain . . .*

*. . . hers shall be the breathing balm,*
*And hers the silence and the calm*
*Of mute insensate things.*

*. . . and she shall lean her ear*
*In many a secret place*
*Where rivulets dance their wayward round,*
*And beauty born of murmuring sound*
*Shall pass into her face . . .*

*Thus Nature spake—The work was done—*
*How soon my Lucy's race was run!*
*She died, and left to me*
*This heath, this calm and quiet scene;*
*The memory of what has been,*
*And never more will be.*

In "A Slumber Did My Spirit Seal," he used words of physical science to describe the girl after her death.

*No motion has she now, no force;*
*She neither hears nor sees;*
*Rolled round in earth's diurnal course,*
*With rocks, and stones, and trees.*

One father who saw the death of a child as a natural happening found it all the more difficult to accept for that. Alfred Tennyson's first child was stillborn on Easter Sunday 1851, three days before Annie died. At the birth Tennyson "heard the great roll of the organ, of the uplifted psalm" in the chapel next door to their home.

*Little bosom not yet cold,*
*Noble forehead made for thought,*
*Little hands of mighty mould*
*Clenched as in the fight which they had fought.*
*He had done battle to be born,*
*But some brute force of Nature had prevailed*
*And the little warrior failed.*

Emma and Charles's family and friends, knowing her faith and probably having some idea of his doubts and the difficulties between the two, offered what consolations they could, but spoke with care. In all the letters of sympathy that Emma and Charles kept, only his sister Catherine made one brief reference to the idea that Annie had entered a better life in Heaven. She wrote to Charles: "No little dear could have had a happier life, except her health, until her little innocent spirit was called above." She wrote to Emma: "I can only hope that God may comfort you in your great sorrow," but did not suggest how He might.

There were two suggestions that Christian belief might be a consolation in other ways. Catherine wrote that Emma's prayers might comfort her, and a close family friend, Ellen Tollet, believed there was "a mysterious consolation in the act of meekly submitting our will to His." In writing those words, she was echoing Emma's own hope, which she had declared so tentatively to Fanny, that she would be able to "attain some feeling of submission to the will of Heaven."

A few consolations depended on other things than Christian faith. Catherine and Emma's sister Charlotte Langton suggested that Emma and Charles could take comfort in the thought that they had done "everything in the world to make her happy." Catherine also wrote: "It is an infinite mercy that your darling did not suffer. That would have been such an addition." She hoped that Emma's own "wonderful love and thought for others" might support her through the "grievous trial." And Caroline suggested that Emma's approaching confinement was perhaps "the best thing for her."

Everyone agreed that time and memory were the best hope for recovery. Catherine wrote to Emma that there was "no comfort but time." Erasmus wrote: "It is all bitterness to you now, but tender memories will survive and it will not be all loss." Fanny wrote of pain and pleasure. "It will be long before you will be able to conquer that feeling of longing which is so bitter, and yet there will be something on the other side, the recollection of *herself* as she always was, bright, happy and loving. How difficult and how sad are all these comparisons, and yet one cannot help perpetually having them before one."

These various beliefs and uncertainties were the background for Charles and Emma's thinking and understanding with each other. When they wrote to each other during the last few days of Annie's illness, the couple often used words which suggested that God had foreseen and possibly in some way influenced the outcome. Emma wrote at one point "How I do thank God" for her improvement. On another occasion she pleaded "God grant that dreadful sickness may keep off."

And after the end, "I do feel very grateful to God that our dear darling was apparently spared all suffering." Just as Erasmus had written to Fanny Wedgwood when Dr. Darwin died in 1849, Charles also now referred to God as a kind figure overlooking human life and mitigating pain. He wrote: "God only knows the issue," and "God only knows what miseries would have been in store for Annie had she lived." At other times, he thanked God that she did not suffer, that she was not worse, that she had not been sick, and that he had hardly ever cast a disapproving eye on her. He said "God bless" Emma and Fanny; he exclaimed "God help us"; he wished that God would preserve and cherish Emma, and he said that if diarrhea did not come, "I trust in God we are nearly safe."

After Annie's death, he wrote to his cousin Fox, who was an Anglican clergyman, "Thank God she suffered hardly at all, and expired as tranquilly as a little angel. Our only consolation is that she passed a short though joyous life . . . Poor dear little soul. Well, it is all over."

Emma hoped that Annie would go to Heaven and that she would join her there, but she could not fathom God's purpose in taking her child from her. Charles, on the other hand, had no belief that there was any Divine purpose behind such events. For Emma as a devout Christian, death was inextricably bound up with sin, but for Charles there was no connection. Since writing his essay on evolution in 1844, he had held to the view he had reached then. Death was a purely natural process. Medicine might eventually find natural causes and work out treatments, but there were no explanations in religion for the loss of a loved child.

Charles found one consolation in an idea which Rousseau had mentioned in *Émile.* For too many, Rousseau suggested, childhood passed "amidst tears, punishments, threats, and slavery." He asked fathers, "Do you know the moment when death awaits your children? Do not prepare regrets for yourself in depriving them of the few instants nature gives them. As soon as they can sense the pleasure of being, arrange it so that they can enjoy it; arrange it so that at whatever hour God summons them they do not die without having tasted life." Charles thanked God that he hardly ever cast a look of displeasure on Annie, and wrote many years later that it was his "greatest comfort" that he had never spoken a harsh word to her.

As for the hope that time would heal the wound, Charles and Emma both knew it would be a balance of remembering and forgetting. Each wanted to keep certain memories, purged of pain, but they dealt with them in different ways. Emma was as reticent in her recollections as she was about any deep feelings, but she kept precious letters and objects, and used them to help her remember the people she cared for. When her sister Fanny had died in 1832, Emma did not write about her as some would have done, but made a small packet of her housekeeping memoranda and lists. When Etty saw them many years later, she was struck by the triviality of the scraps, and yet knew how her mother had cared for them. "There is something strangely pathetic in finding these simple records thus carefully kept for sixty years or more after they were written."

After Annie's death, Emma made a few notes about her recollections of her, but they were only brief prompts for her memory. She found words for her feelings in two poems by Hartley Coleridge, son of Samuel Taylor Coleridge, about the death of a young woman. The first asked "Where dwells she now?" and suggested that she no longer had any part in this life. The second poem in reply argued that the world was a "record sad of ceaseless change." Emma copied out three stanzas.

*She passed away, like morning dew*
*Before the sun was high;*
*So brief her time, she scarcely knew*
*The meaning of a sigh.*

*As round the rose its soft perfume,*
*Sweet love around her floated;*
*Beloved she grew—while mortal doom*
*Crept on, unfeared, unnoted.*

*Love was her guardian angel here,*
*But love to death resigned her;*
*Though love was kind, why should we fear?*
*But holy death is kinder.*

The poem made no reference to the afterlife or reunion with loved ones, only the kindness of "holy death."

Emma gave Annie's small embroidery case to Effie. She wrote to Fanny Wedgwood that she thought Effie would "like to possess some little keepsake out of poor Annie's treasures. She was always the one Annie loved best." Emma gave some other playthings of Annie's to Etty, but took her writing case for herself. She gathered some of Annie's letters, a piece of her embroidery and one or two of her trinkets. As she placed the things together in the box, their simple meanings added to each other.

The quills, steel nibs, paper, sealing wax and seals had all been kept in the writing case, and called Annie to mind as she had sat writing letters to cousins and sealing the envelopes for Emma to send. Of the other things that Emma put in the box, the silk needle case, the thimble and the needlework also reflected Annie's neatness and concentration. The ribbon, the glass beads and the pendant were small reminders of her love of trinkets and treasures. Annie's pocketbook for 1848 recalled her seventh birthday. Emma had inscribed it, "Anne Elizabeth Darwin March 1848 From her Mamma."

Annie's letter to Sarah Thorley brought to mind one of the friendships outside the family circle which Annie had been learning how to take forward as a polite young lady. The notebook Brodie had made and embroidered for Annie spoke of the nurse's devotion to her during the last months. Emma found the piece of paper with Charles's daily notes on his care of Annie and her condition. Charles's words, brief and frank, showed his private understanding of Annie's distress, unsoftened by the glossings of hope for recovery with which he would have spoken to her and others. Putting the folded paper in the box with Annie's "childish things," Emma kept the memory of her and Charles's hidden concern and anxieties during Annie's lingering sickness.

Etty's letters to Emma and George in the first days at Malvern pictured Annie buying oranges in the village and riding a donkey on the hills above. She had been up and about then, and the letters were a sharp reminder of how sudden and unexpected was the final crisis. The lock of Annie's hair had been cut in the hours after her death. Emma wrote the date on the paper, closed up the writing case and put it away in a private place.

Charles kept nothing of Annie's for himself, but chose instead to write a piece about her. It was a common practice at the time for a bereaved father or mother to write a private "memorial" of a loved child. Most wanted to bring their experience of loss into key with the comforting idea of the "good death." They would describe the child's moral state and approach to death; they would explain the consolation that had been found, and look forward to the next life. Charles, by contrast, had no wish to dwell on Annie's illness and nothing to say about the meaning of her death or a life beyond. His aim was to preserve his memories of the living child and what she had meant to him and Emma when she was happy and well.

Charles had always had clear recollections of things he loved, and liked to dwell on them. During the *Beagle* voyage he had found that cherished memories rose "the more vividly in my imagination" because of his remoteness from the people and places remembered. Etty once saw him reaching for a memory of his father when they visited the family home in Shrewsbury some years after his father's death. The occupant showed them round, and as they were leaving, Charles said regretfully: "If I could have been left alone in that greenhouse for five minutes, I know I should have been able to see my father in his wheel-chair as vividly as if he had been there before me." When Charles's cousin, the anthropologist Francis Galton, asked him about his visual memory some time later, he explained with a telling detail: "I remember the faces of persons formerly well-known vividly, and can make them do anything I like."

Charles was, though, unhappy about one missing figure. His mother had died when he was eight, after a painful illness. She had been confined to her bedroom for some time beforehand; Charles had been looked after by his elder sisters, and was kept away from her room during the last days. When he came to write about his childhood memories in 1838 and again in his last years, he mentioned that he could scarcely recall anything of her except being sent for when she died, going into her room, and "my Father meeting us crying afterwards." Almost at the end of his life, after his brother Erasmus's death, he discovered a miniature of her among Erasmus's possessions and wrote to his sister Caroline about it. He was glad to learn from her that the picture, with its "most sweet expression," was a good likeness. He was clearly concerned about his inability to remember his mother, and suggested that his "forgetfulness" might be partly accounted for "by none of you being able to endure speaking about so dreadful a loss." When Annie died, Charles may well have recognised the possibility that he and Emma would find it difficult to talk about her with the children. They might then all lose their memories of her in the silence.

A week to the day after Annie's death, Charles took a gathering of special black-bordered mourning paper and prepared a quill. He st[illegible] to write with a degree of detachment, but as the memories came, he wrote more and more freely, dwelling on glimpses of Annie, words and gestures that brought out his deepest feelings. He ended with a direct but unanswerable plea.

> Our poor child, Annie, was born in Gower St on March 2nd 1841 and expired at Malvern at Midday on the 23rd of April 1851. I write these few pages as I think in after years, if we live, the impressions now put down will recall more vividly her chief characteristics. From whatever point I look back at her, the main feature in her disposition which at once rises before me is her buoyant joyousness, tempered by two other characteristics, namely her sensitiveness, which might easily have been overlooked by a stranger, and her strong affection. Her joyousness and animal spirits radiated from her whole countenance and rendered every movement elastic and full of life and vigour. It was delightful and cheerful to behold her. Her dear face now rises before me, as she used sometimes to come running down stairs with a stolen pinch of snuff for me, her whole form radiant with the pleasure of giving pleasure. Even when playing with her cousins when her joyousness almost passed into boisterousness, a single glance of my eye, not of displeasure (for I thank God I hardly ever cast one on her), but of want of sympathy would for some minutes alter her whole countenance. This sensitiveness to the least blame, made her most easy to manage and very good; she hardly ever required to be found fault with, and was never punished in any way whatever. Her sensitiveness appeared extremely early in life; and showed itself in crying bitterly over any story at all melancholy, or on parting with Emma even for the shortest interval. Once when she was very young she exclaimed "Oh Mamma, what should we do, if you were to die?"

The other point in her character, which made her joyousness and spirits so delightful, was her strong affection, which was of a most clinging, fondling nature. When quite a Baby, this showed itself in never being easy without touching Emma, when in bed with her, and quite lately she would, when poorly, fondle for any length of time one of Emma's arms. When very unwell, Emma lying down beside her seemed to soothe her in a manner quite different from what it would have done to any of our other children. So again, she would at almost any time spend half-an-hour in arranging my hair, "making it" as she called it "beautiful," or in smoothing, the poor dear darling, my collar or cuffs, in short in fondling me. She liked being kissed; indeed every expression in her countenance beamed with affection and kindness, and all her habits were influenced by her loving disposition.

Besides her joyousness thus tempered, she was in her manners remarkably cordial, frank, open, straightforward, natural and without any shade of reserve. Her whole mind was pure and transparent. One felt one knew her thoroughly and could trust her: I always thought, that come what might, we should have had in our old age, at least one loving soul, which nothing could have changed. She was generous, handsome and unsuspicious in all her conduct; free from envy and jealousy; good-tempered and never passionate. Hence she was very popular in the whole household, and strangers liked her and soon appreciated her. The very manner in which she shook hands with acquaintances showed her cordiality.

Her figure and appearance were clearly influenced by her character: her eyes sparkled brightly; she often smiled; her step was elastic and firm; she held herself upright, and often threw her head a little backwards, as if she defied the world in her joyousness. For her age she was very tall, not thin, and strong. Her hair was a nice brown and long; her complexion slightly brown; eyes dark grey;

*Jenny the orang-utan at the Zoological Gardens*

*Emma Darwin by George Richmond in 1840*

*Charles Darwin by George Richmond in 1840*

*Ernest Wedgwood by George Richmond in 1840*

*Daguerreotype of Charles and William Darwin in 1842*

*Charles Darwin's study, photographed by Leonard*

*Daguerreotype of William Darwin in 1849*

*Daguerreotype of Anne Darwin in 1849*

*Daguerreotype of Henrietta Darwin in 1849*

*Jessie Brodie, the Darwin children's nurse*

*Dr. Gully,*
*the physician at Malvern*

*Catherine Thorley, the governess, photographed by William Darwin in 1859*

*Anne Darwin's gravestone in the churchyard of Malvern Priory*

13) she said "I quite thank you"; & these, I
believe were the last precious words ever
addressed by her dear lips to me.
But looking back, always the spirit of
joyousness rises before me as her emblem and
characteristic: she seemed formed to live a life
of happiness: her spirits were always held in
check by her sensitiveness lest she should displease
those she loved, & her tender love was never
weary of displaying itself by fondling & all the
other little acts of affection. — We have
lost the joy of the Household, and the
solace of our old age:— she must have
known how we loved her; oh that she
could now know how deeply, how tenderly
we do still & shall ever love her
dear joyous face. Blessings on her. —

April 30. 1851.

*Charles Darwin's memorial of Annie*

*Daguerreotype of Henrietta Darwin in August 1851*

*Charles Darwin by Samuel Laurence in 1853*

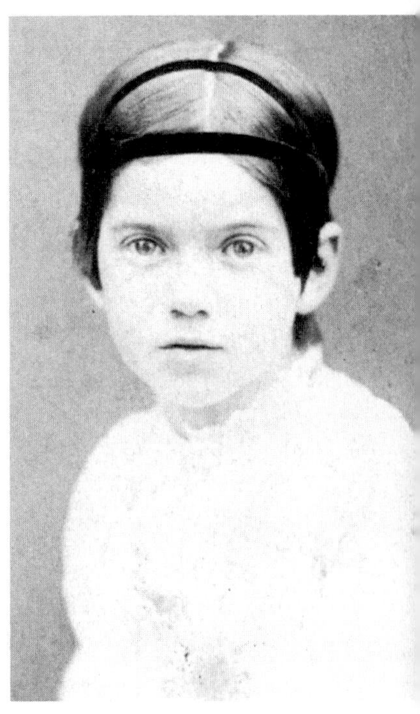

TOP LEFT: *Daguerreotype of George Darwin in August 1851*

TOP RIGHT: *Elizabeth Darwin in the mid-1850s*

BOTTOM: *Emma Darwin with Leonard in 1854*

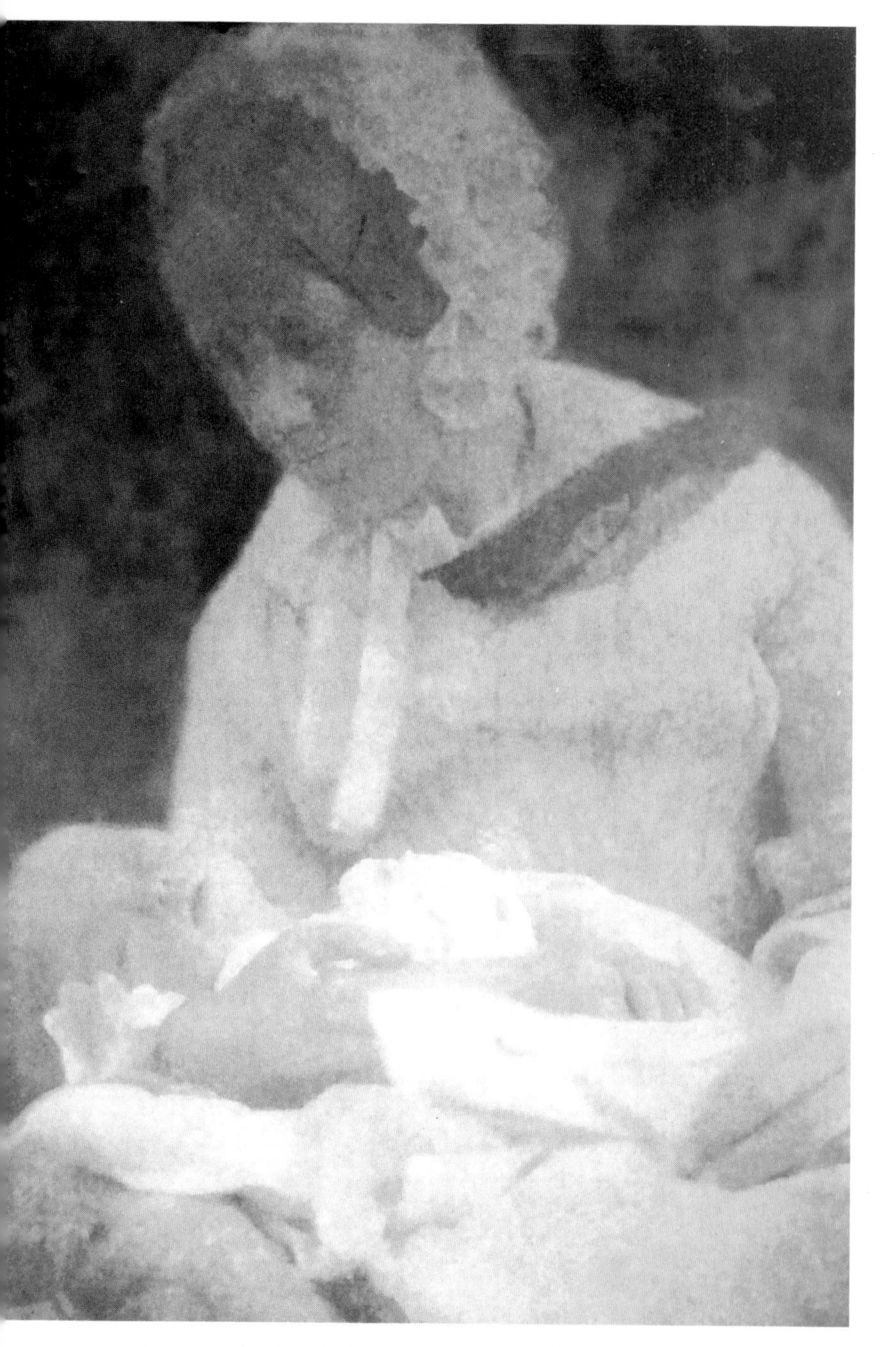

*Emma Darwin with Charles Waring, photographed by William in 1857*

*Charles Darwin, photographed by Leonard in 1878*

*Emma Darwin in her widowhood*

*Down House and the garden by Alfred Parsons in 1883*

her teeth large and white. The daguerreotype is very like her, but fails entirely in expression: having been made two years since, her face had become lengthened and better looking. All her movements were vigorous, active and usually graceful; when going round the Sand-walk with me, although I walked fast, yet she often used to go before pirouetting in the most elegant way, her dear face bright all the time, with the sweetest smiles.

Occasionally she had a pretty coquettish manner towards me, the memory of which is charming: she often used exaggerated language, and when I quizzed her by exaggerating what she had said, how clearly can I now see the little toss of the head and exclamation of "Oh Papa, what a shame of you." She had a truly feminine interest in dress, and was always neat: such undisguised satisfaction, escaping somehow all tinge of conceit and vanity, beamed from her face, when she had got hold of some ribbon or gay handkerchief of her Mamma's. One day she dressed herself up in a silk gown, cap, shawl and gloves of Emma, appearing in figure like a little old woman, but with her heightened colour, sparkling eyes and bridled smiles, she looked, as I thought, quite charming.

She cordially admired the younger children; how often have I heard her emphatically declare "What a little duck Betty is, is not she?"

She was very handy, doing everything neatly with her hands: she learnt music readily, and I am sure from watching her countenance, when listening to others playing, that she had a strong taste for it. She had some turn for drawing, and could copy faces very nicely. She danced well, and was extremely fond of it. She liked reading, but evinced no particular line of taste. She had one singular habit, which, I presume, would ultimately have turned into some pursuit; namely a strong pleasure in looking out words or names in dictionaries, directories, gazetteers, and in this latter case finding out the places in the Map: so also she would take a strange interest in comparing word by word two editions of the same book; and again she would spend hours in comparing the colours of any objects with a book of mine, in which all colours are arranged and named.

Her health failed in a slight degree for about nine months before her last illness; but it only occasionally gave her a day of discomfort: at such times, she was never in the least degree cross, peevish or impatient; and it was wonderful to see, as the discomfort passed, how quickly her elastic spirits brought back her joyousness and happiness. In the last short illness, her conduct in simple truth was angelic; she never once complained; never became fretful; was ever considerate of others; and was thankful in the most gentle, pathetic manner for everything done for her. When so exhausted that she could hardly speak, she praised everything that was given her, and said some tea "was beautifully good." When I gave her some water, she said "I quite thank you"; and these, I believe were the last precious words ever addressed by her dear lips to me.

But looking back, always the spirit of joyousness rises before me as her emblem and characteristic: she seemed formed to live a life of happiness: her spirits were always held in check by her sensitiveness lest she should displease those she loved, and her tender love was never weary of displaying itself by fondling and all the other little acts of affection.

We have lost the joy of the household, and the solace of our old age: she must have known how we loved her; oh that she could now know how deeply, how tenderly we do still and shall ever love her dear joyous face. Blessings on her.

April 30. 1851.

CHAPTER ELEVEN

# THE DESTROYING ANGEL

*Cause of death—Tuberculosis—Fear of the disease—
Charles's understanding—Child mortality—Bacteria*

IT HAD BEEN LEFT TO THE LANDLADY in Malvern to report Annie's death for the official record. Eliza Partington took a note from Dr. Gully to Mr. Dancocks, the Registrar of Births, Deaths and Marriages for the neighbourhood, and he entered the details in his ledger. He recorded the cause of death as "bilious fever with typhoid character, certified." The last word, a touch of procedure, formally acknowledged Dr. Gully's note as the judgement of a qualified physician.

Dr. Gully did not mean typhoid as we know it now. The disease of that name, caught from contaminated water, was recognised as a specific illness only in the 1860s and 1870s. Dr. Gully had not even identified a disease as the cause of death in the way a doctor would now; he was only describing Annie's symptoms. Her "bilious fever" was the vomiting and fever in the days before her death. The fever's "typhoid character" was the delirium she had drifted in and out of as Charles and the others sat beside her. In the medical language of the time, "typhoid" simply meant "like typhus," a separate and well-recognised disease of which delirium was one of the main symptoms.

As far as is known, no one in the Darwin family said anything more than Dr. Gully about the cause of Annie's death. It was enough to be told that she had died of a fever. So little was understood about the many different conditions of which fever was a symptom, and so little could be done to treat any of them, that there was no reason to say more. Dr. Holland had written in 1839: "We can scarcely indeed touch upon this subject of fever . . . without finding in it a bond by which to associate together numerous forms of disease; but withal a knot so intricate, that no research has hitherto succeeded in unravelling it."

Doctors now have many more diagnoses and treatments, and we now want to know what Annie was suffering from and what she died of. A medical history can be drawn up for her from the notes and letters that survive, but the information is difficult to interpret because many of the words used were vague and there is almost no clinical detail. To find out what could be said about the illness, I gave all the clues to four experienced physicians and medical historians for their advice. Looking at the pattern throughout the months after the time when Annie "first failed," they agreed that she probably died of tuberculosis, which was known at the time as consumption, phthisis or simply "decline." The early symptoms were characteristically non-specific and intermittent—discomfort, wakefulness and tiredness, but the change in Annie's health by the end of October 1850 must have been marked as it was worrying enough for Emma to take her twice to London to see Dr. Holland. Annie had a low fever in October and a barking cough from December. The crisis in early April may have been triggered by her 'flu in mid-March, or an infection she caught in Malvern. Tuberculosis then took hold of her body, weakened as it was by the infection. The disease may have struck Annie finally in the abdomen causing tuberculous peritonitis, or in its generalised "miliary" form infecting the blood and leading to meningo-encephalitis, the vomiting and fatal coma. Few who fell victim to tuberculosis before adolescence developed the racking consumptive cough with blood-stained phlegm. Severe damage to the lungs occurred more often in young women and men.

Tuberculosis is caused by a slow-working bacillus, *Mycobacte*...... *tuberculosis,* which can be picked up in infected milk or passed from person to person through the air when the carrier coughs. Many people who acquire the infection never suffer any active disease. Others are rapidly affected, and some develop the symptoms after a period of dormancy. The disease is said to be Protean; besides the commonest form which wrecks the lungs, other forms attack the intestines, the lining of the brain, the lymph glands, the spine, other bones, the kidneys and genitals. The bacillus was first identified as the cause of the disease by the German bacteriologist Dr. Robert Koch in 1882. The most obvious pathological signs had always been the tubercles—white fatty deposits which were often found in areas of infection. Tubercles were also found in slaughtered cows and other animals, especially the monkeys and apes who so often sickened and died in the zoos and menageries.

In the 1840s and 1850s there were deep fears of consumption in almost every household. Dr. Thomas Yeoman, a physician in north London, wrote in 1848: "Consumption, Decline or Phthisis, is the plague-spot of our climate; amongst diseases it is the most frequent and the most fatal; it is the destroying angel who claims a fourth of all who die." He wrote about the fears of the disease, "Does the individual exist who has not some special interest in every attempt to arrest its ravages? Is there a family without anxiety, lest some loved relative or connection should fall a victim to its ruthless arm?" It struck both rich and poor without favour. "Consumption steadily and surely pursues its way, and desolation of heart, of home, of hope, follow in its path."

Nothing was known at the time about the causes of the disease and there was no effective cure. The leading authority on the condition and its treatment was Sir James Clark, who had published his *Treatise on Pulmonary Consumption* in 1835. He was physician to Princess Victoria at the time; Charles had consulted him in 1838 about his illness, and he treated the pianist Frédéric Chopin when he was suffering from consumption in London in 1848. Clark estimated that a third of all deaths in England arose from tuberculous diseases, and noted that a physician at the Hôpital des Enfants Malades in Paris had reported in 1824 that five out of six children who died in the hospital were found at autopsy to be "more or less tuberculous."

Clark described a number of forms in which the illness might appear. The one closest to Annie's symptoms he called "latent or occult." It was particularly common in delicate young persons; it was marked by fever and night sweats, and it could continue without any more specific symptoms for some time. Any cough was rarely accompanied by expectoration with traces of blood until late in the disease. Dr. Yeoman wrote that the earliest symptoms were often "so obscure or doubtful that consumption . . . cannot be detected with certainty. We should always suspect the presence of consumption when we . . . find a cough continuing for some length of time, inducing increasing debility and emaciation." Another physician, Dr. Richard Cotton, wrote: "In childhood . . . the child is peevish, irritable, and indisposed to exertion; and, in general appearance, is evidently labouring under some deep-rooted malady, which, at no very distant time, will exhibit itself either as phthisis or some other form of tubercular disease."

It is a mark of how little was understood about the disease that no one established the critical point that it was infectious or how the condition was passed on. A few doctors had claimed it was infectious, but Clark argued strongly against them, and his view was widely accepted at the time. He and many others believed that the condition could arise in any person. Many suggestions were made about factors that might trigger it. Dr. Gully was among those who believed that it had its origins in poor digestion. There was also general agreement that the condition could be inherited. Dr. Yeoman wrote: "Many persons acquire a predisposition to consumption from their parents, although the latter may attain an advanced age without evincing any symptoms of pulmonary disorder." The parents need not themselves be consumptive. "Bad general health in one generation is frequently converted into tuberculous disease in the succeeding one."

Clark suggested that parents could protect their children from the disease by careful upbringing with fresh air and exercise, but insisted that once the disease had taken hold, it was fatal. "No physician acquainted with the morbid anatomy of Tuberculous consumption, can for a moment indulge the hope that we shall ever be able to cure what is usually termed 'confirmed consumption.'" By the 1840s and 1850s, his view that the disease was fatal was the accepted medical opinion.

Against this background, to say firmly that a child was consumptive was a sentence of death, and the force of the judgement had a deep effect on what people were prepared to admit to themselves and say to others about the possibility. Some young women dwelt on the idea. Harriet Martineau wrote in later life: "I romanced internally about early death till it was too late to die early." For most others, the possibility was a lurking fear to be kept out of mind as long as possible. Describing the early stages of the disease, Dr. Cotton noted a common pattern of denial. "From some cause, for which no good reason can be assigned, there is a slow but marked diminution of bodily vigour, compelling the individual to abandon many of his accustomed pursuits; the spirits, nevertheless, are good, and not only is the idea of consumption never entertained, but any allusion to it is at once ridiculed. So general, indeed, is this hopeful condition,—this almost instinctive blindness to the real cause of distress, that in its absence, however suspicious certain symptoms appear, these may, with much probability of accuracy, be pronounced unconnected with phthisis. The complexion is, at the same time, pallid or sallow; the expression is that of care united with animation; the features are somewhat sharpened; the movements of the body are hurried and anxious; the mental condition is irritated and capricious; whilst every act betrays an effort, sometimes instinctive, and at others voluntary, to conceal the presence of the disease."

Another physician, Dr. Henry Hillier, wrote insistently about the need for extreme caution in giving a clinical opinion. "The extreme prevalence of consumption in this country, the very insidious nature of the symptoms which mark its positive existence, its undoubted hereditary tendency, the various complications with other diseases which attend its development, and the very uncertain means we possess of staying its progress, renders it the imperative duty of every physician, no matter how celebrated his professional reputation, to pause ere he gives a positive opinion that his patient is the subject of consumption, a disease from which there is little chance of recovery, and which must sooner or later prove destructive to life."

Because there was no known cure for the condition, no hospital in London would admit consumptive patients until the Hospital for Consumption and Diseases of the Chest—now the Royal Brompton Hospital—was opened in 1842. It had two aims, to provide care for those poor victims of the disease to whom all other hospitals' doors were "irrevocably closed," and to learn what it could from them about the disease. "Pain and suffering must be alleviated; the agonies of disease must be mitigated and soothed: but we hope for more; we hope that here discerning and patient minds, investigating the progress of a fearful complaint in all its gradations, and narrowly observing the results of different remedies, will discover the best mode of treatment; and that from this Institution rays of light will be dispersed, not only through our country, but we might fain hope through the whole civilised world." An appeal for funds was widely supported, and Charles was one of many people hoping for progress in medical understanding and care who made contributions. The hospital staff were soon treating hundreds of patients, but the disease would not yield its secrets.

Charles Dickens had given a vivid impression of ordinary people's understanding of the slow onset of consumption in his novel *Nicholas Nickleby,* which appeared in 1838–39. "There is a dread disease which so prepares its victim, as it were, for death; which so refines it of its grosser aspect, and throws around familiar looks unearthly indications of the coming change—a dread disease, in which the struggle between soul and body is so gradual, quiet, and solemn, and the result so sure, that day by day and grain by grain, the mortal part wastes and withers away, so that the spirit grows light and sanguine with its lightening load and feeling immortality at hand, deems it but a new term of mortal life—a disease in which death and life are so strangely blended, that death takes the glow and hue of life, and life the gaunt and grisly form of death—a disease which medicine never cured, wealth warded off, or poverty could boast exemption from—which sometimes moves in giant strides, and sometimes at a tardy sluggish pace, but, slow or quick, is ever sure and certain." Dickens was familiar with the pattern of denial. When Nicholas noticed that his friend Smike was growing ill, he took him to a physician "with some faint reference in his own mind to this disorder, though he would by no means admit it, even to himself." The physician avoided commitment. "There was no cause for immediate alarm, he said. There were no present symptoms which could be deemed conclusive."

The *Christian Remembrancer,* an Anglican periodical, felt that Dickens was trying to romanticise the disease and found his approach objectionable. It commented: "We have great doubts about the propriety of this incessant working up our feelings by pictures of consumption. It is hardly fair. The subject is, to half the families of England, too fraught with painful reality to be thus introduced in a work of amusement, and amid dreamy sentiment. It suggests reminiscences at once too agonising and too sacred to make it admissible in fiction. Like the death to which, in all its manifold varieties, it surely, whether slowly and inch by inch, or with impetuous torrent-like rapidity, conducts its prey, consumption is a thing too terribly real to be fitly sported with." The writer urged realism. "Let our thoughts about death be always as practical as death is actual and certain."

But, as the *Christian Remembrancer* itself recognised, there was no avoiding the truth. Everyone had relatives, friends or neighbours who had been killed by tuberculosis. In writing about it in the way that he did, Dickens cannot have believed that his readers would forget what they knew about the fear and wasting pain involved. He and others dramatised the disease because so many of its victims were young people; they could feel it working in their bodies and see their death approaching at an uncertain pace for months or years; they had all too much time to dwell on what lay ahead for them. The "good death" of the evangelical and other traditions was a way to draw something of lasting value from the suffering. There was no denying or avoiding the long-drawn-out and cruel pain of the illness beneath the devout hopes.

Charles and Emma both read *Nicholas Nickleby,* and knew many other books for children and adults in which young people died slowly of consumption. In the books that Charles borrowed from the London Library for Annie in her last winter, there were two set-pieces: the illness and death of Ben the young shepherd in William Howitt's *Boy's Country Book,* and of Lady Eleanor in *The Earl's Daughter* by Elizabeth Sewell. The disease also claimed two of the main characters in Elizabeth Wetherell's *The Wide, Wide World,* the novel about the American child with a writing case which Emma read to her grandchildren. The book is now mocked by many for its sanctimonious tearfulness, and it seems strange at first that Emma, with her extreme reticence and distrust of emotional display, should have read it. But it was a melodrama like those which she had loved as a young woman at the theatre in London. The young heroine, Ellen Montgomery, cried and wept in different ways in different situations, and quite as often it was the suppression or utter lack of tears which expressed her emotion.

Elizabeth Wetherell wrote in rich detail about Ellen's life in a small town in New York State, and dwelt eloquently on her hesitant but growing faith in God, but she never named the illness which killed Ellen's mother and her closest friend Alice, and she wrote nothing about any physical symptoms. Yet the hints and silences left no doubt, and Ellen showed the familiar pattern of denial. When Alice revealed her condition, Ellen could not believe what she had told her. "To her mind it seemed an evil *too great to happen;* it could not be! . . . 'But have you seen somebody?—have you asked somebody?' said Ellen,—'some doctor?' 'I have seen, and I have asked,' said Alice; 'it was not necessary, but I have done both. They think as I do.' 'But those Thirlwall doctors—' 'Not them; I did not apply to them. I saw an excellent physician at Randolph, the last time I went to Ventnor.' 'And he said—' 'As I have told you.' Ellen's countenance fell—fell." Later, Alice's father responded in the same way. "It was impossible at first to make Mr. Humphreys believe that Alice was right in her notion about her health. The greatness of the evil was such that his mind refused to receive it, much as Ellen's had done." In Emma's copy of the book, these passages are marked with a straggling pencil line.

There is no hint in the Darwin family papers of a belief that Annie might have had consumption, but it was almost certainly the main unspoken fear. Emma's brother-in-law, Charles Langton, had lost nine brothers and sisters from the disease, and himself suffered an attack in 1833 which alarmed the Wedgwoods and the Darwins. Charles's father had judged that cousin Allen Wedgwood who had christened Annie at Maer was consumptive; Emma's sister Elizabeth had a severe spinal deformity which may have been due to tuberculosis, and there was a belief in the family that Charles's brother Erasmus had tuberculous damage to one lung. Herbert Mayo's *Philosophy of Living* which Charles read in the late 1830s described a kind of child who was likely to become consumptive. Among the points believed to be characteristic was that the child's mind was "quick, forward, intelligent, [and] touched with a high degree of sensibility and gentleness."

The water treatment which Dr. Gully prescribed for Annie may also have been devised as a regime suitable for the early stages of consumption. Dr. Yeoman wrote: "Water is one of the best prophylactics of disease that beneficent nature has provided for us, and in the malady now under consideration, when judiciously employed, is of considerable utility." Dr. Gully followed Sir James Clark in his belief that there was no cure for "confirmed consumption," but suggested that the water cure might be effective in its early stages. He wrote in *The Water Cure in Chronic Disease* that he was "convinced, that the judicious use of the water treatment reduces the harassing evils of consumption, the hectic, sweatings, bad sleep, and languor, and prolongs existence to some extent." "Is it not gain to be spared even a little of the stupor of opiates, the exhaustion of bad sleep and sweatings, which are the 'heavy day on day' of patients in consumption? . . . these can be in great measure avoided, notwithstanding the unceasing onward progress of the miserable malady."

The nature of all diseases was a mystery to Charles, but he had to take account of death by disease at a number of points in his thinking about natural life and human origins. In 1838, he had noted how hydrophobia, cowpox and many other diseases were shared between man and animals, and saw the point as "proof of common origin of man." He returned to the theme many years later in *The Descent of Man,* mentioning that humans shared consumption with monkeys, and suggesting that the two-way communication of diseases between man and animals "proves close similarities of tissue and blood far more plainly than does comparison under the best microscope."

The strength of heredity was a central strand in his thinking from before his marriage until the end of his life. He wrote about his grandfather Erasmus that he "fully recognised the truth and importance of the principle of inheritance in disease," and he paid particular attention to it himself. In a note he made in 1838, he used the point to illustrate his fundamental insight into the lack of direction or purpose in the processes of variation and inheritance. "It should be observed that transmission bears no relation to utility of change. Hence harelips [are] hereditary, [and] disease." Thirty years later, in *The Variation of Animals and Plants Under Domestication,* he wrote: "Unfortunately it matters not, as far as inheritance is concerned, how injurious a quality or structure may be if compatible with life. No one can read the many treatises on hereditary disease and doubt this." He cited consumption alongside epilepsy, asthma and cancer as examples.

In early 1838, he had read a book on breeding domestic animals and noted without further comment that it contained some "excellent observations of sickly offspring being cut off—so that not propagated by nature." In 1839, he wrote that "the numbers of fatal diseases in mankind" were no doubt due to the "rearing up of every hereditary tendency towards fatal diseases, and such constitutions only being cleared off by fatal diseases." So the incidence of severe illness among humans was an exceptional pattern in the natural world, a cruel side effect of our care for each other and our protection of the weak and infirm from the struggle for existence.

Charles's comment to his cousin Fox in March 1851 that he feared "with grief" that Annie had inherited his "wretched digestion" revealed that he was already then dwelling on the possibility that he was himself in a way responsible for her condition. A year after Annie's death, he wrote to Fox about his surviving children that "My dread is hereditary ill-health." Six months later he wrote again about his obsessive private worries. "The worst of my bugbears is hereditary weakness." Two years further on, in 1854, he revealed his worries indirectly to Hooker in a letter urging him to attend carefully to his troubling stomach condition. "Do reflect and act resolutely. Remember your troubled heart-action formerly plainly told how your constitution was tried. But I will say no more, excepting that a man is mad to risk health, on which everything, including his *children's inherited health,* depends." The comment reflects a significant point in Charles's view of evolution, that while arguing that natural selection of random variations was the main mechanism for adaptation, he also followed the French evolutionist Jean-Baptiste Lamarck in believing that some characteristics acquired during a human or animal's life could somehow be inherited. As he dwelt on this idea, Charles was caught in a web of guilt. In facing Annie's death, he had chosen to disregard the religious idea that it might be a punishment for his sin, but an element in his scientific thinking suggested instead that he was responsible for causing her death by a natural process.

Powerless as they were to provide any effective medical treatment for the most fatal illnesses, some Victorian doctors tried to identify any possible causes in the victims' conditions of life, so that any factors to do with the water supply, say, or poisons at the workplace could be dealt with. Sir James Clark wrote in his *Treatise:* "It is only by convincing the public of the comparative futility of all attempts to cure consumption . . . that physicians can ever hope to produce those beneficial results in improving public health and in preserving and prolonging human life, which it is the distinguishing privilege of their profession to aim at."

When Clark was preparing his work for publication, he was helped by a young medical writer called William Farr. He was the son of a farm labourer in Shropshire and had been working as a dresser in Shrewsbury Infirmary in the 1820s when Charles's father was one of the leading physicians in the town and Charles was studying medicine at Edinburgh. While Farr worked for Clark and was helping him set out the argument that consumption was fatal, his own young wife became ill with the disease and died. Farr had a strong interest in medical statistics and decided to take up Clark's challenge to prevent fatal illnesses by identifying factors in the patterns of death and improving public health. He was appointed Compiler of Abstracts for the Registrar-General of Births, Deaths and Marriages, and produced his *Abstracts of Causes of Death in England and Wales* each year for the rest of his working life. The *Abstracts* and his accompanying reports are the foundations of the modern sciences of medical statistics and epidemiology, and stand among the great achievements of the age.

Farr wrote in his first report: "Diseases are more easily prevented than cured, and the first step to their prevention is the discovery of their exciting causes . . . The registration of the causes of death . . . will give greater precision to the principles of physic. Medicine, like the other natural sciences, is beginning to abandon vague conjecture where facts can be . . . determined by observation; and to substitute numerical expressions for uncertain assertions." Commenting many years later on population growth as a factor in natural selection, he wrote: "The great source of the misery of mankind is not their numbers, but their imperfections, and the want of control over the conditions in which they live. Without embarrassing ourselves with the difficulties the vast theories of life present, there is a definite task before us—to determine from observation, the sources of health, and the direct causes of death in the two sexes at different ages and under different conditions. The exact determination of evils is the first step towards their remedies."

Farr's reports were known as his "ledgers of death," and every year he offered tables of figures showing the patterns of mortality from all causes in each five-year age band. In 1851, five children died in every thousand between the ages of ten and fourteen. Comparable figures have been produced ever since, and in 1996 one child died in five thousand between the same ages. Charles and Emma were therefore over twenty-five times more likely to lose a child of Annie's age than a family is now.

Dickens gave statistical analysis of this kind a bad name in *Hard Times.* His criticism of Mr. Gradgrind, his love of hard fact and his "tabular statements" did not apply to those like Farr who compiled their tables in the hope that they could be used for vital and humane purposes.

Farr made a detailed analysis of certified causes of death. For 1851, he reported that there were 395,396 deaths in all, of which 64,708, or one in six, were from tubercular diseases. Next came pneumonia with 21,983 deaths—only a third as many. Pulmonary consumption alone was fatal to almost three in every thousand of the population, a quite exceptional figure for a single disease in the whole country.

Charles studied Farr's reports and kept notes on the figures for child mortality. In the 1870s, he wrote to him about another issue connected in his mind with Annie's death. Some people at the time believed that the children of marriages between first cousins, as Charles and Emma were, were doomed to deformity and illness which might include consumption. One doctor with extreme views claimed that they might be dwarfish or ill-formed; they might be "cut off by consumption" or "become inmates of a lunatic asylum." In short, consumption "in its most hideous forms, revels in the system of the unhappy child of two persons of one blood. Nature seems to abhor this incestuous compact, and visits on the children the sins of the parents." Another doctor with humane and moderate views wrote that consumption was "for the most part . . . the parent's gift" and marriage of "persons who are too nearly allied in blood" was often the cause. Charles's own findings on the fundamental role of sexual reproduction, variation and outbreeding in the formation of species had prompted him to look at the effects of inbreeding on the partners' offspring, and led him to fear that first-cousin marriages might indeed be harmful. He asked Farr if the matter could be covered in the National Census for 1871. Farr was eager to help and a question was drafted for the census form. It was put forward for parliamentary approval, but rejected as an unwarranted breach of family privacy on a sensitive matter. Charles was left to protest in the conclusion to *The Descent of Man* that "ignorant members of our legislature" had blocked an important inquiry into a matter of great concern for the avoidance of human suffering.

The eventual key to the full understanding of tuberculosis was the discovery by Dr. Koch in the 1880s that the agent was a living organism. It now appears that the disease spread in the nineteenth century, declined in the first part of the twentieth century and is now spreading again, because of changes in the germ's conditions of life in the organisms it infects, and because it is evolving by natural selection to survive those changes. There is now a "natural history of infectious disease" which looks at the relations between parasitic micro-organisms and their hosts in an evolutionary framework. Ironically but with characteristic insight, while Charles had no conception of micro-organisms as carriers of disease, he clearly identified the special features of co-adaptation between parasites and hosts. He noted in *The Origin of Species* how mistletoe as a parasite was "dependent on other organic beings." Writing about epidemics working as a limiting check on populations "independent" of natural selection, he pointed out that some were due to parasitic worms "and here comes in a sort of struggle between the parasite and its prey." However, despite this recognition of the struggle, he found himself acknowledging "beautiful co-adaptations . . . in the humblest parasite which clings to the hair of a quadruped or the feathers of a bird . . . We see beautiful adaptations everywhere and in every part of the organic world." But the adaptations by one of the humblest parasites were lethal for its human hosts.

Many years later, in 1877 as the germ theory of infection was being developed, Charles's friend Professor Ferdinand Cohn, a leading plant physiologist at the University of Breslau in Silesia, sent him a copy of his periodical *Beiträge zur Biologie der Pflanzen*. The issue contained the first photographs ever published of bacteria; they had been taken by Dr. Koch, who was to identify the tuberculosis bacillus five years later. Dr. Koch had come to Professor Cohn with his first microscopic preparations of the bacilli that were responsible for anthrax, and had offered him a paper arguing for the first time that the bacilli were the cause of the disease. Professor Cohn recognised at once the great importance of his findings for medicine and the preservation of human life, and wrote to Charles that the photographs showed "the least but also perhaps the mightiest living beings." Charles replied: "I well remember saying to myself between twenty and thirty years ago, that if ever the origin of any infectious disease could be proved, it would be the greatest triumph to Science; and now I rejoice to have seen the triumph." It was twenty-six years since Annie had died. Bacteria were indeed the least but mightiest beings, living and killing unseen.

CHAPTER TWELVE

# THE ORIGIN OF SPECIES

*Etty's distress—Time and memory—Struggle for life—*
*Last child—The Origin of Species*

In the weeks after Annie's death in April 1851, Emma prepared for the birth of her next child. During the hours when Charles was writing his memorial of Annie, Emma had some hopes her "troubles were beginning," but it was a false alarm. The next day, she wrote to Fanny and Hensleigh: "It will be a very soothing occupation looking after a young baby." She gave birth to Horace a fortnight later, but had difficulties with him. A few days after, Elizabeth, who was still looking after her, wrote a disappointing account to Aunt Fanny Allen. Her letter does not survive, but Aunt Fanny replied that she had looked forward "with so much hope to this time for the healing influence to her sorrow: However we must have patience and wait." A wet-nurse came for Horace and some "artificial nipples" were obtained. Emma stayed in her room, away from the life of the household, while she recovered slowly. She came downstairs in mid-June.

Charles tried to avoid the pain of his grief by concentrating on his species theory. "The only chance of forgetting for short times your dreadful loss," he wrote to Francis many years later after Francis's wife had died in childbirth, lay in "the habit of close mental attention." A week after writing his memorial of Annie, he turned his mind to a problem which, like the design of the eye, posed a serious challenge to his argument about how species were formed. It "pressed so hardly" on the theory, he was to write eventually in *The Origin of Species,* that an effective answer had to be found. Sitting in his study, he wrote that it had often occurred to him that there ought to be "intermediate forms" between different kinds of creature. He took up the idea that the forms that would be found would not lie halfway between, but would be connected with both kinds by separate descent from a common ancestor. "I suggest not halfway between bird and reptile, but some third form equally connected with both." That suggestion lies at the heart of the evolutionary understanding of taxonomic relations, and has been borne out by all subsequent discoveries. The fact that all fossil hominids and other primates, for example, can best be placed on different branches of a family tree is a support for the theory of common descent which is now taken for granted. But as Charles sketched it out that day in May 1851, the idea was also a distraction from his pain.

Etty had been severely shaken by Annie's death. When she looked back on her childhood in her old age, she remembered how she felt that Annie was "the flower of the flock" with a gift for music, more beauty and much more charm than she had. "The maids told me, I well remember, how superior she was to me in all ways, but especially in sweetness of disposition."

In July Charles and Emma took Etty and George to London for the Great Exhibition, which had been drawing huge crowds. They stayed with Charles's brother Erasmus; Charles took an intense interest in the exhibits, and the family went for a second day. Aunt Fanny Allen felt that the children lost their interest after a short time. All the others whom she had seen there looked "wretched victims of *ennui,* and so it would be with these children, except for the sweet cakes and ices" that Uncle Ras bought for them "which I believe would please them better if they had them in the gardens here close at hand." Etty did not think much of the exhibition; she stayed at Erasmus's home on the second day and helped the housemaid scrub the back stairs "as being better fun."

The next day, Charles and Emma took the children to the Zoological Gardens where Obaysch the hippopotamus was still the most popular attraction. Then they went to the Polytechnic Institution in Regent Street, an exhibition hall which rivalled the Adelaide Gallery where the children had sat for their daguerreotypes two years before. A children's book of the time described a boy's visit to the Polytechnic. He was weighed in the patent weighing machine; he saw the diving bell go down; he tried the electrical machine; he saw the glass-blower and brought away a glass ship; and, last of all, he saw "the magnified figures and the dissolving views." Etty and George were taken up to Richard Beard's daguerreotype studio on the roof and sat for their "sun pictures." Taking Etty and George to be photographed, Charles must have been thinking of the daguerreotype of Annie which he was now "so thankful" to have. George, who was just six, was dressed in a small child's tunic, with white cotton drawers, white socks and leather shoes. He sat cross-legged on the studio chair with a composed and determined look. Etty wore a dark dress for half-mourning; she looked straight at the camera as she had done two years before, but her mind was now elsewhere and her face showed her unhappiness.

When they returned to Down, Emma watched Etty with concern, and did what she could to help her. "Etty nearly 8 years old. She appeared for some time to have lost the distressing feelings she used to have on hearing music, but one evening I saw her countenance change when Miss Thorley was singing, and on taking her out of the room, she said, rather distressed, 'But Mamma, where do the women go to, for all the angels are men?' She burst into tears when I asked her if she had been thinking of Annie, but said she had not." The angels Etty was thinking of may have been the "Shining Ones" in John Bunyan's vision of the Holy City in *Pilgrim's Progress,* which Emma read with the children. "Behold the City shone like the sun, the streets also were paved with gold, and in them walked many men with crowns on their heads, palms in their hands, and golden harps to sing praises withal."

One afternoon later in the month, Etty came to Emma "looking very much distressed." She asked: "Mamma, what can I do to be a good girl?" Emma told her several things, "openness and so on," and said she had better pray to God to help her to be good. Etty asked: "Shall I pray to God now?" and Emma led her through a short prayer. At bedtime a few days later, Etty asked: "Will you help me to be good?" Emma replied that Annie was a good child and that she did not think Etty would find it difficult to be as good as she was. She asked Etty "what made her so unhappy when she thought of being good." Etty replied: "I am afraid of going to hell." Emma said she thought Annie was safe in Heaven. "Come to me and I will try to help you as much as I can." Etty replied: "But you are always with somebody."

The following day at bedtime, Etty whispered to her mother, "Do you think I have done anything wrong today?" "No I don't think you have." Emma wrote: "We consulted a little over her prayers. I repeated 'Suffer little children' and so on. It did not seem to be *Pilgrim's Progress* as I had suspected which had alarmed her." Etty asked: "Do you think you shall come to Heaven with me?" "Yes, I hope so, and we shall have Annie." Etty suggested: "And Georgy too I hope." Emma wrote: "The next day she seemed trying to be good all day, and ended at night looking very sweet and happy, and I hope her fears are passed."

Emma read to the children from a Unitarian book of *Stories for Sunday Afternoons.* In the story of Adam and Eve, God told them after the Fall that "they would often be weary and sick, and at last would die. But He did not take *all* his kindness from them, though they had done this wrong thing. He clothed them with skins to keep them warm; and showed them how to get their food in the country whither He sent them forth. And He always watched over them in mercy, as He does over us now."

Some time later Emma found Etty crying "in great distress." Etty said: "Mamma, I used to be a very naughty girl when Annie was alive. Do you think God will forgive me? I used to be very unkind to Annie." The following February, Emma felt that Etty's mind had "developed itself wonderfully in the last few months. She asked me to put some of Annie's hair in her locket." One night, Etty said: "Mamma, I think of Annie when I am in bed." The next night, she added: "Mamma, when I see anything belonging to Annie, it makes me think of her. Sometimes I make believe (but I know it is not true) that she is not quite dead, but will come back again sometime . . . Mamma, I want you to put something in my prayers about not being proud, as well as not being selfish."

The other children quickly regained their high spirits. Emma's Aunt Emma Allen wrote in January 1852 that the six-year-old George had "a laugh so hearty, so merry, she would defy anyone not laughing with him." Emma dwelt on their doings when she wrote to her Aunt Jessie. The letters do not survive, but Aunt Jessie loved them. "You are poetic without knowing it, which is the prettiest poetry of all. The drop of water on the cabbage leaf is delicious."

Emma also touched on her lasting grief for Annie, and Aunt Jessie replied that she believed there was "no cloud without a silver lining." She now set aside the uncertainties she had mentioned after her husband's death in 1842, and gave Emma what consolation she could. Emma would see the silver lining on her dark cloud "when you rejoin the one lost," but who was lost only "for so short a time." Now that Aunt Jessie stood at the end of a long life, "the whole appears to me so short, so fleeting, as if nothing was worth thinking of but the Eternity in which we recover all our earthly loves. Tender and loving mother, what a fountain of love will burst forth there for you, from your many happy loving children."

. . .

In the years that followed, Emma hardly ever spoke about Annie to her children or anyone else, but when she did, Etty felt that "the sense of loss was always there unhealed." Emma's reticence had become ingrained. Aunt Jessie once said: "Why sorrow should make us shy is inexplicable to me, but I am sure it does."

Charles would not talk about Annie to the children either. Etty remembered that he mentioned her hardly twice in the rest of his life. She commented: "I should never have ventured to say her name to him." This was an utter change from his openness when he and Emma were engaged to be married and she felt that his "every word expresses his real thoughts." Bearing in mind what Emma knew of Etty's unhappiness after Annie's death, and how close Etty was to her father, her sense that she could not "venture" to talk with him about Annie was remarkable. He felt deeply for others, but was unable to help them by talking with them about their pain.

We can only guess at the reasons for Charles's reticence. Etty was a very different child from Annie, and yet in the years after Annie's death she became her father's daily companion, sharing his absorption in their fancy pigeons and hothouse plants. Years later, after she had helped him with the text of *The Descent of Man,* he called her "my very dear coadjutor and fellow-labourer" and commented to a friend that she was "the deepest critic I know in the world." She was direct and sharp while Annie had been affectionate and playful, but he loved her for her commitment to the things she cared about, a feeling he knew so well in himself. He may have feared that if they spoke of Annie, she might put the most difficult questions straight to him. Perhaps, as she had asked her mother, "Do you think you shall come to Heaven with me?"

For the family, avoiding difficult subjects became a way of life, and the children acquired their parents' reticence. My American great-grandmother, Maud Du Puy, met George in 1883 when she came to Cambridge from her home in Philadelphia to find a husband. She wrote home, "The Darwin family *are* a nice family together, extremely nice, always cordial and kind together and yet it strikes me that they are like affectionate second cousins more than brothers and sisters. I don't know why it is, unless that when there is a family gathering there is no family talk, no personal talk, it is more about the world at large and everything in general. Each goes his own way, thinks his own thoughts to himself." She was struck that there was never any quarrelling, and found them all so reserved that conversation was sometimes difficult. She had discussed the problem with her sister Ella, who was with her. "They do not talk about their own affairs at all. Ella says you feel like telling them the deepest secrets just for the sake of talking."

A year after Annie died, Charles and Emma's cousin Dr. Henry Holland, who had delivered Annie and to whom Emma had taken her twice in the months before her final illness, published a volume of *Chapters on Mental Physiology.* He was now a leading doctor in London, and had just been appointed Physician Extraordinary to Queen Victoria. He shared Charles's interest in the natural history of mankind, and found the question whether species changed a matter "of deep interest, carrying us by diverse paths into the midst of the most profound questions which can legitimately exercise our reason." Like Charles, he believed strongly in "the great law of continuity which equally governs all mental and material phenomena," and felt that we could gain many insights into human nature by looking at comparable features in animals. He had treated many patients with severe mental disorders, and had seen what we might be able to learn about the workings of the mind in a healthy state from what happened when memory or speech broke down. He saw the brain with its two hemispheres as a double organ and suggested that some forms of derangement of mind were due to a "sort of double dealing with itself." The healthy mind attained a "singleness in all acts of perception, volition and memory" which marked "the intellectual character of man." But beyond and below that state of mind stretched "paths too obscure" for human reason to follow.

Charles found Dr. Holland's chapter on instincts and habits particularly interesting, as it dwelt on all the difficulties in understanding the parallels and differences between human and animal behaviour. When Dr. Holland treated bodily movements linked with emotions, such as laughing and crying, Charles read carefully, thinking back to his ideas about the connections between body and mind when he had watched Jenny the orang and his own young children. Dr. Holland noted one puzzling feature, that the gestures linked with certain feelings "change at different periods of life." "The child cries and sobs from fear or pain; the adult more generally from sudden grief, or warm affection or sympathy with the feelings of others." Charles had known all those feelings with Annie; he marked the passage with a pencil line, and jotted down the page number on the fly-leaf, noting "Sobbing in child."

Another writer who had looked into the recesses of his own mind was Wordsworth's one-time friend, Thomas De Quincey. Charles read his *Confessions of an English Opium Eater* in 1854. Opium was freely available from any chemist and the Darwins and Wedgwoods were familiar with its effects. Charles's grandfather Erasmus had prescribed it to Tom Wedgwood, Charles and Emma's uncle; he had become addicted and his addiction was almost certainly a factor in his suicide. Robert Darwin prescribed the drug freely to his patients, and was "bitterly reproached" by one who became dependent on it. Charles took it occasionally for his sickness; Emma took it for the migraines which tormented her through her life, and Charles's brother Erasmus relied on "his opium," as Charles put it, to help him endure his long-standing depression.

In his *Confessions,* De Quincey gave a vivid and painful account of the mental derangement he suffered from the drug, and the nightmares it induced. In his dreams, memories welled up from parts of his mind below the threshold of awareness. He suggested that "there is no such thing as *forgetting* possible to the mind; a thousand accidents may, and will, interpose a veil between our present consciousness and the secret inscriptions on the mind; accidents of the same sort will also rend away this veil; but alike, whether veiled or unveiled, the inscription remains for ever; just as the stars seem to withdraw before the common light of day, whereas, in fact, we all know that it is the light which is drawn over them as a veil." In one recurring nightmare, "upon the rocking waters of the ocean the human face began to appear: the sea appeared paved with innumerable faces, upturned to the heavens: faces imploring, wrathful, despairing, surged upwards by thousands, by myriads, by generations, by centuries: my agitation was infinite; my mind tossed and surged with the ocean."

Of the faces that haunted De Quincey, one meant as much to him as Annie's meant to Charles. When he had been a young vagrant in London, he had lived for a few weeks in the alleyways of Soho with a fifteen-year-old street-walker called Ann who had a consumptive cough. She once saved him by her compulsive generosity when he collapsed in a doorway from hunger and exhaustion. A short time later, he left her for a journey of a few days, and when he came back, he could not find her. He searched everywhere but eventually had to give up. Afterwards, whenever he walked in the city and heard the barrel-organ tunes they had listened to together, tears came to his eyes as he thought about "the mysterious dispensation which so suddenly and so critically separated us for ever." He "looked into many, many myriads of female faces in the hope of meeting her. I should know her again amongst a thousand if I saw her for a moment; for, though not handsome, she had a sweet expression of countenance, and a peculiar and graceful carriage of the head." She came to him at last in his dreams. "There sat a woman; and I looked; and it was—Ann! She fixed her eyes upon me earnestly; and I said to her at length: 'So then I have found you at last.' I waited: but she answered me not a word. Her face was the same as when I saw it last, and yet again how different! Seventeen years ago, when the lamp-light fell upon her face, as for the last time I kissed her lips (lips, Ann, that to me were not polluted), her eyes were streaming with tears: the tears were now wiped away; she seemed more beautiful than she was at that time, but in all other points the same, and not older . . . In a moment all had vanished; thick darkness came on; and, in the twinkling of an eye, I was . . . by lamp-light in Oxford-street, walking again with Ann—just as we walked seventeen years before, when we were both children."

Charles found the book "very poor." This may have been disquiet and dislike. He may have sensed in his own feelings and memories something of what De Quincey had been through, but found it too painful to follow the young writer in his wild imaginings.

Betty, Charles and Emma's youngest daughter, whom Annie had called her "little duck," was now four. Emma noticed that she had nervous tics and some strange ways of talking. "When telling her a story or if she is observing anything, she has the most curious way of playing with any dangling thing she can get hold of, sometimes twiddling her fingers as Charles used to do." A few months later, she had acquired "a great habit of abstraction, going by herself and talking to herself for an hour. She does not like to be interrupted." Charles felt a special closeness to her because he was acutely aware of his own nervous tics. One, which he hid from people, he realised was remarkably like one of hers, and the similarity gave him evidence of a point about the inheritance of instinctive behaviour which was important for his scientific thinking. Many years later he gave a clinical but carefully veiled account of it in his book on the mechanisms of heredity, *The Variation of Animals and Plants Under Domestication*. It was essential for his theory of evolution by natural selection to show that traits of instinctive behaviour could vary between creatures of a species just like bodily features, and that special features, some of which might turn out to have adaptive value, could then be passed on from one generation to the next. "One instance . . . has fallen under my own observation, and . . . is curious from being a trick associated with a peculiar state of mind, namely, pleasurable emotion. A boy had the singular habit, when pleased, of rapidly moving his fingers parallel to each other, and, when much excited, of raising both hands, with the fingers still moving, to the sides of his face on a level with the eyes; when this boy was almost an old man, he could still hardly resist this trick when much pleased, but from its absurdity concealed it. He had eight children. Of these, a girl, when pleased, at the age of four and a half years, moved her fingers in exactly the same way, and what is still odder, when much excited, she raised both her hands, with her fingers still moving, to the sides of her face, in exactly the same manner as her father had done, and sometimes even still continued to do so when alone. I never heard of any one, excepting this one man and his little daughter, who had this strange habit; and certainly imitation was in this instance out of the question." Charles was fifty-nine when the book appeared. "This boy was almost an old man."

While Emma would not talk of Annie to anyone, Charles wrote briefly about her to people outside the family when he felt it fitting and helpful to do so. In July 1853, his cousin Fox's two-year-old daughter Louisa died of scarlet fever. Charles replied to the news: "I thank you sincerely for writing to me so soon, after your most heavy misfortunes. Your letter affected me much. We both most truly sympathise with you and Mrs. Fox. We too lost, as you may remember, not so very long ago, a most dear child, of whom I can hardly yet bear to think tranquilly." Referring to the death of Fox's first wife in 1842, Charles wrote: "Yet, as you must know from your own most painful experience, time softens and deadens, in a manner truly wonderful, one's feelings and regrets. At first it is indeed bitter. I can only hope that your health and that of poor Mrs. Fox may be preserved; and that time may do its work softly."

In October 1856, Fox went to Malvern for a time with his family, and mentioned Annie's grave in a letter to Charles. Charles replied that the thought of that time was still "most painful" to him. "Poor dear happy little thing." A month before, he had thought of returning for treatment by Dr. Gully, "but I got to feel that old thoughts would revive so vividly that it would not have answered; but I have often wished to see the grave, and I thank you for telling me about it."

The difference between Charles's two letters to Fox is noticeable. In the first, "time softens and deadens," but in the second over three years later, the memory of Annie's death was "yet most painful to me" and Charles feared that "old thoughts" of her last days "would revive so vividly." The comfort he offered in the first letter followed the normal expectation of time's healing power, but the deep wound of the last days in Malvern would not heal.

Leonard believed he saw one glimpse of his father's secret feelings during his childhood in the 1850s. "I went up to my father when strolling about the lawn, and he, after, as I believe, a kindly word or two, turned away as if quite incapable of carrying on any conversation. Then there suddenly shot through my mind the conviction that he wished he was no longer alive."

Charles's feelings about Annie changed gradually as time passed. When his close friend Thomas Huxley's three-year-old child died of scarlet fever in 1860, Charles wrote as soon as he heard, with characteristic directness and brevity. "I was indeed grieved to receive your news this morning. I cannot resist writing, though there is nothing to be said. I know well how intolerable is the bitterness of such grief. Yet believe me, that time, and time alone, acts wonderfully. To this day, though so many years have passed away, I cannot think of one child without tears rising in my eyes; but the grief is become tenderer and I can even call up the smile of our lost darling, with something like pleasure." When he mentioned that he could now "even call up" Annie's smile, he was thinking of his special power to make remembered faces "do anything I like." He was calling up her smile and hoping to find warmth in the recollection.

After Annie's death, Charles set the Christian faith firmly behind him. He did not attend church services with the family; he walked with them to the church door, but left them to enter on their own and stood talking with the village constable or walked along the lanes around the parish. He did, though, still firmly believe in a Divine Creator. But while others had faith in God's infinite goodness, Charles found him a shadowy, inscrutable and ruthless figure. Charles had dealt with the struggle for existence in 1838 and noted the "pain and disease in world" without further comment. When he returned to the theme in his scientific writings during the years after Annie's death, he wrote about it in a new way. He never referred directly to his personal experience; that would have been quite inappropriate. But he made some new points; there was a darkness in the wording of some passages, and others echoed his feelings about human loss.

Charles's first comment was a private and casual aside in 1856. Two months before, Lyell had at last persuaded him to start writing his "big species book," the work that was to become *The Origin of Species.* Charles asked Huxley whether a kind of jellyfish could take in sperm by the mouth, and Huxley replied archly, "The indecency of the process is to a certain extent in favour of its probability, nature becoming very low in all senses amongst these creatures." Reporting Huxley's comment to Hooker, Charles wrote: "What a book a Devil's Chaplain might write on the clumsy, wasteful, blundering low and horridly cruel works of nature!" His remark had special force because he and Hooker both knew that some would claim the book he was writing was the Devil's work. A few months later, Charles put what he had in mind into his draft. "Can the instinct, which leads the female spider savagely to attack and devour the male after pairing with him, be of service to the species? The carcase of her husband no doubt nourishes her; and without some better explanation can be given, we are thus reduced to the grossest utilitarianism compatible, it must be confessed, with the theory of natural selection." This was "low and horridly cruel," but could perhaps be explained by "the grossest utilitarianism." "Blundering," on the other hand, went beyond Huxley's point. It hinted at careless and pointless destruction. In that sense Annie's death had been a blunder.

Charles continued to work on the "laws of life," but was now sharply aware of the elimination of the weak as the fit survived. He wrote powerfully about Nature's ruthless culling. "She cares not for mere external appearance; she may be said to scrutinise with a severe eye, every nerve, vessel and muscle; every habit, instinct, shade of constitution—the whole machinery of the organisation. There will be here no caprice, no favouring: the good will be preserved and the bad rigidly destroyed, for good and bad are all exposed during some period of growth or during some generation, to a severe struggle for life . . . Nature is prodigal of time and can act on thousands of thousands generations: she is prodigal of the forms of life; if the right variation does not occur under changing conditions so as to be selected and profit any one being, that form will be utterly exterminated as myriads have been."

Linking the point to the cruelty and other "imperfections" of some animal instincts, Charles suggested that those traits were best understood not as "specially given by the Creator," but as "very small parts of one general law leading to the advancement of all organic beings. Multiply, vary, let the strongest forms by their strength live and the weakest forms die." Here some might have heard echoes of Tennyson's evil dream of Nature in his poem *In Memoriam*, which had appeared in 1850. "I care for nothing, all shall go."

Writing in this way, Charles was facing up again to the cruelty in natural life. He was thinking of Annie, but also of every other victim. His view was not an exceptional insight at the time; it was shared by Matthew Arnold and others who could at last see through the shallow complacency of natural theology. John Stuart Mill commented on Nature's utter amorality. "Next to the greatness of these cosmic forces, the quality which most forcibly strikes every one who does not avert his eyes from it, is their perfect and absolute recklessness . . . Nearly all the things which men are hanged or imprisoned for doing to one another, are Nature's everyday performances." She never turns "one step from her path to avoid trampling us into destruction"; such are "her dealings with life."

What was special in Charles's view was his unique understanding of the extraordinary outcome of the process of natural selection, the paradox of how all the diversity and beauty of living things was the outcome of the endless suffering and death in the "war of nature," and how each new form might itself be destroyed as others took its place.

Charles decided that he would open his argument for the idea of evolution by natural selection with a reminder of some generally known facts about fancy pigeons. There was wide interest in "the fancy" at the time, and people of all ranks were fascinated by the extraordinary variety of shapes and plumage that the fanciers had managed to produce by careful choice in breeding—fantails like peacocks, pouters with their enormous inflated crops, barbs with bright red eye-wattles, and so-called swallows with fanned feathers on their legs and feet. Charles saw that the range of breeds offered a perfect illustration of the variation that could be achieved by selection for reproduction. He built a pigeon house in the garden at Down for a collection to experiment with. Etty, who was now twelve, joined him in looking after them. He was grateful for her help, and found quickly that the pigeons were "a decided amusement to me, and a delight to Etty." At first, things did not all go well and he complained that "all nature is perverse and will not do as I wish it." But he and Etty soon mastered the tricks of breeding, and he came to love the pigeons so much that he could not bear "to kill and skeletonise them." Etty got to know their characters and judged some sharply, just as she commented on Miss Thorley. A pouter pigeon, she recalled later, was "good-natured but not clever" and one hen Jacobin she considered "rather feeble-minded."

Charles was now working hard on his "big species book," and he was preparing in his mind for the controversy that would follow when it was published. He was also dandling his last son, Charles Waring Darwin, whom Emma had borne in 1856. After the birth of Horace five years before, even though Emma was then forty-three, Charles had wanted another child. Emma carried on noting her periods in her diary. Charles wrote to his cousin Fox in October 1852: "Emma has been very neglectful of late and we have not had a child for more than one whole year." She missed a period some months later and numbered the following weeks in her diary for the "reckoning," but appears to have miscarried. She had another miscarriage in the following year, but eventually conceived again and, after weary months of discomfort, gave birth to Charles Waring when she was forty-eight.

Charles played with his tenth child just as he had done with each of the others, and watched him with the same devoted absorption. "He was small for his age and backward in walking and talking, but intelligent and observant. When crawling naked on the floor, he looked very elegant. He had never been ill, and cried less than any of our babies. He was of a remarkably sweet, placid and joyful disposition, but had not high spirits, and did not laugh much. He often made strange grimaces and shivered, when excited; but did so, also, for a joke and his little eyes used to glisten after pouting out or stretching widely his little lips. He used sometimes to move his mouth as if talking loudly, but making no noise, and this he did when very happy. He was particularly fond of standing on one of my hands, and being tossed in the air; and then he always smiled, and made a little pleased noise. I had just taught him to kiss me with open mouth when I told him. He would lie for a long time placidly on my lap looking with a steady and pleased expression at my face; sometimes trying to poke his poor little fingers into my mouth, or making nice little bubbling noises as I moved his chin."

During Willy's summer holiday in 1857, Charles gave him a set of photographic apparatus and the seventeen-year-old became a keen amateur photographer, walking in and out of the house with his hands blackened by the chemicals he used to prepare his plates. One of his first photographs was a picture of his mother watching over Charles Waring as he lay in her lap. She mounted the picture on a card and kept it until she died. I showed it recently to a consultant paediatrician, and gave him Charles's comments on his child. The paediatrician said that the infant's appearance in the photograph, his placid temperament and Emma's age when he was conceived were all consistent with the condition we now call Down's syndrome.

Charles and Emma had noticed Charles Waring's slow development, but probably did not recognise the signs in his appearance as there was no general awareness of the condition at the time. Few sufferers were ever seen in ordinary life, and the appearance was first identified tentatively with the disability in 1866 by Dr. John Langdon Down, the superintendent of a mental asylum, who corresponded with Charles in later years about facial features. Dr. Down had a positive approach to caring for the "feeble in mind," and looked attentively for any patterns among his patients' symptoms in the hope that suitable treatments could be found for any groups he could identify. From his experience with children with the distinctive "Mongolian" appearance, he suggested that they often responded better to treatment than many others with otherwise similar disabilities. The first step for all, though, was "to rescue the feeble one from the solitary life, to give him the companionship of his peers, to place him in a condition where all the machinery shall move for his benefit and where he shall be surrounded by influences both of art and nature, calculated to make his life joyous, to arouse his observation, and to quicken his power of thought."

In June 1858, Charles received the letter from Alfred Russel Wallace, a young naturalist in Malaya, in which he revealed that he had come quite independently to the idea of natural selection and sought Charles's views. Charles suddenly faced the possibility of losing credit for the theory which it had become his life's work to build. He wrote in desperation to Lyell and Hooker (now assistant director of the Royal Botanic Gardens at Kew) and waited for their advice on how to reply. As the days passed, a wave of scarlet fever was spreading from household to household through the village. Charles Waring caught the infection and died five days later in extreme distress. When Hooker's letter came, Charles was thinking only about his son. The day after the infant died, Charles wrote to Hooker that he hoped to God he had not suffered "as much as he appeared." "Thank God he will never suffer more in this world." The last thirty-six hours had been "miserable beyond expression." But "In the sleep of Death he resumed his placid looks." Charles added about Wallace's letter and the species theory: "I cannot think now on [the] subject but soon will."

In the event Hooker and Lyell presented a joint paper by Charles and Wallace to the Linnean Society in London. The historic announcement of their joint theory of evolution by natural selection was made at a meeting on the first day of July; Hooker and Lyell were both present, but Charles was not. That day he had followed Charles Waring's small coffin to the village churchyard and watched as it was laid in the family tomb next to Mary.

After the announcement to the Linnean Society, Charles accepted that he must publish his full argument without further delay, and set to work on a shortened version of the "big species book." It appeared a year later as *The Origin of Species.* His thoughts about Annie and Charles Waring coloured two pictures of the "face of Nature" at the heart of his argument about natural selection in the struggle for life. In his early accounts of his ideas he had written how difficult it was when seeing "the contented face of Nature," "to believe in the dreadful but quiet war of organic beings going on in the peaceful woods and smiling fields." In his draft for the "big species book," he had given a full chapter to the "Struggle for existence." "All Nature . . . is at war. When one views the contented face of a bright landscape or a tropical forest glowing with life, one may well doubt this . . . Nevertheless the doctrine that all nature is at war is most true. The struggle very often falls on the egg and seed, or on the seedling, larva and young; but fall it must sometime in the life of each individual."

In the final text of *The Origin of Species,* Charles sharpened the wording in ways that echoed his memories of Annie as a child and the unexpected pain of her loss. He had written in his memorial of her that "from whatever point" he looked back on her, he saw the joyousness which "radiated from her whole countenance . . . Her dear face now rises before me . . . her eyes sparkled brightly." And he had ended: "Oh that she could now know how deeply, how tenderly we do still and shall ever love her dear joyous face." He now wrote: "Nothing is easier than to admit in words the truth of the universal struggle for life, or more difficult—at least I have found it so—than constantly to bear this conclusion in mind. Yet unless it be thoroughly engrained in the mind, I am convinced that the whole economy of nature, with every fact on distribution, rarity, abundance, extinction, and variation, will be dimly seen or quite misunderstood. We behold the face of nature bright with gladness . . . we do not see, or we forget, that the birds which are idly singing around us mostly live on insects or seeds, and are thus constantly destroying life; or we forget how largely these songsters, or their eggs, or their nestlings, are destroyed by birds or beasts of prey." Charles had found that when loved children were healthy and lively, pain and death were easily forgotten. In Annie's last days, he had watched as her face was changed beyond recognition by the emaciation of her fatal illness. You could understand the true conditions of life only if you held on to a sense of the ruthlessness of the natural forces that could waste the bright surface, of how Tennyson's "brute force of Nature" might prevail.

The second picture of the face of Nature had also changed as Charles developed his ideas. In 1838, he had written: "There is a force like a hundred thousand wedges trying to force every kind of adapted structure into the gaps of the economy of Nature, or rather forming gaps by thrusting out weaker ones." He chose wedges for his simile to convey a critical point about the mechanism of natural selection, because when you drove a wedge in, it displaced whatever was on either side. The image gave a sense of the struggle for survival, and how alongside every survivor there was a victim. Now, in *The Origin of Species,* he suddenly linked the two images of the glad face and the ten thousand wedges in an extraordinary and violent way. He wrote that each creature "lives by a struggle at some period of its life" and "heavy destruction inevitably falls either on the young or old." The face of Nature, which he had just described as a human face "bright with gladness," might be compared to "a yielding surface, with ten thousand sharp wedges packed close together and driven inwards by incessant blows, sometimes one wedge being struck, and then another with greater force." The cruelty of the image was too painful, though, and he cut the sentence from all editions after the first.

In *The Origin of Species,* Charles ended his account of the struggle for existence with a last attempt at reassurance. He insisted again that we must "keep steadily in mind, that each organic being is striving to increase at a geometrical ratio; that each at some period of its life, during some season of the year, during each generation or at intervals, has to struggle for life, and to suffer great destruction." But now, "When we reflect on this struggle, we may console ourselves with the full belief, that the war of nature is not incessant, that no fear is felt, that death is generally prompt, and that the vigorous, the healthy, and the happy survive and multiply." He had reflected on the struggle. Consolation was needed. He had a consolation to offer, but it was hollow for him after the deaths of his two sick children. "The vigorous, the healthy, and the happy survive," but we continue to care for the others.

Charles held to his wonder at the richness of natural life; it remained the inspiration of his theory, evident in page after page of *The Origin of Species,* and when he came to the conclusion, he returned to the idea of evolution as an endless unfolding which he had first sketched in his notebook and then developed in each reworking of the theory. He now gave it its final form in an imagined corner of the countryside he walked through every day on his own or with his children—the Sandwalk copse perhaps, or Orchis Bank.

"It is interesting to contemplate an entangled bank, clothed with many plants of many kinds, with birds singing on the bushes, with various insects flitting about, and with worms crawling through the damp and to reflect that these elaborately constructed forms, so different from each other, and dependent upon each other in so complex a manner, have all been produced by laws acting around us . . . Thus, from the war of nature, from famine and death, the most exalted object which we are capable of conceiving, namely, the production of the higher animals, directly follows. There is a grandeur in this view of life, with its several powers, having been originally breathed into a few forms or into one; and that, whilst this planet has gone cycling on according to the fixed law of gravity, from so simple a beginning endless forms most beautiful and most wonderful have been, and are being, evolved."

The words became famous. In them, Charles offered his riddle to the world, how the "endless forms most beautiful" evolved "from the war of nature, from famine and death."

CHAPTER THIRTEEN

# GOING THE WHOLE ORANG

*Apes and humans—Lyell and Queen Victoria—The Water Babies—Sympathy—Workings of the mind*

ETTY SUGGESTED AFTER HER father's death that "the habit of looking at man as an animal had become so present to him, that even when discussing spiritual life, the higher life kept slipping away." Emma had lived with his "habit of looking at man as an animal" since she joked with him before their marriage that he would watch her as a specimen of the ape "genus." Etty was right to suggest that this habit undermined his thinking about "the higher life"; he was developing his own ideas about human nature at the same time, deep rather than high, to put in place of the claims of Christianity. His eventual view of human origins was a humble one, but after comparing his children with Jenny the orang, he did not see our link with animals as demeaning. Others, when they wanted to emphasise animals' utter inferiority to mankind, called them "brutes" or the "brute creation"; Charles seldom used the word, and never with that contemptuous sense.

When Charles had first developed his species theory in 1838, mankind was at the centre of his thinking. When, twenty years later, before the publication of *The Origin of Species,* Alfred Russel Wallace asked him if he would discuss human origins, he acknowledged that it was the "highest and most interesting problem for the naturalist." But "I think I shall avoid the whole subject, as so surrounded with prejudices."

Few apes had been seen in England after the second Jenny died in 1844, but Wombwell's Menagerie had a male orang in the 1850s which they took in their cavalcade from town to town in England and Scotland. They claimed in their handbills that it was "beyond doubt or dispute the SECOND LINK in the chain of the Animal Creation." The citizens of Aberdeen were told that the orang exhibited "sagacity little inferior to that of a human being," and so close were its links to our species that "we are literally lost in wonderment, and almost doubt the class to which it immediately belongs." The Darwins and their friends went often to the Crystal Palace at Sydenham, which had a stuffed chimpanzee in a tableau of African life. The official guide commented that the living ape "exhibits an intelligence that presses with rather uncomfortable nearness upon the pride of the sole rational animal."

Zoologists were also learning about the gorilla from skulls and other bones brought from West Africa by travellers. Professor Owen was now widely recognised as a leading authority on the comparative anatomy of mammals. In 1853, he obtained the first complete gorilla skeleton for the Royal College of Surgeons' museum, and in 1858, another was displayed in the Mammalian Gallery of the British Museum alongside a human skeleton. The juxtaposition was so striking that the British Museum had a photograph made of the two together, and it may have been among the pictures that were sold at the photographer's stall in the entrance hall of the museum.

The year before, Professor Owen had tried to break the evident close links between man and ape by taxonomic sleight of hand. He proposed making *Homo sapiens* the sole member of a subclass of mammals, quite separate from all other primates. He argued that the structure of the brain should be the key to the main distinctions in the class of mammals; that the human brain was unique, and that one of its special features, a small swelling at its base known as the hippocampus minor, warranted the separation of humans from all other mammals. When

Charles read Professor Owen's paper, he told Hooker that he had enjoyed Owen's argument but could not swallow the placing of man in a position as far removed from a chimpanzee as a horse was from a platypus. With his old impulsive empathy he put himself in the ape's position looking at man, and wrote: "I wonder what a chimpanzee would say to this."

Charles did not enter the debate because he had no standing as a mammalian taxonomist and, as he put it later to Hooker, "knew nothing about the brain." Their friend Huxley, though, could argue confidently about the anatomical issues. In 1858, a few months before the announcement of the theory of evolution at the meeting of the Linnean Society, Huxley suggested in a lecture at the Royal Institution that if we compared man, gorilla and baboon, we should find little greater interval "as animals" between man and gorilla than between gorilla and baboon. Echoing Dr. Holland's point in his *Chapters on Mental Physiology* about the continuity between man and animals, he went on: "I believe that the mental and moral faculties are essentially and fundamentally the same in kind in animals and ourselves. I can see no line of demarcation between an instinctive and a reasonable action." Speech had led to our "unlimited intellectual progress" but "to the very root and foundation of his nature, man is at one with the rest of the organic world."

No one in England had yet seen a gorilla in the flesh, but in September 1858 a corpse of a young adult male, which had been killed in Gabon, was brought to the British Museum in a cask of spirits. When the cask was opened, the gorilla was found to be partly decomposed and smelt dreadfully. Professor Owen examined it and gave a detailed description of its anatomy to the Zoological Society, but could say little about its behaviour because he had only travellers' reports from "Gorilla-land" to go on. Mindful perhaps of the specimen's wretched semi-human look when the spirits were poured away, he noted two "redeeming qualities . . . the male's care for his family and the female's devotion to her young."

Professor Owen entrusted the specimen to Abraham Bartlett, a leading taxidermist who knew Charles and was working at the time on the animal displays at the Crystal Palace. Bartlett prepared the body and was allowed to display it for a few months at the Crystal Palace, where the gorilla was portrayed in *The Illustrated London News* as a terrifying monster. "The strength of the adult male being prodigious, and the teeth heavy and powerful, it is said to watch, concealed in the thick branches of the forest trees, the approach of any of the human species, and as they pass under the tree, let down its terrible hind foot, furnished with an enormous thumb, grasp its victim round the throat, lift him from the earth, and, finally, drop him on the ground dead. Sheer malignity prompts the animal to this course, for it does not eat the dead man's flesh, but finds a fiendish gratification in the mere act of killing." The force of the emphasis on the gorilla's wilful and "fiendish" evil is remarkable. It helps explain why, when Charles later suggested that the apes were our cousins, people were reluctant to accept the link.

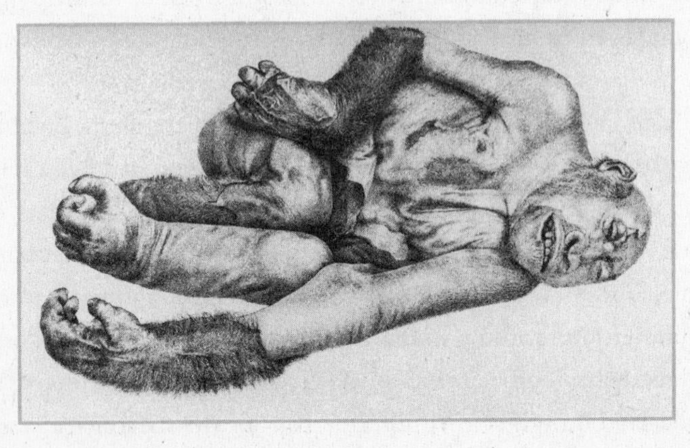

*The young gorilla*

. . .

Charles did not deal with human origins in the main argument of *The Origin of Species,* but declared his belief twice in the conclusion, writing later that he had done so "in order that no honourable man should accuse me of concealing my views." He said about one familiar example of our bodily similarities with other mammals, that "The framework of bones being the same in the hand of a man, wing of a bat, fin of the porpoise, and leg of the horse . . . at once explain themselves on the theory of descent with slow and slight and successive modifications." And when he came to sketch the "considerable revolution in natural history" which he believed would follow if his ideas were accepted, he suggested that "psychology will be based on a new foundation, that of the necessary acquirement of each mental power and capacity by gradation. Light will be thrown on the origin of man and his history." Charles thus placed the final emphasis, not only on bodily structure in which most of the obvious likenesses lay, but on psychology which was the science of the human soul or mind. Here the links were far less clear, and most people denied that there were any, but Charles was confident that they would prove to be the key to the understanding of human nature.

A few days after *The Origin* was published, Charles revealed privately to Lyell that he was confident that the theory of natural selection would explain fully how man had evolved as a thinking being. True to Lyell's uniformitarian principles, he saw a gradation of mental powers between an orang-utan and a savage, and wrote: "To show how minds graduate, just reflect how impossible everyone has yet found it to define [the] difference in mind of man and the lower animals; the latter seem to have [the] very same attributes in [a] much lower stage of perfection than the lowest savage." He rejected the idea that God had miraculously inserted the human soul in an animal body, and set a challenge for his argument. "I would give absolutely nothing for [the] theory of natural selection if it required miraculous additions at any one stage of descent."

*The Origin of Species* opened the way to an evolutionary view of human origins, both body and mind. Charles Kingsley, a country rector, Christian Socialist and chaplain to Queen Victoria, suggested to Charles that he had set a "villainous shifty fox of an argument" running, and people should follow it into "whatsoever unexpected bogs and brakes" it might lead them. That "shifty fox" quickly became the hunters' quarry in the controversy about the theory. Bishop Samuel Wilberforce wrote in his anonymous review of *The Origin* in the *Quarterly Review* that Mr. Darwin clearly declared "that he applies his scheme of the action of the principle of natural selection to Man himself, as well as to the animals round him." He insisted that "the degrading notion of the brute origin of him who was created in the image of God" must be rejected because it was utterly irreconcilable with mankind's supremacy over all other creatures. During the notorious debate on evolution at Oxford in 1860, Wilberforce jokingly asked Huxley whether it was on his grandfather's or grandmother's side that he was descended from an ape. Huxley replied that he would prefer to be descended from an ape than from a person who treated serious issues with the bishop's frivolity. It is not clear exactly what was said and heard on the day, but the exchange in Huxley's telling was quickly identified by Charles's friends and supporters as the defining moment of the whole controversy about evolution, because it harped on man and animal, bishop and baboon.

The general distaste at the idea that humans were cousins to gorillas was heightened in 1861 when Paul Du Chaillu, a young American explorer, published his *Explorations and Adventures in Equatorial Africa*. He had hunted the "monstrous and ferocious ape" in the mountain forests of Congo, and gave a vivid and almost evil account of the living creature. Just as the writers on chimpanzees and orangs in the 1830s had done, he played heavily on the animal's links with man. "Suddenly I was startled by a strange, discordant, half-human, devilish cry, and beheld four young gorillas, running toward the deep forests . . . their whole appearance like men running for their lives. Take with this their awful cry, which, fierce and animal as it is, has yet something human in its discordance, and you will cease to wonder that the natives have the wildest superstitions about these 'wild men of the woods.'" A gorilla, when shot, "dies as easily as a man." His hideous death cry "tingles" the hunter's ears "with a dreadful note of human agony." "It is this lurking reminiscence of humanity, indeed, which makes one of the chief ingredients of the hunter's excitement in his attack."

Struck by the likenesses with humans, people nevertheless insisted that the beast must be alien. In 1863, the Literary Institute in Bromley, the Darwins' nearest town in Kent, heard a talk by a visiting lecturer on the gorilla as compared and contrasted with man, and the *Bromley Record* made a dark hint to Charles, referring to his grandfather Josiah Wedgwood's anti-slavery medallion. "One thing we may as well state, not that some we could name, who ought to have been present, deserve to know it . . . the lecturer clearly proved that the gorilla is not 'a Man and a Brother.'"

The public were fascinated by the issue of human origins, and Huxley and Owen entertained them with a fierce and highly personal public argument about the hippocampus minor, which Huxley managed to show apes had as well as *Homo sapiens.* Charles, meanwhile, experimented on plants and insects, feeling that the best contribution he could now make to the case for evolution was to develop his ideas about inheritance and cross-fertilisation, both relatively uncontroversial topics. In 1862, he commented to Kingsley that the genealogy of man was "a grand and almost awful question." He had "long attended" to the subject and had "materials for a curious essay on human expression, and a little on the relation in mind of man to the lower animals." But he would keep them to himself for the time being. "How I should be abused if I were to publish such an essay!"

In 1863, Huxley published his lectures on the anatomical relation of man to the lower animals as a short book, *Evidence as to Man's Place in Nature.* (Charles wrote to him: "Hurrah, the monkey book has come!") He showed clearly the close similarities in anatomical structure between the human brain and those of the man-like apes, and repeated the case for a close natural link. On our mental powers, he emphasised the differences between human and animal minds, and how far humans had progressed from their animal origins. He insisted that humans had their origins in the natural world, but then said: "no one is more strongly convinced than I of the vastness of the gulf between civilised man and the brutes." "Intelligible and rational speech" was mankind's unique possession. The capacity had a natural origin, but with it, man was "transfigured from his grosser nature by reflecting, here and there, a ray from the infinite source of truth." Those were not words that Charles would have used. Whether they were Huxley's true ideas, or whether he was paying lip service to the pious presumptions of the age, is difficult to say.

Everyone was waiting for Lyell to declare his view, but although he had encouraged Charles to publish, he was troubled by the implications of the theory and unwilling to commit himself. He was a respected champion of the method of science, believing that "truth is the highest aim"; he was committed to the principle of continuity in nature, and to uniformitarian explanations. At the same time he was a devout and scrupulous Unitarian, and thought long and hard about the difficulties he faced in reconciling Charles's idea of evolution with his Christian faith. He had no fear that the idea would diminish our awe of the Creator, because nothing could detract from God's grandeur. The dignity of man was Lyell's worry. In the fortnight after the appearance of *The Origin of Species,* he had noted in his private journal how the laws of nature and continuity pointed clearly to human development by descent from an animal ancestor, but he had an "apprehension lurking in the depth of the soul," "fear lest the dignity of Man . . . should be lowered by establishing a nearer link of union between him and the inferior animals," a threat to "his conscious feeling of superiority." Lyell did not avoid the problem, but wrote bravely: "Let us look it steadily in the face. Now before it is proved. When it is only possible, perhaps probable, that it will be established."

By 1863, Lyell had concluded privately that he must "go the whole orang" with Charles and the others; he told Charles, but he could not bring himself to say so in public. He produced a book, *The Geological Evidences of the Antiquity of Man,* which marshalled the evidence that humans had lived with ancient mammals that are now extinct. The book gave mankind a history reaching far back into the geological past, with ample time for humans to have evolved with the apes from a common ancestor. Lyell admitted the possibility that we were descended from animals, but firmly rejected Charles's idea that human nature had developed from an animal condition by "gradation," small steps in a natural process with no sudden leaps to break the continuity with our ancestors. He chose to suggest instead that mankind might have at some point in the remote past "cleared at one bound the space which separated the highest stage of the unprogressive intelligence of the inferior animals from the first and lowest form of improvable reason manifested by man."

Charles was bitterly disappointed by Lyell's comment, and it was a cruel irony for him that the eminent geologist from whom he had learnt the uniformitarian approach to the explanation of natural processes had to introduce a discontinuity or leap in his conjecture about human origins. Charles cared deeply about his species theory and all that it suggested about the secrets of human nature, and was upset by the unwillingness of others to explore its implications as freely and boldly as he wanted to. During the days after first reading Lyell's *Antiquity of Man,* he suffered "much sickness and weakness" and had to put off a visit by him. The day after the Lyells' next stay, Emma noted in her diary that Charles had "bad hysteria and sickness." Some time later, Charles told his doctor that one of his recurring symptoms was "vomiting preceded by shivering, hysterical crying, dying sensations or half faint." With his comment to his doctor, there can be no doubt about the intensity of the distress that Emma witnessed.

Queen Victoria had by now been mourning Prince Albert's death for two years, but she was interested in the debate about human origins, and summoned Lyell for an audience at her summer residence on the Isle of Wight. He wrote to his wife: "She asked me a good deal about the Darwinian theory as well as *Antiquity of Man.* She has a clear understanding, and thinks quite fearlessly for herself."

One of the most positive if peculiar responses to *The Origin of Species* was Charles Kingsley's story for children, *The Water Babies,* which also appeared in 1863. The *Spectator* wrote: "The purpose of this tale—and it was a fine one—seems to have been to adapt Mr. Darwin's theory of the natural selection of species to the understanding of children, by giving it an individual, moral and religious as well as a mere specific and scientific application." Kingsley, an eclectic and mercurial thinker, saw the presence of God in all natural life, but a God at once loving and ruthless whose character was "consistent with all the facts of nature, not only those that are pleasant and beautiful." Among "all the facts" in Kingsley's natural theology were some absurdities. No human could have invented "anything so curious and so ridiculous" as the lobster.

*The Water Babies* is a surreal fantasy in which two children who die meet strange characters in a parallel water world and act out a satirical allegory of selfishness and moral transformation. Kingsley pictured God the Creator as Mother Carey, a white marble lady sitting on a marble throne at the world's end. She created all living things, but not each separately. Instead, she explained, "I sit here and make them make themselves." Kingsley welcomed Charles's idea of evolution and felt it should be offered to children in this way because it pointed to our links with other living creatures; it explained human degeneration as well as progress, and he believed that "your soul makes your body, just as a snail makes his shell."

When, towards the end of the book, Tom, the drowned chimney sweep, reached the Other-End-of-Nowhere, he found an island where everyone was "running for their lives day and night . . . and entreating not to be told they didn't know what." They were being chased by "a poor, lean, seedy, hard-worked old giant" who "had a heart, though it was considerably overgrown with brains." He was a naturalist "made up principally of fish bones and parchment, put together with wire and Canada balsam." He smelt strongly of spirits for preserving specimens; he had "a butterfly-net in one hand, and a geological hammer in the other; and was hung all over with pockets, full of collecting boxes, bottles, microscopes, telescopes, barometers, ordnance maps, scalpels, forceps, photographic apparatus, and all other tackle for finding out everything about everything, and a little more too."

Tom told the giant that he had seen Mother Carey, and the giant was delighted to find someone who might be able to "tell him what he did not know before." He said "quite simply—for he was the simplest, pleasantest, honestest, kindliest old . . . Sampson of a giant that ever turned the world upside down without intending it . . . . 'If I had only been where you have been, to see what you have seen!' " Tom asked the giant why he ran after the people, and he replied that it was they who had been running after him. All he wanted was " 'to be friends with them, and to tell them something to their advantage . . . only somehow they are so strangely afraid of hearing it. But, I suppose I am not a man of the world, and have no tact.' 'But why don't you stop, and let them come up to you?' 'Why, my dear, only think. If I did, all the butterflies and . . . birds would fly past me, and then I should catch no more new species, and should grow rusty and mouldy, and die. And I don't intend to do that, my dear; for I have a destiny before me, they say.' "

Earlier in the book, Kingsley had ridiculed Professor Owen's egotistical clash with Professor Huxley over the hippocampus minor. When he wanted to show science in a different way, as an approach to knowledge and understanding which could reveal the secrets of natural life for the good of mankind, he portrayed it as a tall, kind, unworldly figure whose heart was "considerably overgrown with brains." All who knew Charles would have recognised him in the picture.

At last in 1864, Alfred Russel Wallace took up the argument about mankind and carried it a step forward. He read a paper to the Anthropological Society on "The origin of human races and the antiquity of man deduced from the theory of 'Natural Selection.' " He suggested that man was "social and sympathetic" by nature, and that early in the development of human societies the capacity for cooperation and "sympathy which leads all in turn to assist each other" benefited each community and was favoured by natural selection. "From the time, therefore, when the social and sympathetic feelings come into active operation, and the intellectual and moral faculties became fairly developed, man would cease to be influenced by natural selection in his physical form and structure . . . Every slight variation in his mental and moral nature which should enable him better to guard against adverse circumstances, and combine for mutual comfort and protection, would be preserved and accumulated." That was "the true grandeur and dignity of man." Wallace asserted that if his conclusions were just, it must inevitably follow that "higher" communities, "the more intellectual and moral," must displace the lower and more degraded ones.

Charles warmly welcomed Wallace's argument, with its suggestion that our moral sense might have developed naturally from our social instincts by evolution and reflection. Wallace's emphasis on the human power of sympathy also struck a chord. Charles had not taken his own ideas further since his private notes in 1839, but the novelist George Eliot had made sympathy the key to the "religion of humanity" which she wrote about in her essays and stories from the early 1850s, believing that "Our moral progress may be measured by the degree in which we sympathise with individual suffering and individual joy." She had taken up Wordsworth's idea of the Romantic imagination which Charles had read about in 1838, and made it the focus of her artistic purpose, writing that "The greatest benefit we owe to the artist, whether painter, poet or novelist, is the extension of our sympathies." She developed the theme in her first full-length novel, *Adam Bede,* which appeared in 1859 a few months before the publication of *The Origin of Species.* She embodied the idea most clearly in the figure of Dinah Morris, the young Methodist lay preacher, and her "passionate pity" for Hetty Sorrel waiting in her prison cell to be hanged for the murder of her illegitimate child. At a point of great strain in preparing *The Origin* for publication, Charles had to take a week's rest to "drive" the species theory "out of my head." He read *Adam Bede* with Emma and it did him "a world of good." It remained their favourite of George Eliot's novels, and he referred to a passage from it in one of his last works, *The Expression of the Emotions in Man and Animals,* as an example for a scientific point about infants' feelings.

. . .

Charles was still working on his plant experiments when he read Wallace's paper in 1864. He hoped Wallace would take the lead in further work on human nature and wrote to him: "I have collected a few notes on man, but I do not suppose I shall ever use them. Do you intend to follow out your views? and if so, would you like at some future time to have my few references and notes? I am sure I hardly know whether they are of any value, and they are at present in a state of chaos. There is much more that I should like to write, but I have not strength."

Wallace did not take up Charles's offer, and wrote nothing more at the time. He returned to the theme five years later, but by then he had changed his view, to Charles's intense disappointment. Wallace had attended a séance in 1865 and was intrigued as a man of science by the possibilities of spiritualism. He experimented with paid mediums, and was soon convinced of the reality of spiritual forces. In 1869, he set out his changed thinking about human origins in a review of a new edition of Lyell's *Principles of Geology.* He argued that savages had mental powers beyond their needs for survival. Primitive man could not, therefore, have evolved by the workings of natural selection alone, and we must "admit the possibility that in the development of the human race, a Higher Intelligence has guided the same laws for nobler ends." He was now arguing for just the kind of "miraculous addition" to explain human evolution that Charles had declared to Lyell he would never accept. Charles felt betrayed and wrote unhappily to him about their theory. "I hope you have not murdered too completely your own and my child." He said later: "I differ greatly from you, and I feel very sorry for it."

Meanwhile, Charles's acquaintance Professor W. B. Carpenter of University College London was developing an organic view of the workings of the mind in his *Principles of Human Physiology.* Following previous writers, he suggested that memory, on which our feeling of personal identity depended, was rooted in the structure of the brain, and was "essentially an automatic form of mental activity." Recollection was the "volitional exercise" of that power, and had special importance for mankind since reflection and reasoning by conscious will were held to be the unique powers which set us apart from animals. Carpenter emphasised how much mental and emotional activity took place below the threshold of awareness, and his approach soon influenced other writers. In 1865, Charles read William Lecky's *History of Rationalism* which had a long footnote about the power of deep-rooted prejudices. Following Carpenter, Lecky suggested that the mind was "perpetually acting, pursuing trains of thought automatically, of which we have no consciousness." "Opinions, modes of thought and emotions belonging to a former stage of our intellectual history" often reappeared through "the automatic action of the mind when volition is altogether suspended." As a result, "the origin of most of those opinions we attribute to pure reasoning is more composite than we suppose."

Another person working in the area whom Charles knew was the psychiatrist Henry Maudsley, who argued and worked for the humane treatment of the mentally ill. He was then an ambitious young physician who believed strongly in an organic rather than spiritual approach to the human mind. Knowing all he did from his medical practice about the patterns of breakdown in disturbed and demented patients, he could not see human awareness and reason as a God-given faculty independent of organic life. In his *Physiology and Pathology of Mind,* published in 1867, Maudsley aimed to bring scientific method and biology to bear on mental activity, and to explore what light disease might throw on the workings of the brain. He was impatient with philosophers' reliance on introspective examination of their own minds. They gave no account of the workings of the minds of children, uneducated adults or the insane; nor did they cover the influence of the body on the mind, or the large field of unconscious mental action. The mind was not, as so many assumed, "a wondrous entity, the independent source of power and the self-sufficient cause of causes." It was, instead, "the most dependent of natural forces," highly developed, but still strongly influenced by all the natural factors in its makeup. Man was "a sort of compendium of animal nature," carrying his history with him, and only by looking at the evolution of human nature through time could we see the underlying relations.

Charles read *The Physiology and Pathology of Mind* carefully, and noted some passages about expressions and other involuntary actions linked with sharp and deep feelings. "The beatings of the heart, the movements of respiration, the expressions of the countenance, the pallor of fear or the flush of anger, and the effects upon all the secretions and upon intuition—all these evince with certainty that the organic life participates essentially in the manifestation of emotion."

CHAPTER FOURTEEN

# GOD'S SHARP KNIFE

*Providence, design and suffering—Return to Malvern—Emma's faith—Spiritualism*

THROUGHOUT THE 1860S, Charles thought again and again about the puzzle of suffering and the sense of order in the natural world, and the mystery only deepened. He first explained his growing doubts to Asa Gray, an American naturalist and friend. Commenting on the "theological view" of his theory, he said: "This is always painful to me. I am bewildered." He held firmly to his long-standing belief that God took no special interest in the fate of individuals. "The lightning kills a man, whether a good one or bad one, owing to the excessively complex action of natural laws." Did Gray believe "that God *designedly* killed this man? Many or most persons do . . . I can't and don't." Instead, he was inclined to look at everything as resulting from the operation of God's "designed laws, with the details, whether good or bad, left to the working out of what we may call chance." He wrote later that it was "more satisfactory to attribute pain and suffering to the natural sequence of events." He was prepared to allow that when God set the laws, he foresaw every eventuality. But he found that notion so all-embracing that it was valueless. "It may be said that when you kick a stone, or when a leaf falls from a tree, that it was ordained before the foundations of the world were laid exactly where that stone or leaf should lie. In this sense the subject has no interest for me."

Charles had always shared the common wish to see "evidence of design and beneficence" in the natural world. He had rejected the claim of natural theology that every species had been separately contrived by an all-powerful Creator, but commented to Gray and Hooker that he could not "view this wonderful universe, and especially the nature of man" "as the result of blind chance" or "brute force." Yet still, he could not see the evidence of design and beneficence "as plainly as others do." "There seems to me too much misery in the world." He gave examples from nature like the huge numbers of ichneumons, wasp-like insects which feed on the living bodies of caterpillars, but his word "misery" clearly applied to human suffering and grief.

Annie's cousin Snow Wedgwood had grown into a serious young woman with a strong interest in the harmonies and discords between her uncle's theory and her own deeply felt Christian morality. She set out the problem in an essay which was published in *Macmillan's Magazine* in 1861. When Charles was shown a copy, he found that she gave a correct account of his argument, and that, he told her gratefully, was a "rare event" with his critics. She was concerned that in his account of natural selection, "the work of creation" was carried out by natural forces which had elements of "what in man would constitute sin." His theory seemed to carry us back to the point when "God saw everything that he had made, and behold, it was very good," but the theory then revealed a "scene of strife, of bloodshed, of suffering." "Surely it was not on *this* that the Creator pronounced a blessing! Surely the command 'Be fruitful and multiply' did not mean 'Let every creature engage in an unremitting warfare with its fellows for the means of subsistence'!" Looking throughout nature, she felt it was difficult to avoid the feeling that "something is amiss; something is the work of an evil power." And we must ask ourselves what it would mean for human nature if we accepted that "man is the result of the predominance among his ancestors of those tendencies which in him are sinful."

Snow suggested that man and nature both bore the "impress of imperfection," and that the task of reconciling that with our belief in Divine omnipotence lay beyond human reason. Trying to make sense of how God had placed mankind in this imperfect world, she suggested that nature bore a lesson at every turn, "that failure, suffering, and strife, and even death, are but the steps by which [man] has been raised to the height at which he finds himself . . . What a depth of meaning do we find in such a view of creation as this, of such mighty changes accomplished through such faint and dim gradations, such innumerable failures for one success, such a slow and such an unpausing movement in the stream of creation, widening towards the mighty ocean!"

Charles wrote to Snow that her conclusions had "several times vaguely crossed" his mind. But when he tried to think the points through, "the result has been with me a maze." He could not imagine that the universe had not been designed; yet the closer he looked where one would most expect to find design, "in the structure of a sentient being," the less proof he could see. As he thought about the problem, his sense of order in nature fluctuated, and he found he could not reconcile it with his understanding of the extent of animal and human pain. He could not see how a truly beneficent and omniscient God could have created the order out of that boundless weight of suffering. "I feel most deeply that the whole subject is too profound for the human intellect. A dog might as well speculate on the mind of Newton."

Around the same time, Charles told Herschel he was "in a complete jumble on the point," and commented to Gray: "I am in thick mud . . . yet I cannot keep out of the question." Some years later, he suggested to Hooker that thinking about the riddle was a waste of time, but he still could not stop. "How difficult it is not to speculate."

Charles's memories of Annie softened and changed as the years passed, but he still felt for her as he always had since her childhood. The lasting sense of her loss and the fresh pain of Charles Waring's death deepened his fear of the hurt he would suffer if any of his other children were to die.

Infectious diseases continued to kill many children of both rich and poor throughout the 1850s and 1860s. William Farr's "ledgers of death" showed that for years after its peak in the 1840s, child mortality remained almost as high. In 1857, a new disease, diphtheria, spread from northern France to southeast England, and caused public alarm. Etty suffered an attack with dreadful inflammation of her throat, and as soon as the worst of the crisis was over, Charles wrote to Hooker that she had been "very seriously ill with Dipterithes (or some such name)." In 1862, Leonard almost died of scarlet fever caught at his boarding school in Clapham. For Charles and Emma the "misery of having illness among one's children" grew worse. When Etty was ill, she could sometimes hardly bear it when her father came to see her, because his concern and emotion were "too agitating."

Charles revealed his fears and stress most clearly in 1863 when he and Emma returned at last to Malvern and saw Annie's grave. His sickness had recurred through the years after the publication of *The Origin of Species.* He talked often with Emma about returning to Malvern for treatment by Dr. Gully, but was torn between the hope of relief from his illness and the fear of reviving his memories of Annie's last days. In the first months of 1863, Horace developed a chronic stomach complaint; Charles was vomiting frequently, and in June he told Hooker he was "languid and bedevilled."

Charles found it difficult to work, but watched a wild cucumber plant growing in a pot in his study. A neighbour's gardener, for whom he had great respect as an observer, believed the tendrils could see, because wherever he placed the plant, its tendrils found any stick quickly. Charles had long had a special interest in plant shoots. In the Brazilian forests in 1832 he had seen them growing, and had watched entranced. One day he wrote in his pocket notebook: "Twiners entwining twiners—tresses like hair—beautiful Lepidoptera—silence." Now looking at the plant in his study from hour to hour, Charles spotted the circular sweeping of the tendrils, now clockwise and now anti-clockwise, as they searched for an object to attach themselves to. He suggested to Hooker that the tendrils had "some sense, for they do not grasp each other when young."

Emma urged Charles to take the family to Malvern for a month or two to see if Dr. Gully's treatment could relieve his symptoms, and after a fortnight of sickness in August, he agreed. Emma travelled ahead, and took a house for their stay. She went at once to the churchyard to find Annie's grave, but looked from headstone to headstone in vain. The sexton told her that the churchyard had been altered a few years before and the stone might have been stolen. When Charles arrived, he wrote at once to Fox to ask if he could remember from his visit in 1856 where the grave was. "We want, of course, to put another stone." Fox replied immediately that it was "among several tombs which have shrubs and trees thickly planted round them . . . It was a good strong upright stone, and I remember well 'To a good and dear child'." Charles and Emma then sought out Eliza Partington, who was still at Montreal House, and with Fox's information and her help, they found the headstone. It was shaded by trees and looked so green and old that Emma had not thought it could be the one. She wrote to Fox: "This has been a great relief."

Finding the stone and reading the words he had chosen for Annie, now patched by lichen after twelve years, seems to have helped Charles in coming to terms with his memories of her death, but only to a point. As the weeks passed at Malvern, the water treatment had an effect and he made some progress, but he was not at ease with himself. He had a day of "languor," and suffered bouts of "sinking," "swimming," giddiness and distress in the night.

Then, in the chill of early autumn, Hooker wrote to him from Kew in an agony of pain. "My darling little second girl died here an hour ago, and I think of you more in my grief than of any other friend." Maria was six. Charles wrote back at once, but the following day he suffered "much swimming in head." Hooker's next letter opened: "Dear old Darwin, I have just buried my darling little girl and read your kind note." Hooker wrote about his devotion to his daughter, "the companion of my walks, the first of my children who has shown any love for music and flowers, and the sweetest tempered affectionate little thing that ever I knew. It will be long before I cease to hear her voice in my ears or feel her little hand stealing into mine by the fireside and in the garden. Wherever I go she is there."

Charles replied with a restrained effort at consolation, as he had tried long before with Fox. "I understand well your words 'Wherever I go, she is there.' I am so deeply glad that she did not suffer so much, as I feared was inevitable. This was to us with poor Annie the one great comfort. Trust to me that time will do wonders, and without causing forgetfulness of your darling." These two comforts were the consolations that he and Emma had relied on, but behind his calm words, he was deeply shaken by the sense of what his friend was going through, and was overcome by another wave of feeling. "I am very weak and can write little . . . My head swims badly, so no more."

Charles and Emma stayed on at Malvern for another two weeks, but his head kept swimming; he grew very weak, and eventually he "could not walk a step but from one room into another." Watching him with concern, Dr. Gully decided that he was too ill for the water treatment, and the family returned to Down. It was a relief to be back at home and Charles tried hard to recover, walking a little further every day. In the last week of October, he "accomplished twice round the Sand-walk," but he was unable to work or write letters to anyone. Hooker, not aware of his state of mind, wrote frankly to him again about his grief for Maria. "I am very well, but it will be long before I get over this craving for my child, or the bitterness of that last night. To nurse grief I hold is a deadly sin, but I shall never cease to wish my child back in my arms as long as I live." In a postscript he added that he had just learnt that his son William had scarlet fever but was recovering.

Hooker's letter shook Charles deeply. In the few days after receiving it, he tried to reply but could not manage to write anything. Eventually, at the end of the month, he penned a brief note which was unlike any other he ever sent. "My dear old friend, I must just have pleasure of saying this. Yours affect[ionatel]y C. Darwin." He added a short postscript that he had a letter from a geologist. "I do not know whether you would care to see it. It has something on spreading of European plants." He could not focus on what the letter said. His own message, lamely inconsequential as it was, was a reaching out to his closest friend in the memory of his own pain.

During the following days, he tried again many times to write to Hooker, but still could not find words to put on the page. A physician came from London to examine him and suggested that "a little headwork" might help. Charles's mind kept swimming, but a week later he managed to write to Hooker about a botanical matter, explaining: "I have tried many days to write to you, but could not." He hoped he could recover. "Unless I can, enough to work a little, I hope my life may be very short, for to lie on sofa all day and do nothing but to give trouble to the best and kindest of wives and good dear children is dreadful."

As the days passed, Charles could not concentrate on any work or practical matters, but sat in his study watching another climbing plant growing in a pot on the table next to the sofa. It was a wax flower from Queensland which Hooker had sent him from Kew. In the third week of November, he jotted scrappy notes as a shoot circled near an arm of the sofa, and then back past a bell glass, a copy of Lyell's *Principles of*

*Geology* next to it, and the *Post Office Directory.* The day after the tendril reached the *Directory,* he was at last able to write again to Hooker, but he had to ask him to "excuse my jumping from subject to subject." "The more I look at plants, the higher they rise in my mind; really the tendril-bearers are higher organised, as far as adapted sensitivity goes, than the lower animals." He had looked again at Hooker's letters about his daughter Maria. "How well I remember your feeling, when we lost Annie, that it was my greatest comfort that I had never spoken a harsh word to her. Your grief has made me shed a few tears over our poor darling; but believe me that these tears have lost that unutterable bitterness of former days." He was "so glad" to hear that Hooker's son William was recovering from scarlet fever, and ended: "Goodbye. I am tired."

In the last week of November, Hooker wrote to say that William had relapsed. Charles wrote back at once. "I grieve to hear about the scarlet fever." He had learnt that his own sister Susan was now very ill with the disease. Thinking again of Hooker's worry for his child, he ended his letter with a short string of remarks in which he voiced his feelings about the strength and pain of human affection with utter simplicity and directness. "I shall be glad to hear sometime about your boy, whom you love so. Much love, much trial, but what an utter desert is life without love. God bless you. C.D."

Writing to Hooker in the first week of December, Charles explained why he cared so much. He hoped that Hooker would have "some lull in anxiety and fear" for his children. "Nothing is so dreadful in this life as fear: it still sickens me when I cannot help remembering some of the many illnesses our children have endured." Not "when I remember" but "when I cannot help remembering." The sickening memory of the fear of loss struck again and again.

In the next weeks, Charles experienced "very bad sinkings" and his head swam. He stayed upstairs in his bedroom with one or two climb-

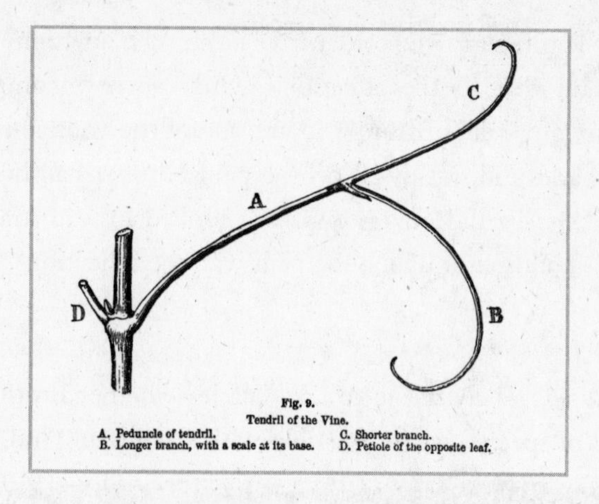


On the Movements and Habits of
Climbing Plants *(1865)*

ing plants for company, watching their tendrils reaching, touching and curling round any object they found. As the year turned and the January days passed, he recovered slowly, but fears for children were still eddying round his mind. He wrote to Hooker in the last week of the month: "As I do nothing all day, I often get fidgety and I now fancy that Charlie or some of your family ill. When you have time let me have a short note to say how you all are." Charlie was one of Hooker's young sons. This remark was revealing. There was no reason for Charles to worry about Charlie's health; it was nothing more than a "fancy," but the idea was clearly preying on Charles's mind.

Charles's collapse during October and November 1863, with the days when he could not even put words together for a message to his closest friend, was one of the most severe in his forty years of illness. In the years that followed, Hooker looked to Charles again and again for sympathy when his father, mother and wife died, and on the anniversary of Maria's death. Charles offered companionship and said what he could to help. On one occasion he wrote: "You have sometimes spoken if you felt growing old: I have never seen any signs of this, and I am certain that in the affections, which form incomparably the noblest part of a man's nature, you are one of the youngest men that I know." Charles did not dwell on the pain Hooker had lived through any more than he did on his own, but looked instead straight to the love for his children of which the pain was one reflection.

In 1866, when a lady wrote to ask Charles whether his theory about the origin of species was compatible with a belief in God, he replied: "It has always appeared to me more satisfactory to look at the immense amount of pain and suffering in this world as the inevitable result of the natural sequence of events, i.e. general laws, rather than from the direct intervention of God." It was easier to come to terms with pain and suffering if there was no question of a Divine purpose governing the life and death of individuals you cared for.

A few years later, a Dutch writer asked for Charles's views on the grounds for belief in God. He replied that "the impossibility of conceiving that this grand and wondrous universe, with our conscious selves, arose through chance, seems to me the chief argument for the existence of God," but if we assumed a First Cause, "the mind still craves to know whence it came, and how it arose." He went on: "Nor can I overlook the difficulty from the immense amount of suffering through the world . . . The safest conclusion seems to me that the whole subject is beyond the scope of man's intellect; but man can do his duty."

Emma devoted her life to caring for Charles and running the household so that he could work without distraction. She claimed to take no interest in his work, but helped with experiments and read proofs of his books when he needed her help. Francis remembered that when he was ill or in distress, "he depended entirely on her presence to make his discomfort bearable. And she would often sit drumming on his head as he lay down."

Emma kept her deep thoughts and feelings to herself, but broke her reserve on one occasion. Sometime before or in June 1861, she looked after Charles while he suffered a few weeks of acute sickness. She wrote a note to him. "My heart has often been too full to speak or take any notice . . . I find the only relief to my own mind is to take it as from God's hand, and to try to believe that all suffering and illness is meant to help us to exalt our minds and to look forward with hope to a future state." When she saw his patience and sensed his gratitude to her, she could not help "longing that these precious feelings should be offered to Heaven for the sake of your daily happiness." But she found it difficult enough for herself. "I often think of the words 'Thou shalt keep him in perfect peace whose mind is stayed on thee.' It is feeling and not reasoning that drives one to prayer." She ended the note with the same diffidence she had shown in her two letters to Charles over twenty years beforehand. She felt "presumptuous" in writing to him. "I shall keep this by me till I feel cheerful and comfortable again about you, but it has passed through my mind often lately, so I thought I would write it, partly to relieve my own mind."

Emma did, though, manage to give the note to Charles, and he wrote on it: "God Bless you. C. D. June 1861."

When Etty was ill in the early 1860s, Emma would read to her. William Cowper, the evangelical poet of the 1780s, was one of her favourite authors, and Etty remembered her reading from his "Winter Walk at Noon." In one insistent passage Cowper argued against the Deist view of a remote Creator which Charles had adopted, asserting instead that God was everywhere and in everything.

*The Lord of all, himself through all diffused,*
*Sustains and is the life of all that lives.*

*Nature is but a name for an effect*
*Whose cause is God . . .*

For those like Emma who hoped for the Last Judgement and a life after death, Cowper offered a quiet approach to their final destiny.

*The groans of nature in this nether world*
*Which heaven has heard for ages, have an end . . .*
*The time of rest, the promised sabbath comes.*
*Six thousand years of sorrow have well-nigh*
*Fulfilled their tardy and disastrous course*
*Over a sinful world. And what remains*
*Of this tempestuous state of human things,*
*Is merely as the workings of a sea*
*Before a calm, that rocks itself to rest.*
*For he whose car the winds are . . .*
*Shall visit earth in mercy; shall descend*
*Propitious, in his chariot paved with love,*
*And what his storms have blasted and defaced*
*For man's revolt, shall with a smile repair . . .*
*Thus heavenward all things tend. For all were once*
*Perfect, and all must be at length restored.*
*So God has greatly purposed . . .*

A few years later, Emma read two books by Ashton Oxenden, an Anglican clergyman whose devotional writings were popular for their plain and simple language. In *Words of Peace; or the Blessings and Trials of Sickness* he wrote: "What are God's reasons for afflicting us? Is it to *punish?* Sometimes it is; but not, I think, usually . . . There must be *another and truer reason* why the Lord chastens. It is because *He desires to do you some great good.* The gardener cuts and prunes his tree, to make it grow better, and bear more precious fruit; and God often uses His sharp knife for some gracious purpose . . . *God cannot afflict wrongly.* He never makes mistakes . . . Before then you go a step further, ask God to convince you of this precious truth—It is my Father who corrects me, even He who loves me." In *Fervent Prayer,* Oxenden wrote about the "spirit of unbelief which is ever creeping into the hearts of God's people, tempting them to feel that the Lord is far off, that their prayers will not be heard, and that it is useless to seek Him. Who has not felt something of this?" We should constantly implore God "to give us that faith, which is not in us naturally, but which comes from Him. 'Lord, I believe; help thou my unbelief.' "

During these years, Charles and Emma were as close to each other as ever in their marriage, but the "painful void" which Emma had spoken of just before their wedding still lay between them. Unable to share her faith with Charles, Emma felt it weaken. Etty wrote that "As years went by her beliefs must have greatly changed, but she kept a sorrowful wish to believe more, and I know that it was an abiding sadness to her that her faith was less vivid than it had been in her youth."

Emma's feelings for Annie remained as deep as ever. In April 1875, Aunt Fanny Allen, who had been with her at Down when Annie died, was ninety-four and the last survivor of Emma and Charles's parents' generation. Emma wrote to her as she approached death. The letter does not survive, but Aunt Fanny replied that Emma's "grateful remembrance of the sad April days of '51 makes my heart beat with gratitude to you for its recollection, coupled as it was by the memory of your grief for your darling." Taking up what Emma had said, Aunt Fanny went on: "It is true gaps can never be filled up, and I do not think we should wish them to be filled other ways than as our memory fills them."

. . .

While Charles and Emma lived with their private thoughts and worries, they presented a picture of ease and contentment to the world. In 1869, Henry James, then a young American visitor to London and as yet unknown as a writer, accompanied a friend to lunch at Down. He wrote to his family that the Darwins' carriage met them at Bromley Station, and they "rolled quietly along through a lovely landscape, between springing hedges and ivy-crowned walls . . . ineffably verdurous meads and tender-bursting copses . . . fine old seats and villas." "Darwin's house is a quiet old place . . . We lunched and spent an hour and a half seeing the old man, his wife and his daughter. Darwin is the sweetest, simplest, gentlest old Englishman you ever saw . . . He said nothing wonderful and was wonderful in no way but in not being so."

Within the privacy of the household, Charles revealed a growing dislike of established religion. While Emma was reading *Fervent Prayer,* he subscribed to *The Index,* a newspaper produced by a group of disaffected American Unitarians and philosophical unbelievers. The paper advocated a spirit of reform "without deference to authority of Bible, Church or Christ." It argued for rejection of the Christian confession, and proposed in its place a humanistic "Free Religion" in which "lies the only hope of the spiritual perfection of the individual and the spiritual unity of the race." Charles allowed the editor to print in each issue a comment by him endorsing these views, and pressed the newspaper's claims in conversation with his sons and daughters. Francis remembered how his father would tell the family "the most extraordinary facts" from the newspaper and was indignant with anyone who doubted their complete accuracy.

In his published writings Charles paid lip service to Christian belief, but his words were carefully chosen and non-committal, as when he wrote in *The Descent of Man* that "the question whether there exists a Creator and Ruler of the universe . . . has been answered in the affirmative by some of the highest intellects." He made occasional tongue-in-cheek comments, suggesting, for instance, that there might be a link between religious devotion and "the deep love of a dog for his master, associated with complete submission, some fear, and perhaps other feelings." He commented on the primitive origins, or "natural history," of religious belief. The idea that "natural objects and agencies are animated by spiritual or living essences, is perhaps illustrated by a little fact which I once noticed: my dog, a full-grown and very sensible animal, was lying on the lawn during a hot and still day; but at a little distance a slight breeze occasionally moved an open parasol, which would have been wholly disregarded by the dog, had any one stood near it. As it was, every time that the parasol slightly moved, the dog growled fiercely and barked. He must, I think, have reasoned to himself in a rapid and unconscious manner, that movement without any apparent cause indicated the presence of some strange living agent, and that no stranger had a right to be on his territory." In *The Expression of the Emotions in Man and Animals* he wrote about the torment of eternal punishment: "There is said to be 'gnashing of teeth' in hell; and I have plainly heard the grinding of the molar teeth of a cow which was suffering acutely from inflammation of the bowels."

Emma and Charles talked together in the 1870s about another form of spiritual life: "manifestations," the other world and messages from the departed. At a time of intense concern about the natural or supernatural reality of spiritual forces and life beyond death, mediums were coming from America with parlour performances claiming to confirm both. Some men of science took an interest in the phenomena, hoping they could find sound evidence to resolve the issues. Among Charles's acquaintances, Dr. Gully was a firm believer and Alfred Russel Wallace had been converted in the mid-1860s. Charles took an interest because he wanted his theory to explain as much as could be explained by the regular working of observable and purely natural forces, and if he had to accept the reality of a separate spiritual realm, human life would have to be approached in a different way.

The mediums played boldly on people's feelings about loved ones who had been taken from them by death. One produced a book, *Heaven Opened; or Messages for the Bereaved from our Little Ones in Glory.* Charles's acquaintance Robert Chambers, the publisher, amateur geologist and secret author of *Vestiges of the Natural History of Creation,* which had argued for evolution many years before *The Origin of Species* appeared, started attending séances in 1853 in a cool and critical frame of mind, watching carefully for fraud. In 1857, he met Daniel Dunglas Home, a charismatic showman from America, and was greatly impressed by his performances. He took copious notes of a number of sessions and decided that the phenomena compelled a reasonable man to believe in a "spiritual agency," immortality and the hereafter, as no other explanation would fit what he had witnessed. At a séance in 1860, an accordion was provided and Chambers called for his deceased father to play his favourite Scottish and English tunes. "Ye Banks and Braes" and "The Last Rose of Summer" were sounded as if by magic, and Chambers told the company that they had indeed been his father's favourite airs. Home was always seeking public endorsements from men of science and other prominent figures, but Chambers was not prepared to give him one, and Home's friends kept working on him quietly.

Chambers had lost two daughters and was greatly affected by their deaths. In 1866, Home formed a society called the Spiritual Athenaeum with Dr. Gully and some other friends, one of whom knew Chambers's family. At two meetings, Home saw the spirits of the two daughters; one gave a message to be passed on to their father, and the other gave the last words she had spoken to him, "Pa, love," to prove their identity. The society wrote to Chambers and he confirmed that those were his second daughter's last words. The message was then sent to him; he acted on it and was convinced it was genuine. He now openly declared his support for Home, and wrote to Wallace: "My idea is that the term 'supernatural' is a gross mistake. We have only to enlarge our conceptions of the natural and all will be right." It is strange to realise that if Home and his friends had thought the notorious Mr. Darwin might welcome a message from a loved daughter beyond death, and Dr. Gully had told them about Annie, Home might have received a table-tapping from her.

An ambitious chemist, William Crookes, who discovered the element thallium in 1861 and later experimented with cathode rays, launched a personal inquiry into "psychic force" in the early 1870s. Like Chambers, he was alert to the possibility of fraud; he kept careful notes through a series of Home's séances, and was impressed by what he saw. In April 1871, Home held a séance for Crookes and a group of friends. Crookes recorded that "At first we had very rough manifestations, chairs knocked about, the table floated about six inches from the ground and then dashed down, loud and unpleasant noises bawling in our ears and altogether phenomena of a lower class. After a time it was suggested that we should sing, and as the only thing known to all the company, we struck up 'For he's a jolly good fellow' . . . After that D.D. Home gave us a solo—rather a sacred piece—and almost before a dozen words were uttered Mr. Herne was carried right up, floated across the table and dropped with a crash of pictures and ornaments at the other end of the room."

Charles was interested in Crookes's inquiry, and his cousin Francis Galton reported to him about a séance he had attended with Crookes in April 1872. The table moved while Galton was sitting under it checking for deception, and "beautiful sacred music" was played on the accordion. He wrote to Charles about another occasion that "The absurdity on the one hand, and the extraordinary character of the thing on the other, quite staggers me; wondering what I shall yet see and learn, I remain quite passive with my eyes and ears open."

In 1874, Charles's brother Erasmus arranged a séance with a paid medium at his house in Mayfair. Charles, Emma and Etty attended, together with George Eliot and her partner G. H. Lewes, the journalist and philosopher. Mr. Lewes was firmly and openly doubtful, and Erasmus was almost certainly hoping for quiet entertainment, watching in the darkness. Etty remembered that Mr. Lewes "was troublesome and inclined to make jokes and not play the game fairly." The usual manifestations occurred, "sparks, wind-blowing, and some rappings and movings of furniture." Emma watched and kept an open mind. Charles, on the other hand, found it so hot and tiring that he went away, as he wrote to Hooker, "before all these astounding miracles, or jugglery, took place." "The Lord have mercy on us all, if we have to believe in such rubbish."

Charles and Emma's niece Snow discussed the séance with Emma when she was staying at Down a few months later. Snow wrote to a friend that Emma thought Charles had "quite made up his mind he *won't* believe it, he dislikes the thought of it so much." Referring to a remark Charles had once made to her, Snow asked Emma if he did not say it was a great weakness to allow wish to influence belief. Emma replied: "Yes, but he does not act up to his principles." Was that not bigotry? Emma replied with heavy irony: "Oh yes, he is a regular bigot." The sharpness of her comment suggests she may have felt it was more difficult for her to keep her open mind and her hope of salvation, than it was for him to reject religion as he did in the name of scientific reason.

CHAPTER FIFTEEN

# THE DESCENT OF MAN

*Descent of Man—Expression of the Emotions—*
*Biographical Sketch of an Infant*

IN 1869, CHARLES DECIDED AT LAST that he must tackle the issue of human origins himself. He explained to a friend: "I am thinking of writing a little essay on the Origin of Mankind, as I have been taunted with concealing my opinions." After completing *The Variation of Animals and Plants Under Domestication,* the last part of his main argument about the origin of species, he began work on *The Descent of Man.* The book, which eventually appeared in 1871, dealt first with the animal ancestry of mankind and how we became human, and then tackled the controversial question of human race, introducing the idea of evolution by sexual selection to explain racial differences.

On the first theme, Charles thought back to Jenny the orang; he remembered his ideas about the natural origins of the moral sense, and he read David Hume's moral philosophy again. He took up Hume's suggestion that the "social virtues" were part of our instinctive makeup rather than the product of reasoning from abstract principles, since they had a natural appeal to "uninstructed mankind" long before we had received any "precept or education." He thought again about his "natural history of babies," and remembered all he had learnt from Annie and after her death about the strength of a parent's love and how memory lasted. He wove his observations as a naturalist together with his own experiences into a view of human nature which looked beyond received ideas.

The book was widely read, and was not criticised as fiercely as Charles had feared. The times were changing and Charles was encouraged to follow up the book with two more works on aspects of human nature, *The Expression of the Emotions in Man and Animals* and "A Biographical Sketch of an Infant." As things worked out, few of his themes were taken further in his lifetime, but they have been since. The subjects are still as bedevilled by controversy and prejudice as they were in Darwin's day, but that is hardly surprising when it is the buried history of our own nature that we are arguing about.

Charles was thinking about *The Descent of Man* in early 1870 when Etty, now a young lady of twenty-seven, was wintering in Cannes. Emma wrote to her that Charles was working on his new book. "I think it will be very interesting, but that I shall dislike it very much as again putting God further off." She was to repeat that phrase a year later in a comment to a friend when the book was published. Her offhand remarks touched again on the "painful void" between Charles and herself as he held to his idea of a remote and mysterious Deity while she tried to maintain her faith in Cowper's loving God who was present everywhere and was "the life of all."

Charles sent Etty a chapter of *The Descent of Man,* explaining that its object was "simply comparison of mind in men and animals." He feared that "parts are too like a sermon; who would ever have thought that I should turn a parson?" He asked Etty for "deep criticism" and any corrections of style. He knew that she was close to him in much of her thinking. She believed in the essential unity of mind and matter; she saw disease and death as purely natural processes; she had doubts about life after death, and one of her few convictions was "the worship of humanity."

With a characteristic preference for things unconnected with

Charles's "stiff" ideas, Emma added a wry footnote to his letter. Etty had written about the hotel in Cannes, and her mother replied: "How very odd the meat being so bad. One would have thought with a population of rich invalids, that would have been the first thing to attend to."

In *The Descent of Man,* Charles emphasised our animal origins with a new force and sharpness. He still felt a strong need to puncture human arrogance, but he was now more hopeful that others would agree. It was only our natural prejudice, and "that arrogance which made our forefathers declare that they were descended from demigods," which led people to reject the idea of common descent. "But the time will before long come, when it will be thought wonderful that naturalists, who were well acquainted with the comparative structure and development of man, and other mammals, should have believed that each was the work of a separate act of creation." He offered his insistently humble view of man's place in nature with a deadpan sharpness. He repeated the point he had made about the human frame in *The Origin of Species.* "It is notorious that man is constructed on the same general type or model as other mammals. All the bones in his skeleton can be compared with corresponding bones in a monkey, bat, or seal." He now added a number of other points by which the link was shown, though few readers would have been happy to recognise them. Wild baboons like beer, get drunk and are hung over; we share syphilis, cholera and herpes with animals, and we are infested with many of the same parasites. Charles took up a point made by Huxley in his "monkey book." Charles wrote: "As some of my readers may never have seen a drawing of an embryo, I have given one of man and another of a dog, at about the same early stage of development, carefully copied from two works of undoubted accuracy." And he drove the point home with a twist. Anyone who rejected with scorn the belief that the shape of his canine teeth was due to his early ancestors having been given them as weapons, would "probably reveal, by sneering, the

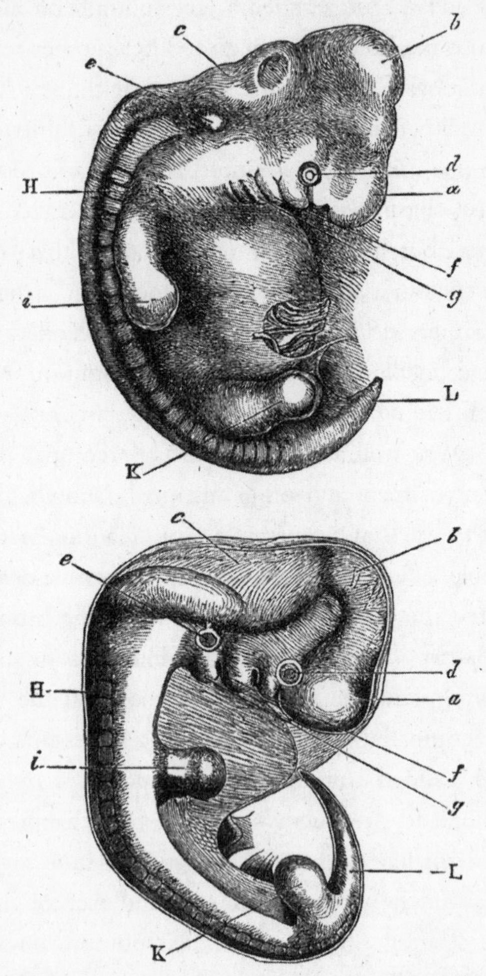


Fig. 1. Upper figure human embryo, from Ecker. Lower figure that of a dog, from Bischoff.

*a*. Fore-brain, cerebral hemispheres, &c.
*b*. Mid-brain, corpora quadrigemina.
*c*. Hind-brain, cerebellum, medulla oblongata.
*d*. Eye.
*e*. Ear.
*f*. First visceral arch.
*g*. Second visceral arch.
H. Vertebral columns and muscles in process of development.
*i*. Anterior } extremities.
K. Posterior } extremities.
L. Tail or os coccyx.

The Descent of Man *(1871)*

line of his descent. For though he no longer intends, or has the
to use these teeth as weapons, he will unconsciously retract his 'snarling muscles' . . . so as to expose them ready for action, like a dog prepared to fight."

Charles gave man "a pedigree of prodigious length, but not, it may be said, of noble quality." He echoed Huxley and Wallace in suggesting that human progress was a matter for admiration and gave hope for the future, but his tone was slightly different from their triumphalism. "Man may be excused for feeling some pride at having risen, though not through his own exertions, to the very summit of the organic scale; and the fact of his having thus risen, instead of having been aboriginally placed there, may give him hope for a still higher destiny in the distant future." He went on to insist, though, that "we are not here concerned with hopes or fears, only with the truth as far as our reason permits us to discover it; and I have given the evidence to the best of my ability."

Charles was now at last prepared to reveal to the world the ideas about the animal roots of human nature which he had explored so freely and boldly in his private notes thirty years before. He worked them up with the fuller understanding of ties and feelings in a close family that he had gained in the thirty years of his life with Emma, their surviving children, and the three who had died. He pointed first to the basic instincts of survival and affection that we share with the lower animals, "self-preservation, sexual love, the love of the mother for her new-born offspring, the desire possessed by the latter to suck, and so forth." The last two he had watched with devoted attention in Emma and their newborn children.

Taking up the ideas that Wallace had put forward in 1864, Charles suggested that "the parental and filial affections, which apparently lie at the base of the social instincts" were probably developed through natural selection. With the bravado of a person who is confident he is right but does not expect to be believed, he added that "Parental affection, or some feeling which replaces it, has been developed in certain animals extremely low in the scale, for example in star-fishes and spiders. It is also occasionally present in a few members alone in a whole group of animals, as in the genus Forficula, or earwigs." It was Etty who had suggested the point; she had always found small creatures and their families absorbing, looking with Annie at her ladybirds in their "little box" in Malvern, watching over the farmyard cats with their kittens in her shed at Down, and caring for the fancy pigeons she had bred with her father.

Moving on to the more complex emotions, Charles suggested that most were also common to the higher animals and ourselves. All humans and other primates "have the same senses, intuitions and sensations, similar passions, affections and emotions, even the more complex ones such as jealousy, suspicion, emulation, gratitude and magnanimity; they practise deceit and are revengeful; they are sometimes susceptible to ridicule, and even have a sense of humour; they feel wonder and curiosity; they possess the same faculties of imitation, attention, deliberation, choice, memory, imagination, the association of ideas, and reason, though in very different degrees." Charles then worked through the attributes which different writers had argued were unique to mankind: Lyell's improvable reason, the fashioning of tools which the Duke of Argyll had suggested was distinctively human, abstract thought, self-awareness, language, the sense of beauty and the belief in unknown spiritual agencies. In each case he pointed to features of animal behaviour that might be reckoned to be rudimentary forms of the human capability.

One aspect of his approach was criticised at the time by G. H. Lewes and others. He indulged freely in anthropomorphism, making guesses about the mental processes of animals and describing them in terms of human experience. He was aware of the dangers, and acknowledged that as animals could not speak to humans, we could never understand their feelings in the way in which we understand each other's. He made the point in a striking image from one of his country walks. "Who can say what cows feel when they surround and stare intently on a dying or dead companion?" One could, though, make inferences. He based his suggestions about animals' feelings and reasoning on a few simple rules which he was willing to explain. He was also always equally interested to determine where animals' mental processes differed from humans', as to identify where there might be similarities. The important point for him was to explore the possibilities either way, and to find where any boundaries could be traced.

Charles's approach had another special feature which he did not recognise and others did not challenge. It was generally agreed that one essential source of knowledge in the science of man was the thinker's awareness and understanding of his or her own mental experience. Charles accordingly made free use of introspection. He assumed that his own feelings and reflections were shared by others, but some of his generalisations from his own experience were open to question. He wrote on memory that "A man cannot prevent past impressions often repassing through his mind." And "Man, from the activity of his mental faculties, cannot avoid reflection: past impressions and images are incessantly and clearly passing through his mind." That was his experience, but his words, "incessantly and clearly" for example, suggest that memories came to him more insistently and vividly than to most people.

Charles also had an acute sense of how others might be judging him, and presumed that other people shared his feelings. A man's "early knowledge of what others consider as praiseworthy or blameable . . . cannot be banished from his mind, and from instinctive sympathy is esteemed of great moment." He repeated the point in a slightly different form. "Even when we are quite alone, how often do we think with pleasure or pain of what others think of us, of their imagined approbation or disapprobation; and this all follows from sympathy, a fundamental element of the social instincts." He felt the same way about rules on their own. "We recognise the same influence in the burning sense of shame which most of us have felt, even after the interval of years, when calling to mind some accidental breach of a trifling, though fixed, rule of etiquette." Yet again, this sensitivity was not a general truth of human nature, but a special feature of Charles's intense and highly strung sensibility. Most other people have more control over their thoughts, and manage better to overcome, avoid or ignore feelings of shame or guilt.

Charles's account of human awareness was, in this respect, a reading of his own experience. He made the instinctive feeling of sympathy a key notion. He saw it as distinct from love, since "a mother may passionately love her sleeping and passive infant, but she can hardly at such times be said to feel sympathy for it." However, there was a close link; a loved adult or child always received special sympathy, and Charles could never forget his feelings for Annie in her fretfulness and distress throughout her last illness. He saw elements of sympathy in memories charged with feelings, as when he wrote in *The Expression of the Emotions in Man and Animals,* "The vivid recollection of our former home, or of long past happy days, readily causes the eyes to be suffused with tears; but here . . . the thought naturally occurs that these days will never return. In such cases we may be said to sympathise with ourselves in our present, in comparison with our former, state."

Charles saw that humans were essentially social animals like the man-like apes, and the early ape-like progenitors of man were probably social too; they were all likely to have "retained from an extremely remote period some degree of instinctive love and sympathy" for their fellows. In emphasising the instinctive nature of their sympathy, Charles agreed with Wallace that it had probably developed through natural selection, "for those communities which included the greatest number of the most sympathetic members would flourish best and rear the greatest number of offspring."

Moving on to the moral sense, Charles acknowledged that it was the most noble of all human attributes, and "by far the most important of all the differences between man and the lower animals." Recognising the issue to be a second critical test for his theory of evolution alongside the design of the eye, he opened a fresh chapter of his book with Kant's "great question" about the origin of the sense of duty. He wrote that many others far more able than he had discussed it at length, but he would offer his own answer because he could not avoid the issue, and "as far as I know, no one has approached it exclusively from the side of natural history."

Charles took up the argument he had written out in 1839, and suggested that we developed our moral sense when, as part of a natural process, early man first achieved self-awareness, remembered his past actions and reflected on his feelings about them. Charles believed that "The social instincts . . . will from the first have given to [man] some wish to aid his fellows, some feeling of sympathy, and have compelled him to regard their approbation and disapprobation. Such impulses will have served him at a very early period as a rude rule of right and wrong. But as man gradually advanced in intellectual power, and was enabled to trace the more remote consequences of his actions; as he acquired sufficient knowledge to reject baneful customs and superstitions; as he regarded more and more, not only the welfare, but the happiness of his fellow men; as from habit, following on beneficial experiences, instruction and example, his sympathies became more tender and widely diffused, extending to men of all races, to the imbecile, maimed, and other useless members of society, and finally to the lower animals, so would the standard of his morality rise higher and higher."

When he set out his view of the sources of human nature at the end of *The Descent of Man,* Charles explained his feelings about the value of memory. They were points of moral sensibility, not science, and one of the experiences from which they stemmed was the importance to him of his lasting feelings for Annie. Charles now suggested that "The moral faculties are generally and justly esteemed as of higher value than the intellectual powers. But we should bear in mind that the activity of the mind in vividly recalling past impressions is one of the fundamental . . . bases of conscience. This affords the strongest argument for educating and stimulating in all possible ways the intellectual faculties of every human being . . . Whatever renders the imagination more vivid, and strengthens the habit of recalling and comparing past impressions, will make the conscience more sensitive." As he had shown twenty years before when he wrote about Annie after her death, he wanted to remember, even when pain came with the memory.

Charles's ideas echoed themes he and Emma had found in George Eliot's writings of the 1850s and 1860s. Following Wordsworth's suggestions about the Romantic imagination, George Eliot saw memory and feeling, self and other, as bound closely together. In *Scenes of Clerical Life,* she had written: "Sympathy is but a living again through our own past in a new form." In *Adam Bede,* Charles and Emma's favourite of her novels, she commented on Dinah Morris's vivid imaginings of Hetty Sorrel's suffering that "It was in this way that Dinah's imagination and sympathy acted and reacted habitually, each heightening the other." At the end of the book, she linked experience and memory with sympathy in another way when she described how Adam never outlived his sorrow for Hetty. "Do any of us? God forbid. It would be a poor result of all our anguish and our wrestling, if we won nothing but our old selves at the end of it—if we could return to the same blind loves . . . the same feeble sense of that Unknown towards which we have sent forth irrepressible cries in our loneliness. Let us rather be thankful that our sorrow lives in us as an indestructible force, only changing its form, as forces do, and passing from pain into sympathy—

the one poor word which includes all our best insight and our best love." She offered another view in Maggie Tulliver's words in *The Mill on the Floss.* "Love is natural, but surely pity, and faithfulness, and memory are natural too. And they would live in me still, and punish me if I did not obey them."

Despite his deeply held view that morality was rooted in the human affections and sympathy, Charles maintained his unflinching sense of the ruthlessness of natural selection as a force shaping instincts, and he also kept the strong sense of the *ad hoc* and imperfect nature of human instincts that he had first expressed when he wrote in his notebook about the "Devil under form of baboon" being our grandfather. He extended the point to morality, and illustrated it deftly by comparing humans with insects. "If . . . men were reared under precisely the same conditions as hive-bees, there can hardly be a doubt that our unmarried females would, like the worker-bees, think it a sacred duty to kill their brothers, and mothers would strive to kill their fertile daughters; and no one would think of interfering." The sharpness of this passage is breathtaking, as it contradicted the moral assumptions of the age.

Charles believed that morality had been perverted repeatedly throughout human history by religious beliefs. Taking the ideas David Hume had offered in his *Natural History of Religion,* Charles wrote that "The same high mental faculties which first led man to believe in unseen spiritual agencies, then in fetishism, polytheism, and ultimately in monotheism, would infallibly lead him, as long as his reasoning powers remained poorly developed, to various strange superstitions and customs. Many of these are terrible to think of, such as the sacrifice of human beings to a blood-loving god, the trial of innocent persons by the ordeal of poison or fire, witchcraft, etcetera. Yet it is well occasionally to reflect on these superstitions, for they show us what an infinite debt of gratitude we owe to the improvement of our reason, to science, and to our accumulated knowledge . . . These miserable and indirect consequences of our highest faculties may be compared with the incidental and occasional mistakes of the instincts of the lower animals."

A child Annie had played with in Malvern had become a tragic and sensational victim of the shortcomings of human morality and affections. Marian Marsden, daughter of James Marsden, the water doctor in Malvern, had been Annie's age. She was as unlucky with her parents and carers as Annie was lucky with hers. Her mother had died when Marian was six; her father fell in love with a young patient and married her, and he then paid his children's French governess Celestine Doudet to take them to Paris. Dr. Marsden wrote to Mademoiselle Doudet about discipline that "Morals are more important than everything else." When Marian fell ill in Paris, a group of neighbours wrote to Dr. Marsden claiming that the children were being ill-treated. He asked John Rashdall, the vicar who had conducted Annie's burial service, to visit them in Paris. Mr. Rashdall found them "as well as could be expected" and reported that when he asked them about Mademoiselle Doudet, they praised her. Shortly afterwards Marian died and a post mortem revealed a fracture in her skull. Dr. Marsden went to Paris and found the other children starved and bruised. He removed them from Mademoiselle Doudet's care, but Marian's elder sister died shortly after, "crying out in her delirium that Mademoiselle Doudet had sworn to pursue her, even in death, if she revealed what had gone on in Paris." She was buried in Malvern churchyard, and her gravestone stands near Annie's.

Mademoiselle Doudet was tried in Paris for cruelty to the Marsden children and the murder of Marian. The case was reported at great length in a popular periodical, *Les Causes célèbres de tous les peuples,* and the trial was also covered in some English newspapers. One feature of the proceedings was how some actions by Mademoiselle Doudet, seen

*Mademoiselle Doudet with the Marsden children*

by the accusers as murderous cruelty, were claimed by her and other witnesses to be sound discipline, and therefore moral. She was found guilty, but *Les Causes célèbres* voiced widespread sympathy for her. The case pointed obliquely to the relative nature of accepted moral thinking, as it revealed how close some forms of righteous conduct were to evildoing.

When Charles had first thought about the moral sense in 1838, he had suspected that it was "an hereditary compound passion." He now had a notion of its make-up. "Ultimately our moral sense or conscience becomes a highly complex sentiment—originating in the social instincts, largely guided by the approbation of our fellow-men, ruled by reason, self-interest, and in later times by deep religious feelings, and confirmed by instruction and habit." Just as geology had given him the vast time-frame needed for evolution to work in, so philosophy and psychology pointed to mental forces and links operating below personal awareness. The new science of man that he envisaged would not simply trace the complexes of feeling and belief down to one or two supposedly primary factors; it would try to understand the interplay of instinct and conscious thought in order to fathom their workings with each other.

Charles's view of the moral sense prompted him to think again about the involvement of mankind in the struggle for existence. In one comment linked with Annie's death, he contradicted a conclusion that many people had drawn from his ideas about the survival of the fittest. Before the appearance of *The Origin of Species,* some commentators had based a theory of social progress on Malthus's view of perpetual competition in human life. The idea fitted in with the laissez-faire attitude towards the "undeserving poor" which was widely held among prosperous people. When Charles explained his theory of natural selection in *The Origin,* some saw it as further justification for their approach, and applied the idea to the physically unfit. Charles was always respectful towards Herbert Spencer, the social philosopher linked with the ideas which became known as "social Darwinism," but he often felt that his writings were too abstract, and admitted that he did not understand them.

Charles was particularly unhappy with the argument linking social progress with harsh treatment of people who were "unfit" to survive in the struggle for life, and used an opportunity in *The Descent of Man* to make his point. He wrote: "With savages, the weak in body and mind are soon eliminated; and those that survive commonly exhibit a vigorous state of health. We civilised men, on the other hand, do our utmost to check the process of elimination; we build asylums for the imbecile, the maimed, and the sick; we institute poor-laws; and our medical men exert their utmost skill to save the life of every one to the last moment." Thinking perhaps of himself and his chronic illness, he suggested: "Thus the weak members of civilised societies propagate their kind." He then argued that what prompted the aid we "feel impelled to give to the helpless is . . . the instinct of sympathy, which was originally acquired as part of the social instincts, but [was] subsequently rendered . . . more tender and more widely diffused." Mindful of his own experience with Annie and others, he went on with the force of his own conviction: "Nor could we check our sympathy, even at the urging of hard reason, without deterioration in the noblest part of our nature. The surgeon may harden himself whilst performing an operation, for he knows that he is acting for the good of his patient; but if we were intentionally to neglect the weak and helpless, it could only be for a contingent benefit, with an overwhelming present evil."

In the year when *The Descent of Man* appeared, Henry Maudsley the psychiatrist lectured on "Body and Mind" at the Royal College of Physicians. Charles noted his suggestion that our moral sense was a recent inheritance, and the link he made with the observation that "a perversion or destruction of the moral sense" was often one of the earliest symptoms of mental derangement. "As the latest and most exquisite product of mental organisation," the moral sense was "the first to testify to disorder of the mind-centres." The point echoed Charles's comment in 1838 about Emma's mother and how her affections had been destroyed when she became demented. Maudsley now shared Charles's view that human feelings were faculties of our organism that needed to be understood as elements in a complex and obscure mechanism which had developed over time and could break down in a pattern. He described strange animal-like traits in the behaviour of "idiots" and asked whether they might be due to the reappearance of primitive instincts, "an echo from a far-distant past, testifying to a kinship which man has almost outgrown."

Maudsley visited Charles at Down, and gave more of his darkening view of the human mind in his next book, *Responsibility in Mental Disease.* One passage showed clearly the sea change in thinking about human nature which Charles had helped to bring about, reversing the theologians' former proud notions of man as the sole rational being. Maudsley opened his last chapter on "The prevention of insanity" with a comment that undermined the complacency of the age about human reason. "Most persons who have suffered from the malady of thought must at one period or other of their lives have had a feeling that it would not be a hard matter to become insane, that in fact something of an effort was required to preserve their sanity." With his languor, his swimming head, his hysterical crying and the sleepless nights when he could not get a painful idea out of his mind, Charles knew the malady.

Charles had started his next book, *The Expression of the Emotions in Man and Animals,* as a chapter of *The Descent of Man,* but took it out because he found he had more to say than would fit in *The Descent.* One theme was how our common nature with animals extends from body to mind, to our feelings and their expression. He aimed to refute a suggestion by a previous writer, Sir Charles Bell, that the Creator had given us our facial muscles and expressions to enable us to show our feelings to each other for spiritual purposes. He grouped expressions in a number of kinds, and offered explanations of how each had developed by the natural workings of the body. Some he suggested were inherited versions of "serviceable associated habits." Others he explained by a principle of antithesis whereby an opposite emotion to one with a set response would trigger opposite behaviour, and a third group he believed were due to an excess of nervous energy spilling over into other channels. Nowadays, most explanations focus on communication, but some refer to habitually associated actions.

Charles described many careful observations he had made of human emotions and their expressions. He dwelt again on intense feelings, and the obscure links between mind and body. He returned to his point in *The Descent of Man* about humans baring their canine teeth. "With mankind some expressions, such as bristling of the hair under the influence of extreme terror, or the uncovering of the teeth under that of furious rage, can hardly be understood, except on the belief that man once existed in a much lower and animal-like condition . . . No doubt as long as man and all other animals are viewed as independent creations, an effectual stop is put to our natural desire to investigate as far as possible the causes of expression." Again, he wanted to look into the depths of our nature, and he wanted to set aside the obstacles in the way. "He who admits on general grounds that the structure and habits of all animals have been gradually evolved, will look at the whole subject of expression in a new and interesting light."

In exploring emotions and how they are expressed, Charles gathered anecdotes; he corresponded with doctors in charge of lunatic asylums; he looked again inside himself, and he thought of Emma and the children. From his own experience, he wrote: "A strong desire to touch the beloved person is commonly felt; and love is expressed by this means more plainly than by any other. Hence we long to clasp in our arms those whom we tenderly love. We probably owe this desire to inherited habit, in association with the nursing and tending of our children, and with the mutual caresses of lovers." Something of the feeling that lay behind this was caught in a recollection by Francis. "I used to like to hear him admire the beauty of a flower; it was a kind of gratitude to the flower itself, and a personal love for its delicate form and colour. I seem to remember him gently touching a flower he delighted in."

On parental affection, Charles suggested that "an emotion may be very strong, but it will have little tendency to induce movement of any kind, if it has not commonly led to voluntary action for its relief or gratification." Thinking perhaps of Emma with her reserve, he wrote: "No emotion is stronger than maternal love; but a mother may feel deepest love for her helpless infant, and yet not show it by any outward sign; or only by slight caressing movements, with a gentle smile and tender eyes." William's photograph of his mother watching Charles Waring in her lap again captures what his father had in mind.

Charles's comments on infants and young children were remarkable for the focus of his interest and the detail of his recollections almost twenty years after his last child went away to school. His phrasing was now light, now heavy. He wrote about "the art of screaming" which infants "finely developed from the first days" because it was "of service" to them. Then, "When an infant is uncomfortable or unwell, little frowns . . . may be seen incessantly passing like shadows over its face; these being generally, but not always, followed sooner or later by a crying fit." Later, "With very young children it is difficult to distinguish between fear and shyness; but this latter feeling with them has often seemed to me to partake of the character of the wildness of an untamed animal." Charles had clear memories of his children in high spirits. "Under a transport of joy or of vivid pleasure, there is a strong tendency to various purposeless movements, and to the utterance of various sounds. We see this in our young children, in their loud laughter, clapping of hands, and jumping for joy." He remembered a remark by Leonard. "I heard a child, a little under four years old, when asked what was meant by being in good spirits, answer, 'It is laughing, talking, and kissing.' It would be difficult to give a truer and more practical definition."

Returning to his old interest in emotions and their expression, he took up the mystery of the links between the mind and the body, and things we do not understand about our feelings. "The feelings which are called tender are difficult to analyse; they seem to be compounded of affection, joy, and especially of sympathy. These feelings are in themselves of a pleasurable nature, excepting when pity is too deep . . . They are remarkable under our present point of view from so readily exciting the secretion of tears." His eyes still moistened when he thought of Annie. Why?

Charles described how once on a railway journey he had watched "an old lady with a comfortable but absorbed expression" sitting opposite him in the carriage. As he was looking at her, he saw that her *depressores anguli oris,* the muscles that pulled down the corners of her mouth, "became very slightly, yet decidedly, contracted; but as her countenance remained as placid as ever, I reflected how meaningless was this contraction, and how easily one might be deceived. The thought had hardly occurred to me when I saw that her eyes suddenly became suffused with tears almost to overflowing, and her whole countenance fell. There could now be no doubt that some painful recollection, perhaps that of a long-lost child, was passing through her mind. As soon as her sensorium was thus affected, certain nerve-cells from long habit instantly transmitted an order to all the respiratory muscles, and to those round the mouth, to prepare for a fit of crying. But the order was countermanded by the will, or rather by a later acquired habit, and all the muscles were obedient, excepting in a slight degree the *depressores anguli oris.* The mouth was not even opened; the respiration was not hurried; and no muscle was affected except those which drew down the corners of the mouth . . . In this case, as well as in many others, the links are indeed wonderful which connect cause and effect in giving rise to various expressions on the human countenance; and they explain to us the meaning of certain movements, which we involuntarily and unconsciously perform, whenever certain transitory emotions pass through our minds." Charles watched the lady opposite him with clinical attention to her *depressores anguli oris,* and guessing at once when he saw her eyes moisten that she was thinking of a long-lost child, as he did so often himself. This curiosity and compassion, the detached observation sharpened and deepened by his own feeling, was the essence of his approach to the science of man.

Some psychologists were now looking to find a new basis for their science in human evolution, just as Charles had hoped at the end of *The Origin of Species.* In 1876, George Croom Robertson, a young philosopher at University College London, started a periodical called *Mind: A Quarterly Review of Psychology and Philosophy.* The prospectus declared that psychology, "while drawing its fundamental data from subjective consciousness," would be understood in the widest sense, as covering all related lines of objective inquiry including physiology of the nervous system, anthropology, comparative psychology and "mind as exhibited in animals generally." Croom Robertson wrote to Charles about the plan; Charles was interested and offered his support.

In the third issue, Frederick Pollock, a young philosopher of law, wrote on evolution and ethics. "We are not content with saying that the [moral] faculty came from somewhere; we must seek to understand where it came from, and the nature of the process by which it was developed: and this is the knowledge of which Mr. Darwin has laid the foundations in his work on the *Descent of Man* . . . The theory of evolution furnishes us with a far more complete account than we had before of the whole genesis of the feelings which go to make up the Ethical Sanction, and leads to an explanation of one important set of the elements concerned, namely the sympathetic and social instincts, of which there was formerly no explanation at all."

Charles read through the first five issues without registering any special interest. In the sixth, he found an article by Hippolyte Taine, the

French historian and critic who had proposed a method for the scientific study of human personality in his treatise *On Intelligence.* Taine described the stages by which an infant girl developed and learnt to speak. He drew an analogy between the child's successive states and the phases of primitive civilisation, referring to the idea that development of the individual "recapitulated" the evolution of the species. Charles thought at once of Willy and Annie in Macaw Cottage; he found his old white vellum notebook and looked again through his observations of their first years. His chief interest at the time had been in expression, but he now read through his notes and drew out details for a sketch of the development of "the several faculties." He sent it to Croom Robertson, aware that his personal feelings might have influenced his judgement of the paper's value. "I hope that you will read it in an extra critical spirit, as I cannot judge whether it is worth publishing from having been so much interested in watching the dawn of the several faculties in my own infant." Croom Robertson decided it was worth publishing and put it into the eighth issue as "A Biographical Sketch of an Infant."

Charles first described the reflex actions he had noted during Willy and Annie's first weeks. Sneezing, hiccuping, yawning, stretching "and, of course, sucking and screaming" were well performed during the first seven days. Charles commented on the immediate perfection of Willy's reflex movements and the extreme imperfection of his voluntary ones in the first few days. He described Emma's offering her breast to Willy in precise detail. "At the age of thirty two days he perceived his mother's bosom when three or four inches from it, as was shown by the protrusion of his lips and his eyes becoming fixed . . . he certainly had not touched the bosom. Whether he was guided through smell or the sensation of warmth or through association with the position in which he was held, I do not at all know."

Thirty-five years after Annie's first smile at eight weeks, Charles recalled it clearly and suggested that it was a "true smile, indicative of pleasure," because her "eyes brightened" and her eyelids were slightly closed. The smile arose when Annie looked at her mother, and was "therefore probably of mental origin." On the power of reasoning, Charles felt that the facility with which Willy linked ideas was by far the most strongly marked of all the human infant's distinctions from animals. "What a contrast does the mind of an infant present to that of the pike, described by Professor Mobius, who during three whole months dashed and stunned himself against a glass partition which separated him from some minnows." But Charles was happy to give Annie second place to Jenny the orang. "Another of my infants, a little girl, when exactly a year old . . . seemed quite perplexed at the image of a person in a mirror approaching her from behind. The higher apes which I tried with a small looking glass behaved differently; they placed their hands behind the glass, and in doing so showed their sense, but far from taking pleasure in looking at themselves they got angry and would look no more."

Charles had put the underlying question of the "Biographical Sketch" in *The Descent of Man*. "At what age does the new-born infant possess the power of abstraction, or become self-conscious, or reflect on its own existence?" We cannot answer for the infant, nor can we answer for different animals on the "ascending organic scale." While Taine had suggested links between the development of a human infant and primitive societies, Charles in his sketch covered all human awareness. He set out "the probable steps and means by which the several mental and moral faculties of man have been gradually evolved." This evolution of mankind must at least be possible, he argued, "for we daily see these faculties developing in every infant." Reading his "natural history of babies" again, remembering how his first two infants had grown into small children by imperceptible steps, Charles was tracing the pattern every parent watches, wondering how it comes about. Watching Willy and Annie in their first years, he had seen the emergence of human nature—how our ancestors became what we are.

CHAPTER SIXTEEN

# TOUCHING HUMBLE THINGS

*Memories—Wordsworth—Man and ape—Earthworms—Anne Ritchie —Charles's death—Emma's widowhood—Annie's writing case*

IN 1876 CHARLES WROTE A MEMOIR for his children "as if I were a dead man in another world looking back at my own life." He thought back to their childhood. "When you were very young it was my delight to play with you all, and I think with a sigh that such days can never return." He touched briefly on their "one very severe grief" in Annie's death. He would say nothing more about her because he had written a "short sketch" of her character after her death, but "Tears still sometimes come into my eyes, when I think of her sweet ways."

In the years since *The Origin of Species,* Charles had written a sequence of books each identifying new features of pattern and process in natural life, and offering powerful explanations in terms of natural selection, but many of the ideas he put forward raised more questions. In *The Effects of Cross and Self Fertilisation in the Vegetable Kingdom,* which appeared in 1876, he described some features of fertility and sterility in plants for which he could not conceive any explanation. "And so it is with many other facts, which are so obscure that we stand in awe before the mystery of life." He was still perplexed by the emergence of order and beauty out of the struggle for existence. With his own experience of Annie's death, his understanding of what loss meant to others and his awareness of struggle and pain throughout the natural world, he could not make sense even of the idea of a remote God working through universal laws.

In his *Autobiography*, Charles tried to decide whether there was "more of misery or of happiness" in the life of all sentient beings, "whether the world as a whole is a good or a bad one." His reply to the question was halting and flat, and eloquent in its weakness. "According to my judgement, happiness decidedly prevails, though this would be very difficult to prove." He suggested that if every member of a species were to suffer greatly and habitually, they would "neglect to propagate their kind; but we have no reason to believe that this has ever or at least often occurred." Pain and pleasure were both motives to action which would contribute to the survival of the species, but while repeated pain depressed the victim, pleasure was a stimulus. In this way, natural selection had made pleasure the main guide to behaviour, as for instance with the feelings "derived from sociability and from loving our families." He concluded: "The sum of such pleasures as these, which are habitual or frequently recurrent, give, as I can hardly doubt, to most sentient beings an excess of happiness over misery, although many occasionally suffer much."

This argument was obviously of little use in supporting a belief in an all-powerful and beneficent God. Thinking perhaps of Snow Wedgwood's article about his theory, and Emma's effort to persuade herself that "all suffering and illness is meant to help us exalt our minds," Charles wrote that some had tried to explain suffering "in reference to man by imagining that it serves for his moral improvement." But, he insisted, even if suffering had a moral purpose for mankind, it had no value for other creatures. "The number of men in the world is as nothing compared with that of all other sentient beings, and these often suffer greatly without any moral improvement." Charles wrote with sudden vehemence: "It revolts our understanding" to suppose that God denied his benevolence to animals. "What advantage can there be in the sufferings of the millions of lower animals throughout almost endless time?'

Charles acknowledged that the argument was a very old one, but felt it was strong. David Hume had put the point in his *Dialogues Concerning Natural Religion.* His character Philo said: "Epicurus's old questions are yet unanswered. Is [God] willing to prevent evil, but not able? Then is he impotent. Is he able, but not willing? Then is he malevolent. Is he both able and willing? Whence then is evil?" Philo had also anticipated Charles's concern about the natural world. "Look round this universe. What an immense profusion of beings, animated and organised, sensible and active! You admire this prodigious variety and fecundity. But inspect a little more narrowly these living existences, the only beings worth regarding. How hostile and destructive to each other! How insufficient all of them for their own happiness! How contemptible or odious to the spectator! The whole presents nothing but the idea of a blind nature, impregnated by a great vivifying principle, and pouring forth from her lap, without discernment or parental care, her maimed and abortive children."

Charles's final suggestion on the matter returned to his starting point. While the incidence of pain throughout natural life could not be reconciled with any claim that God was universally benevolent, "such suffering is quite compatible with the belief in Natural Selection, which is not perfect in its action, but tends only to render each species as successful as possible in the battle for life with other species, in wonderfully complex and changing circumstances."

He came back also to an underlying question—in a sense the most radical of all those that he asked himself—whether humans could hope ever to understand these deepest issues. David Hume had suggested in his *Dialogues Concerning Natural Religion* that we might believe the world to have been created by a being with intelligence and a purpose simply because that happens to be how we as humans act and understand each other's actions. This might then be another "anthropomorphic" guess, like our presumption that animals had human feelings.

Spiders, on the other hand, might believe that an "infinite spider" had spun the world from his bowels. "Why an orderly system may not be spun from the belly as well as from the brain, it will be difficult . . . to give a satisfactory reason." Charles had felt in the 1860s that the issue of order in the natural world was "too profound for the human intellect," and "a dog might as well speculate on the mind of Newton." He now took up Hume's radical concern; he cast it in his own terms of human descent from animal origins, and applied it to the fundamental issue. "Can the mind of man, which has, as I fully believe, been developed from a mind as low as that possessed by the lowest animal, be trusted when it draws such grand conclusions? May not these be the result of the connection between cause and effect which strikes us as a necessary one, but probably depends merely on inherited experience?"

In the late summer of 1876, Francis's young wife Amy bore him a son, but she died a week later of puerperal fever. Charles and Emma had been looking forward eagerly to the birth of their first grandchild, but were deeply shaken by Amy's death. Charles turned at once to Hooker as Hooker had done before to him. "I saw her expire at 7 o'clock this morning . . . My dear old friend, I know that you will forgive my pouring out my grief. Yours affectionately, Ch. Darwin." Francis came to live with his parents, and Emma, now aged sixty-nine, found herself in charge of the baby, who was called Bernard. Charles took to him at once, and Emma wrote to Etty: "We think he is a sort of Grand Lama, he is so solemn."

A seventeen-year-old girl, Harriet Irvine, came to the household as a wet-nurse for the baby. She was the daughter of a farmworker in a village nearby; she had gone into service, been seduced by her employer, and given birth to an illegitimate daughter just before Bernard was born. She came with "a good breast of milk" as the advertisements offered. Most wet-nurses stayed only as long as their milk was needed, but the family became attached to Harriet, and she remained with them for the rest of her working life. Charles and Emma's granddaughter Gwen remembered "her rich voice and lovely laugh and strong Kentish accent." "She knew she was beautiful, I am sure, for she wore a black velvet ribbon round her neck, like any duchess." Bernard recalled her liveliness and her "laugh not to be quelled, that rang through the house, sometimes penetrating the dining room at inappropriate moments when solemn people had come to luncheon."

By the late 1870s, Charles was widely respected in scientific circles, but many people in the wider world could not accept what his ideas meant for human nature. They joked about them in the popular press, and he was caricatured again and again as a man-monkey. When Cambridge University gave him an honorary degree in November 1877, he was greatly pleased, but he was also embarrassed when undergraduates in the Senate House turned the ceremony into a music-hall spectacle. My great-grandfather John Neville Keynes was then a young fellow of Pembroke College, and wrote in his diary: "The building was crammed, floor and galleries, the undergraduates being chiefly in the galleries; and it was of course an occasion on which undergraduate wit felt bound to distinguish itself. The chief pleasantry consisted of a monkey swung across by strings from gallery to gallery." There was a "ceaseless fire of interruptions (chiefly feeble)" during the long Latin oration, and when the public orator used the Latin word "apes" "the cheering was enormous . . . Darwin bore himself in a rather trying position with remarkable dignity, but I heard afterwards that his hand shook so much when he was signing the register that his signature was scarcely legible." Charles yearned for public approval, and for people to take up and follow through his arguments wherever they might

*Caricature in* The Hornet

yield insights. The tremor in his signature is like the trace of a recording device, a reading of his distress that many people still would not accept his ideas about man's common nature with animals, and chose to avoid their implications by caricature and clowning. The deep truth that he had had the nerve and imagination to recognise in his children and himself, many others could only confront with ridicule.

Tituli Graduum Honoris Causa

George Mackenzie Bacon (5 Nov. 1877) A.M.

Charles Robert Darwin (Christi Coll.) (17 Nov. 1877) L.L.D.

John Morris M.A. 6 June 1878

*Cambridge University Register of Honorary Degrees*

People wrote repeatedly for his views on the great questions about life, death and the afterlife. In May 1879, a Russian diplomat asked for his advice as a man of science on Christ and immortality. Charles spoke firmly about the Christian Revelation. "Science has nothing to do with Christ, except in so far as the habit of scientific research makes a man cautious in admitting evidence. For myself, I do not believe that there has ever been any Revelation." He was non-committal on a future life, but commented that every man must judge for himself between "conflicting vague probabilities." He replied to another inquirer: "My judgement often fluctuates." But he had never denied the existence of a God. "I think that generally, and more and more as I grow older, but not always, that an agnostic would be the most correct description of my state of mind."

. . .

Charles had mentioned to Hooker in 1869 that he would very much like to hear Handel's *Messiah* again, but he was afraid to try it. "I dare say I should find my soul too dried up to appreciate it as in old days; and then I should feel very flat." He felt he was "a withered leaf for every subject except Science" and added: "It sometimes makes me hate Science, though God knows I ought to be thankful for such a perennial interest." This unhappiness about the loss of his former pleasures deepened as the years passed. Etty remembered that in his last years he used to say often that "attending so much to one subject had dried up his soul." He wrote in his *Autobiography* of his "curious and lamentable loss of the higher aesthetic tastes"; he found that he "could not endure to read a line of poetry," and felt that his mind had become "a kind of machine for grinding general laws out of large collections of facts." He often praised George Eliot's novel *Silas Marner* to friends, and may have noticed her comment on Silas the weaver as he had been before the foundling Eppie came into his home. "His life had reduced itself to the functions of weaving and hoarding . . . The same sort of process has perhaps been undergone by wiser men, when they have been cut off from faith and love—only instead of a loom and a heap of guineas, they have had some erudite research, some ingenious project, or some well-knit theory."

Charles wrote in his *Autobiography:* "If I had to live my life again, I would have made a rule to read some poetry and listen to some music at least once every week." If he had done that, "Perhaps the parts of my brain now atrophied could thus have been kept active through use." Reflecting Wordsworth's view of the value of imaginative writing for sympathy and moral understanding, he explained that the loss of the taste for poetry was "a loss of happiness, and may possibly be injurious to the intellect, and more probably to the moral character, by enfeebling the emotional part of our nature."

In the autumn of 1879, Emma persuaded Charles to leave the settled routine of their life at Down and take a holiday with the family in the Lake District. Before they went, he decided that he would try reading some poetry again while he was there. "The place will be propitious." Visiting the landscapes he had known in Wordsworth's poems lifted his spirits and the holiday was a great success. Etty, who was with her parents, remembered their expedition to Grasmere where Wordsworth had lived. "A perfect day and his state of vivid enjoyment and flow of spirits is a picture in my mind . . . He could hardly sit still in the carriage for turning round and getting up to admire the view from each fresh point." He still had his old Wordsworth "marked with his notes as to what to skip and what he cared for." He reread *The Excursion* during the stay but "found parts of it preachy."

One poem which Charles had marked when he first read it in 1841 now had an echo in his own feelings. Wordsworth's daughter Catherine, the one who at three was "tractable, though wild," had died suddenly shortly afterwards. Some time later Wordsworth wrote a sonnet about thinking of her.

*Surprised by joy—impatient as the Wind*
*I turned to share the transport—Oh! With whom*
*But Thee, deep buried in the silent tomb,*
*That spot which no vicissitude can find?*
*Love, faithful love, recalled thee to my mind—*
*But how could I forget thee? Through what power,*
*Even for the least division of an hour,*
*Have I been so beguiled as to be blind*
*To my most grievous loss?—That thought's return*
*Was the worst pang that sorrow ever bore,*
*Save one, one only, when I stood forlorn,*
*Knowing my heart's best treasure was no more;*

*That neither present time, nor years unborn*
*Could to my sight that heavenly face restore.*

Again and again, as Charles talked and read, he was reminded of Annie. In January 1880, he received the first volume of the French entomologist Jean-Henri Fabre's *Souvenirs entomologiques.* Fabre's life as a teacher in Provence had been impoverished and hard-working, but his accounts of insects' lives were recognised as masterpieces of minute observation. Fabre was a devout Catholic and had strong religious objections to Charles's suggestion that man and animals shared a common nature. In the book, he objected to Charles's theory of common descent, and criticised a claim by Charles's grandfather Erasmus in his *Zoonomia* that a wasp he had observed dismembering a bee's carcase had shown intelligence rather than instinct. Charles wrote to Fabre, praising him for his observations but giving his grounds for believing that the wasp's actions had shown signs of reasoning.

Fabre had been helped in his work for many years by his son, but the son had died before the book was published. Fabre ended the last chapter with a moving expression of his loss and a dedication to his son's memory. Charles wrote at the end of his letter to Fabre: "Permit me to add, that when I read the last sentence in your book, I sympathised deeply with you."

For his final contribution to natural science, Charles returned to a subject he had first tackled over forty years before. While staying with the Wedgwoods at Maer Hall in Staffordshire shortly after his return from the *Beagle* voyage, he had been fascinated to see how earthworms worked the soil. He had written a paper "on the formation of mould," and paid attention to worms ever since. In 1877, with his son Horace's help, he designed a "wormograph" to measure the effect of worms' excavations on the level of soil in the garden. For many months he kept pots filled with earth in his study and placed worms in them to observe their behaviour. In 1880, Emma wrote to Leonard that he had "taken to taming earthworms," but he "does not make much progress as they can neither see nor hear." Charles set out the results of his work in *The Formation of Vegetable Mould Through the Action of Worms,* which was published the next year. With its reflections on the "mental qualities" of worms and their unrecognised role in the history of the world, the book was his last flourish of insight and irony about human arrogance and the value of creatures other people despised. When it appeared, it sold thousands of copies and was widely and warmly praised, but after he had finished it, Charles wrote to Hooker: "I am rather despondent about myself . . . Idleness is downright misery to me . . . I have not the heart or strength at my age to begin any investigation lasting years, which is the only thing which I enjoy, and I have no little jobs which I can do. So I must look forward to Downe graveyard as the sweetest place on this earth."

Charles continued to puzzle about order and struggle in the natural world and the idea of a First Cause. He could not make up his mind, and did not pretend to anyone that he had an answer. During a visit to London he called on the Duke of Argyll, the leading Liberal politician and scientist who had argued against the evolution of man and for Divine Creation. Charles found him modest and easy to talk to, even though they disagreed on most subjects. After talking about politics and foreign affairs, the duke drew him back to the question of God's providence in creation, suggesting that it was impossible to look on some of the wonderful contrivances in nature without seeing that they were the effect and expression of a creative mind. The duke wrote later that he would never forget Darwin's answer to his point. "He looked at me very hard, and said 'Well, that often comes over me with overwhelming force; but at other times,' and he shook his head vaguely, adding 'it seems to go away.' " Charles had as full an understanding as anyone then living of "the wonderful contrivances" in the natural world; he often felt that they must be "the effect and expression of mind," but then "he shook his head vaguely . . ." The sense of a mind behind the order of life faded like the grin of the Cheshire Cat.

Shortly afterwards, Charles mentioned to another acquaintance the uncertainty he shared with David Hume about human understanding. He wrote of his "inward conviction . . . that the Universe is not the result of chance." "But then with me the horrid doubt always arises whether the convictions of man's mind, which has been developed from the mind of the lower animals, are of any value or at all trustworthy. Would anyone trust in the convictions of a monkey's mind, if there are any convictions in such a mind?"

Later in the year, Charles's brother Erasmus died and his body was brought to the graveyard at Downe for a family burial by the eighty-five-year-old Allen Wedgwood, who had married Charles and Emma at Maer and then christened Annie there so many years before. With Emma beside him at the graveside, Charles saw his own end ahead. He understood Emma's hopes for reunion with her loved ones in a life after death, but had no such hopes himself. Francis remembered his father "standing in a long black funeral cloak in the scattering of snow that fell, with a grave look of sad reverie." Hooker wrote to Charles shortly afterwards expressing "heartfelt sympathy." He suggested it was better to lose a young person than an old one because with an old person you knew better the value of what you had lost. Charles thanked him for his note but said he could not quite agree. Touching carefully on the losses he and Hooker had shared over the years, he wrote that the death of a young person, "when there is a bright future ahead, causes grief never to be wholly obliterated."

He remained preoccupied by the "horrid doubt." When Snow Wedgwood came to stay for the last time before his death, he came up to her in the dining room "quite abruptly," as she remembered later, "and began without any preface, in a way as if the subject had been much in his mind. 'The reason that I can never give in to the belief that we are all naturally inclined to, of a First Cause . . . is [that] I look on all human feeling as traceable to some germ in the animals.'" Faced with the paradox of purposeless pain at the heart of all life, and looking forward to his own death, Charles saw again the force of Hume's philosophical doubt about the ability of human thought to encompass such issues, and decided that the solution of the riddle must lie beyond our mental reach. His words as Snow remembered them were emphatic. We are "all naturally inclined to" the belief in a First Cause, but he could "never give in" to it. Coming to the end of the long chain of thought that he had followed from his first speculations about human nature over forty years before, he recognised again that the human brain was not a perfect instrument for finding essential truths. His theory suggested that, like other organs of the body, it has been evolved for human survival through a long history of piecemeal adaptations. Our understanding is built out of our animal past and serves only to meet the needs of the species as conditions change. Human reason is a powerful tool and should be used to the full to reveal all it can about the hidden workings of natural life, but we find it has limits, and when we reach the boundaries of present understanding, we should venture beyond only with great care.

In September 1881, the German materialist Ludwig Büchner visited Charles at Down with his young British supporter, Dr. Edward Aveling, whom George Bernard Shaw described as having "a voice like a euphonium" and "the face and eyes of a lizard." Aveling was struck by Charles's easy, frank and unassuming manner, and wrote of the meeting that Darwin was "a man intensely human whose being near you made your own life more intense." Charles asked Büchner and Aveling about their declared atheism and said that he preferred the word "agnostic."

After explanations and comment, Charles suggested that much energy was wasted in argument about the idea of God and the supernatural. "Man had so little time, so much strength at his disposal." While there was work to be done on earth and for humanity, while nature still held so many of her secrets, the effort devoted to aims other than natural could be put to better use.

In its Almanack for 1882, *Punch* portrayed Mr. Darwin to its readers as the culmination of evolution with the caption "Man is but a worm." The caricature showed Charles the change his ideas were leading to in people's thinking about human nature. The "Lord of Creation" had been set down; humans could now acknowledge their ancestors among the lowest orders of animal life.

The following January, Charles and Emma were reminded of the life Annie might have lived when her nurse Brodie's "other Anny," William Makepeace Thackeray's daughter, came to stay. The Darwins and Thackerays had kept in touch in the 1850s and 1860s when Brodie came from Scotland to visit both families. Anny Thackeray had married a civil servant in the India Office and was now Mrs. Anne Thackeray Ritchie, a novelist and essayist and a close friend of Tennyson, Browning, Edward Fitzgerald and Thomas Carlyle.

The Ritchies had made a mistake about the invitation and they came a week before they were expected. "We drove to the door; the butler hospitably said 'Mr. and Mrs. Darwin are sure to want you to remain; pray don't go.' " Charles and Emma invited them in, and Charles said: "You're as welcome as can be, and you must forgive me for laughing. I can't for the life of me help laughing." A dinner was scraped together and Mrs. Evans, the cook who had been a servant in the household for almost forty years, "thought it quite a providence that she had a pigeon pie."

During the Ritchies' stay, they talked with Charles and Emma about Brodie, who had lived to the age of eighty-three in a tenement in Aberdeen. They also spoke about fairy tales. Anne Ritchie had published her own tellings of *Sleeping Beauty, Cinderella, Beauty and the Beast, Little Red Riding Hood* and *Jack the Giant Killer,* and Emma showed her *The Bird Talisman* from Annie and Etty's childhood. Anne Ritchie read it through and admired it so much that Emma decided to have it printed again for the family. They also talked about children's games and the Darwin staircase slide. When the Ritchies returned home to Wimbledon, they found that Charles had had one made and delivered as a present for their small son and daughter.

If Anne Ritchie talked with Emma and Charles about her childhood with Brodie, she may have mentioned the family home in Kensington when it was still a village outside London. "I can remember the tortoise belonging to the boys next door crawling along the top of the wall where they had set it, and making its way between the jessamine sprigs . . . I liked the top schoolroom the best of the rooms in the dear old house; the sky was in it and the evening bells used to ring into it across the garden and seemed to come in dancing and clanging with the sunset . . . We kept our dolls, our bricks and our books, our baby houses in the top room, and most of our stupid little fancies. My sister had a menagerie of snails and flies in the sunny window sill; these latter chiefly invalids rescued out of milk jugs, lay upon rose leaves in various little pots and receptacles." She also remembered her waking dreams. "I used to dream a great deal when I was a little child, and then wake up in my creaking wooden bed, and stare at the dim floating night-light like a little ship on its sea of oil. Then from the dark corners of the room there used to come all sorts of strange things sailing up upon the darkness. I could see them all looking like painted pictures. There were flowers, birds, dolls, toys, shining things of every description."

Charles told her about his travels with Captain FitzRoy on their own "little ship," HMS *Beagle.* "He told us about birds, he told us about fishes, how the little hen starlings lead off and seem to know the way when the time of migrations arrives, and he told us about the tortoises in the Island of Ascension hatched from the eggs in the sand and starting off and plunging into the sea. 'And by Jove,' says he, 'the little tortoises without compass or experience sail straight across by nearest way to Algiers; it's perfectly wonderful.' "

Emma and Charles also shared one of Anne Ritchie's loves, for the letters of Madame de Sévigné in Louis XIV's Paris. In 1881, Mrs. Ritchie had published a short life of the writer with many of her letters. Charles and Emma had enjoyed reading it and he told Mrs. Ritchie at breakfast that when they were all young, they knew Madame de Sévigné's world so well that they gave all their friends the names of her characters. Charles had once written to Emma: "I am in love with Madame de Sévigné. She only shams a little virtue."

During the morning, the Ritchies walked with Charles in the garden and "he showed us his worms which had just been turned out of the study after a course of french horn." The year before Charles had wanted to establish whether earthworms possessed a sense of hearing. They had taken "not the least notice of the shrill notes from a metal whistle which was repeatedly sounded near them; nor did they of the deepest and loudest tones of a bassoon." They were "indifferent to shouts" and when a piano was played "as loudly as possible, they remained perfectly quiet." From which, Charles concluded: "Worms do not possess any sense of hearing."

Charles enjoyed the Ritchies' company and warmed to his Annie's namesake. Anne Ritchie felt his affection and returned it eagerly. Emma wrote to Etty: "Mrs. Ritchie was so effusive when she went away with the tears in her eyes, that I felt I could not properly respond."

A few weeks later, a schoolteacher in Liverpool wrote to Charles asking for his views on the origin of life and its bearing on the existence of God. The schoolteacher suggested that "If we deny the derivation of life from inorganic matter . . . the most probable alternative is the idea of an eternal or ever-living being filling all immensity with his presence, and breathing into the first animal the breath of life." Despite his weariness and depression, Charles wrote back as willingly and frankly as ever. "I hardly know what to say. Though no evidence worth anything has as yet, in my opinion, been advanced in favour of a living being being developed from inorganic matter, yet I cannot avoid believing the possibility of this will be proved some day in accordance with the law of continuity . . . If it is ever found that life can originate on this world, the vital phenomena will come under some general law of nature. Whether the existence of a conscious God can be proved from the existence of the so-called laws of nature . . . is a perplexing subject, on which I have often thought, but cannot see my way clearly."

One day in early March, Charles had an attack of angina in the Sand-walk. Etty wrote that "He was utterly ill after this, hardly sleeping all night, and for the next few days excessively depressed. My mother said he felt as if he had his death blow and that he did not expect to work again."

Etty came to stay with a friend from London. Laura Forster, aunt of the young E. M. Forster, was recovering from an illness; Etty remembered her father's "feeling anxious look as he came out of the study door to learn how she had borne the journey, and his warm sympathy and delight at our having got through it so well." A fortnight later, they were having "the most heavenly spring weather I ever knew in my life." Laura's health improved, and Charles was cheered by her progress. They took short walks in the garden, with Emma, now weakened by the pain of arthritis, in her bath chair. Laura wrote later: "Till I got there I did not realise how much he was changed in strength. Some of the days were sadly weary for him. I remember him coming into the drawing room one afternoon and saying 'The clocks go so dreadfully slowly, I have come in here to see if this one gets over the hours any quicker than the study one does.'"

Charles was still trying to experiment, looking at the effect of animal poisons on insect-eating plants. He wrote at the end of March to a physician who was an expert on the venomous snakes of India. "You will perhaps remember that you gave me some years ago a little cobra-poison for experimenting on *Drosera* [sundew]—Can you redouble your kindness by giving me ever so little of this or any other snake poison? Half a grain and even a quarter of a grain would probably suffice for an experiment which I am anxious to try." He wrote to a botanist who had sent him some insect-eating plants to experiment on. "I have roughly tried the effects of [carbonate] of ammonia on the chlorophyll grain, but I find stooping over the microscope affects my heart." The botanist had commented in his letter on another species, and Charles replied with almost his last words as a scientist. "The facts which you relate about the distribution of the *Mitella* are very curious; and how little we know about the life of any other plant or animal!"

Charles was now seventy-three and his heart was failing. On the night before he died, he woke Emma, saying: "I have got the pain, and I shall feel better, or bear it better, if you are awake." After another attack, he lost consciousness. When he awoke, Emma was with him, but she told Etty later that she was "not sure in how much suffering he was, as she thinks he felt it was his death pain and that he was resolved to bear it." He asked her to "Tell all my children to remember how good they have always been to me." He told her: "I am not the least afraid of death." After another acute attack, he said: "I was so sorry for you, but I could not help you." Emma told Etty later that she "hardly could say anything to him, she felt it so awful; only press his hand." She "felt as if she might break down utterly, but she externally kept her self-control completely." Charles was in great pain and Emma told Etty: "He was longing to die."

Etty and Francis arrived at the house during the morning. Charles was now suffering waves of nausea and fits of retching. He pleaded again and again, "If I could but die." Etty "gave him his salts or rubbed him, and once gave him a little pure whisky by his own desire. His hands were deathly cold and clammy, and Francis could not feel his pulse at all." They sat with him through the early afternoon. As Etty remembered, "He kept lifting his hands to hold his rope, and then they dropped off with a feeble quivering motion, and many times he called out 'Oh God,' 'Oh Lord God.' But only as exclamations of distress I think." Etty kept looking at the clock and felt that the hands never moved.

At about twenty-five minutes past three, Charles said he felt faint. They called for Emma and she came at once to find him grey and cold. He was soon unconscious and "there was the heavy stertorous breathing which precedes death. It was all over before four o'clock."

The next morning, Etty wrote to George: "Mother is very calm, but she has cried a little. You will come at once." She later wrote: "My mother was wonderfully calm from the very first, and perfectly natural. She came down to the drawing room to tea, and let herself be amused at some little thing, and smiled, almost laughed for a moment, as she would on any other day. To us, who knew how she had lived in his life, how she had shared almost every moment as it passed, her calmness and self-possession seemed wonderful then and are wonderful now to look back upon. She lived through her desolation alone, and she wished not to be thought about or considered, but to be left to rebuild her life as best she could and to think over her precious past."

Anne Thackeray Ritchie wrote to Etty when she heard. "No one can have any words—only hearts very full of love and reverence, and sympathy, that must flow from every home to yours . . . When I think of Mr. Darwin, so great, so gentle, so humorous, so beyond all little things and by his kind genius touching humble things and making them great, I feel that he is not gone, and will never die while there are men to look to him . . . Here is one great man to love indeed without fear, and to teach our children to look to and to live towards."

One evening a few weeks later, after the burial in Westminster Abbey, which Emma did not attend, she was wheeled in her bath chair to the Sand-walk to see the bluebells. She wrote to Etty that "it was all so pretty and bright, it gave me the saddest mixture of feelings, and I felt a sort of self-reproach that I could in a measure enjoy it." She had been reading over Charles's old letters. "I have not many, we were so seldom apart, and never I think for the last fifteen or twenty years." She called her small collection her "precious packet" and took it with her wherever she went. Among the letters were the ones he had written to her before their marriage when he was arranging Macaw Cottage and she was spending her last weeks with her parents at Maer; his letters from Shrewsbury when he was there with Willy in the first months of Annie's life; his daily accounts of the young children at Down when Emma was at Maer, and his letters from Malvern during the last week before Annie's death.

As she had done for Annie, Emma found other people's words to express her feelings. She kept a small notebook, her "book of extracts," into which she copied verses that she cared for. One poem she liked to read was Tennyson's *In Memoriam*. The poem was at once fraught and lyrical in its exploration of how faith is weakened by grief and doubt, and its reaching for consolation in hopes which lie beyond reason. Emma chose three verses.

*I know that this was Life, the track*
*Whereon with equal feet we fared;*
*And then, as now, the day prepared*
*The daily burden for the back.*

*But this it was that made me move*
*As light as carrier-birds in air;*
*I loved the weight I had to bear,*
*Because it needed help of Love:*

*Nor could I weary, heart or limb,*
*When mighty Love would cleave in twain*

*The lading of a single pain,*
*And part it, giving half to him.*

When Emma copied in the third verse with its words about the parting of "a single pain," she can only have been thinking of how she had shared with Charles their grief for Annie.

Emma took a house in Cambridge to be near George, Francis and Horace, who were all working there and starting families. Etty wrote that while her mother was in Cambridge, "Down and the past was always in the back of her mind, though she was happy in the present. She rejoiced in all old associations, even caring for the 'dear old azaleas' brought from Down, saying, 'I know their faces so well.' " Emma came back to Down every year in April or May, always wanting to be there "before the trees have become dark and summerlike." One year, she wrote to Etty: "It was a dismal black day on my arrival, but I was glad to wander about alone before the others came."

April, when both Annie and Charles had died, mixed "memory and desire" as T. S. Eliot was to write in *The Waste Land* thirty-five years later. Emma wrote to Etty in 1887 about Charles's death, "I do not find that the day of the month makes the anniversary with me, but the look out of doors, the flowers, and the sort of weather." In 1893, she was reading Tennyson's *In Memoriam* again.

*Is it, then, regret for buried time*
*That keenlier in sweet April wakes,*
*And meets the year, and gives and takes*
*The colours of the crescent prime?*

*Not all: the songs, the stirring air,*
*The life re-orient out of dust,*

*Cry thro' the sense to hearten trust*
*In that which made the world so fair.*

Emma copied the next two verses into her book of extracts.

*Not all regret: the face will shine*
*Upon me, while I muse alone;*
*And that dear voice I once have known,*
*Still speak to me of me and mine:*

*Yet less of sorrow lives in me*
*For days of happy commune dead;*
*Less yearning for the friendship fled,*
*Than some strong bond which is to be.*

Here in the last line were the words for the hope which meant everything to her.

In her last years, Emma went often to places in the neighbourhood where she had been with Charles and the children. She saw the gnarled beeches in a wood near Orchis Bank where the children used to climb, and went often to the terrace in the valley below the Sand-walk copse. She had a special feeling about the terrace with its undergrowth of sloes, traveller's joy, little yellow rock-roses, ladies' fingers and harebells. When she sat there, she remembered how Charles would pace to and fro, and she would sit on the dry chalky bank waiting for him.

George brought his American wife Maud and their young children from Cambridge every summer. Gwen later remembered that when they came each year, "As soon as the door was opened, we smelt again the unmistakable cool, empty, country smell of the house, and we rushed all over the big, under-furnished rooms in an ecstasy of joy." After their breakfast in the nursery, the children would pay a round of calls on the grown-ups who were having breakfast in bed. They played on their grandmother's bed "with little tin pots and pans, called Pottikins and Pannikins, and then she gave us bits of liquorice out of her work-basket, cut up with her work-scissors." In June 1890, when Gwen was four, Emma wrote: "George took Gwenny a walk by Cudham Lodge to the Salt-Box and then along that ridge below. I saw her coming home perfectly fresh and laden with flowers and one strawberry. George said she had been in an ecstasy the whole way, and he looked full of enjoyment himself."

My grandmother Margaret, Gwen's younger sister, was a lively child. Once when the elderly Miss Thorley visited the family in Cambridge, Margaret was called into the garden to meet her. Miss Thorley threw up her arms in amazement when she came, and exclaimed how like Annie she was. One July day at Down when Margaret was five, George wrote to Maud in Cambridge: "Yesterday was a 'scorcher' as you say, really too much to do anything. I pottered round a little with the children in the morning and late in the afternoon took all three to Orchis Bank and Hangrove which they enjoyed immensely. We came back through Sand-walk and Henrietta and I (for we met her there) made bryony wreaths for them all . . . Mother seems below par and didn't come downstairs, although she sat in her room and looked at the children playing on the lawn. The children seem as happy as the day is long, and have all gone off to the Sand-walk now. I have promised them a bonfire one day." As Emma sat at her bedroom window watching her grandchildren play on the lawn below, she may have been thinking of her own children fifty years earlier. And she may have seen Annie for a moment, as Miss Thorley did.

In 1896, Emma was eighty-eight, and through the summer months all her children and grandchildren came to Down to be with her. The furniture around the piano was cleared again and she played the "galloping tune" for the young ones with her arthritic fingers. She went out often with the children, and was taken in her bath chair to some of the family's favourite places, which she had not visited for some years. One day in September, John the manservant wheeled her in her chair along the Cudham Lane. She wrote to Etty: "It looked ever so much deeper, with high hedges and trees grown. I came back over the big field and through the Smiths' yard. I felt the sharp wind over the bare field quite like an old friend."

Emma died early in the morning of the first Friday in October. She had been preparing to leave for Cambridge that day, and her death was "quick, quiet and unexpected." The funeral took place the following Wednesday. The family, neighbours and close friends came, and blinds were drawn in the village.

Forty-five years before, Emma had gathered her keepsakes of Annie and put them away for herself. When Etty found Annie's writing case after her mother's death, and saw her sister's things for the first time since their childhood, recollections of Annie came back to her with "strange vividness." The words Charles used in his memorial of Annie to catch his memories of her work in the same way. "She held herself upright, and often threw her head a little backwards, as if she defied the world in her joyousness."

# NOTES

The book is based on Darwin family papers and other items; the manuscripts in the Darwin Archive at Cambridge University Library; English Heritage's Darwin Collection at Down House; the Wedgwood/Mosley Collection in Keele University Library; the memoirs and letters published after Charles and Emma Darwin died; administrative records of the time including the registers of births and deaths and national census returns; local directories, guides and newspapers; the books and periodicals that the Darwins are known to have read; others which reflected the thinking of the time, and the wealth of recent Darwin scholarship.

The notes give the sources of all important quotations, and those for other points of interest which may not be easy to find in obvious places. I have given references for domestic details only where the points may be of particular interest for some reason.

A number of recent books and articles that I found particularly helpful are noted for further reading. The full story of Charles's life has been told in Adrian Desmond and James Moore's *Darwin* (London, 1991). The first part up to 1856 is also covered in Janet Browne's *Charles Darwin: Voyaging* (London, 1995). *Darwin on Evolution: The Development of the Theory of Natural Selection,* edited by Thomas Glick and David Kohn (Indianapolis, 1996), provides an excellent introduction to his writings, with full explanations of the notebooks of 1836–44 and other important texts that were not published in his lifetime.

I have normalised the spelling and punctuation of some quotations where the forms of the original would distract a general reader. In some other passages I have kept the original spellings and punctuation because they are part of the character of the text.

**Abbreviations**

*Autobiography* — *The Autobiography of Charles Darwin,* ed. Nora Barlow (London, 1969).

*Beagle Diary* — *Charles Darwin's Beagle Diary,* ed. Richard Darwin Keynes (Cambridge, 2001).

"Biographical Sketch" — "A biographical sketch of an infant," *Mind: A Quarterly Review of Psychology and Philosophy,* no. 7, July 1877, pp. 285–94. Reprinted in *The Collected Papers of Charles Darwin,* ed. Paul Barrett (London, 1977), pp. 191–200.

BL — British Library.

*Calendar* — *A Calendar of the Correspondence of Charles Darwin, 1821–1882* (Cambridge, 1994).

*CCD* — *The Correspondence of Charles Darwin* (Cambridge University Press, 1985–).

*CFL* (1904) — *Emma Darwin: A Century of Family Letters,* ed. Henrietta Litchfield (Cambridge, 1904). (This edition was privately printed and contains information and comment that was removed from the later edition.)

*CFL* (1915) — *Emma Darwin: A Century of Family Letters,* ed. Henrietta Litchfield (London, 1915).

CUL — Cambridge University Library.

DAR — Manuscript in the Cambridge University Library's Darwin archive.

*Descent* — *The Descent of Man* (London, 1888).

"Essay" — "Charles Darwin's Essay of 1844," in Charles Darwin and Alfred Russel Wallace, *Evolution by Natural Selection,* ed. Sir Gavin de Beer (Cambridge, 1958).

*Expression* — *The Expression of the Emotions in Man and Animals,* ed. Paul Ekman (London, 1998).

*Life and Letters* — *The Life and Letters of Charles Darwin,* ed. Francis Darwin (London, 1887).

*Marginalia* — *Charles Darwin's Marginalia Volume 1,* ed. Mario Di Gregorio and Nick Gill (New York, 1990).

*MLCD* — *More Letters of Charles Darwin,* ed. Francis Darwin and A. C. Seward (London, 1903).

*Natural Selection* — *Charles Darwin's Natural Selection,* ed. R. C. Stauffer (Cambridge, 1975).

*Notebooks* — *Charles Darwin's Notebooks 1836–1844,* ed. Paul Barrett, Peter Gautrey, Sandra Herbert, David Kohn and Sydney Smith (New York, 1987). The references in brackets after the page numbers indicate the notebook in question and the serial number.

| | |
|---|---|
| *Origin* | *The Origin of Species by Means of Natural Selection,* ed. J. W. Burrow (Harmondsworth, 1985). |
| *Voyage* (1839) | *The Voyage of the Beagle,* ed. Janet Browne and Michael Neve (Harmondsworth, 1989). |
| *Voyage* (1845) | *The Voyage of the Beagle,* intro. H. Graham Cannon (London, 1959). |

## Chapter One: Macaw Cottage

"This is the question."—*CCD,* 2.444.

He found a house—R. B. Freeman, *Darwin and Gower Street, An Exhibition in the Flaxman Gallery of the Library, University College London* (London, 1982).

The imposing new buildings—Ian Jenkins, " 'Athens Rising Near the Pole': London, Athens and the Idea of Freedom," in *London: World City 1800–1840,* ed. Celina Fox (London, 1992), pp. 143–54.

"the largest liberality of opinion"—*North London or University College Hospital. Anniversary Dinner in Aid of the Funds . . . April 12, 1864. Charles Dickens Esq. in the Chair* (London, 1864), pp. 4–5.

"Remnants of carpets"—Gordon Chancellor, "Charles Darwin's St Helena Model Notebook," *Bulletin of the British Museum* (*Natural History*), Historical Series, vol. 18, no. 2, (1990), pp. 218–19.

"If you pluck a branch"—John Hogg, *London as It is* (London, 1837), pp. 193–4.

a strange "wailing whistle"—Alan Jackson, *London's Termini* (London, 1985), p. 32.

The surgeon, Robert Liston—W. R. Merrington, *University College Hospital and its Medical School: A History* (London, 1976), pp. 26–35.

The minister, James Tagart—H. S. Perris, *A Sketch of the History of the Little Portland Street Chapel, London* (London, 1900), pp. 8–9. Dr Williams's Library has a copy of the booklet. Charles Dickens was another member of the congregation.

"the endurance of pain"—John Conquest, *Letters to a Mother, on the Management of Herself and her Children in Health and Disease* (London, 1852), p. 50.

11 "no sentiment is more pregnant"—John Conquest, *Letters to a Mother,* p. 39.

11 "In the case of a woman"—James Blundell, *The Principles and Practice of Obstetric Medicine* (London, 1840), Lecture by James Mackintosh at the School of Medicine, Argyle Square, Edinburgh, pp. 103–4.

12 There were four signs—Thomas Bull, *Hints to Mothers for the Management of Health during the Period of Pregnancy* (London, 1837), Chapter 2, "Of the mode by which pregnancy may be determined."

13 He became ill—Ralph Colp, *To be an Invalid: The Illness of Charles Darwin* (Chicago, 1977).

14 "Mrs Darwin is the youngest daughter"—Maria Edgeworth, *Letters from England 1813–1844,* ed. Christina Colvin (Oxford, 1971), p. 571.

15 "It may be called a fixed law of Nature"—Thomas Bull, *The Maternal Management of Children in Health and Disease* (London, 1848), p. 13.

16 "In all inflammatory ailments"—"Receipts and Memoranda" book in the English Heritage Darwin Collection at Down House, printed in Ralph Colp, *To be an Invalid,* Appendix A, pp. 147–67.

17 "They have freedom in their actions"—*CFL* (1915), 1.59.

## Chapter Two: Pterodactyl Pie

20 "Annie born"—*CCD,* 2.434.

21 Paley argued that if you found a watch—William Paley, *Natural Theology,* Chapter 1, in *The Works of William Paley, D. D.* (Edinburgh, 1837), pp. 435–7.

21 "In studying the conformation of fishes"—Isabella Beeton, *The Book of Household Management* (London, 1861), p. 106.

22 "Their name was legion"—D. Landsborough, *A Popular History of British Zoophytes or Corallines* (London, 1852), p. 55.

22 "the appointment of death"—William Buckland, *Geology and Mineralogy Considered with Reference to Natural Theology* (London, 1836), p. 133. Chapter 13, in which the pas-

sage appears, is entitled "Aggregate of animal enjoyment increased, and that of pain diminished, by the existence of carnivorous races."

man "is placed upon this earth"—William Swainson, *Preliminary Discourse on the Study of Natural History* (London, 1834), p. 112.

"Nine hundred species of intestinal worms"—Robert Grant, *An Essay on the Study of the Animal Kingdom . . . Being an Introductory Lecture Delivered in the University of London on the 23rd of October 1828* (London, 1828), p. 8.

"Can you conceive anything"—Benjamin Disraeli, *Vivian Grey* (London, 1826–27), pp. 315–16, quoted by J. M. I. Klaver, *Geology and Religious Sentiment: The Effect of Geological Discoveries on English Society and Literature between 1829 and 1859* (Leiden, 1997).

"the enlarged views both of time and space"—John Meadows Rodwell to Francis Darwin, CUL DAR 112, folio 94.

"What a capital hand is Sedgwick"—John Meadows Rodwell to Francis Darwin, CUL DAR 112, folio 94.

Herschel wrote of man as a "speculative being"—John Herschel, *A Preliminary Discourse on the Study of Natural Philosophy* (London, 1831), pp. 4, 7, 9, 42.

"one huge, dead, immeasurable steam-engine"—Thomas Carlyle, *Sartor Resartus,* ed. Mark Engel and Rodger Tarr (Berkeley, 2000), pp. 124, 189.

a week in August walking with Professor Sedgwick—James Secord, "The Discovery of a Vocation: Darwin's Early Geology," *British Journal for the History of Science,* 41, vol. 24 (1991), pp. 133–57.

Sedgwick discussed his suggestions—Letter from Charles to Professor Hughes in John Clark and Thomas Hughes, *The Life and Letters of the Reverend Adam Sedgwick* (Cambridge, 1890), 1.380–81.

Sedgwick also showed Charles—John Wyatt, *Wordsworth and the Geologists* (Cambridge, 1995), pp. 76–84.

"There is an intense and poetic interest"—Adam Sedgwick, *Addresses Delivered at the Anniversary Meetings of the Geological Society of London* (London, 1831), p. 26.

27 "No one has put forward nobler views"—William Wordsworth, *A Complete Guide to the Lakes, Comprising Minute Directions for the Tourist, with Mr. Wordsworth's Description of the Country &c. and Three Letters upon the Geology of the Lake District, by the Rev. Professor Sedgwick* (Kendal and London, 1842), p. 3.

27 As HMS *Beagle* sailed through the mid-Atlantic—*Beagle Diary,* pp. 22, 23, 42.

29 a "revolution in natural science"—*Origin,* p. 293.

29 During his years on HMS *Beagle*—*Beagle Diary,* pp. 59, 444.

30 Charles responded to the richness and variety—*Beagle Diary,* pp. 309, 444.

30 the "zoology of archipelagoes"—Nora Barlow, "Darwin's ornithological notes," *Bulletin of the British Museum* (*Natural History*), Historical Series, vol. 2 (1963), p. 262.

31 John Edmonston—R. B. Freeman, "Darwin's Negro bird-stuffer," *Notes and Records of the Royal Society,* vol. 33, 1978, pp. 83–6.

31 "All answered 'No.' "—*Autobiography,* p. 74.

32 A few weeks later—*Beagle Diary,* p. 58; *Voyage* (1845), p. 481.

32 The first Fuegians that Charles met—Nick Hazlewood, *Savage: The Life and Times of Jemmy Button* (London, 2000) gives a full account of Captain FitzRoy's taking of the Fuegians in 1830 and all that happened to them in England and back in Tierra del Fuego after then.

33 "without exception the most curious and interesting spectacle"—*Beagle Diary,* p. 122.

33 "an inherent delight in man"—*Beagle Diary,* p. 445.

34 two pet tortoises—Frank Sulloway, "Darwin's conversion: The *Beagle* voyage and its aftermath," *Journal of the History of Biology,* vol. 15, no. 3 (1982), p. 344.

35 "persistence of type" and "law of succession"—Janet Browne, *Charles Darwin: Voyaging* (London, 1995), p. 350.

The ornithologist John Gould—Frank Sulloway, "Darwin and his finches: The evolution of a legend," *Journal of the History of Biology,* vol. 15, no. 1 (1982), p. 21.

Charles kept notebooks—*Notebooks.*

Herschel had himself suggested—Letter to Lyell of 20 February 1836 quoted in Charles Babbage, *The Ninth Bridgewater Treatise* (London, 1838), pp. 203–4.

how he saw mankind in the scheme of things—Sandra Herbert, "The place of man in the development of Darwin's theory of transmutation," Parts I and II, *Journal of the History of Biology,* vol. 7, no. 2 (Fall 1974) and vol. 10, no. 2 (Fall 1977).

he now explored Wordsworth's writings—Edward Manier, *The Young Darwin and his Cultural Circle* (Dordrecht, 1978); Marilyn Gaull, "From Wordsworth to Darwin: 'On to the Fields of Praise,'" *The Wordsworth Circle,* 10 (1979) pp. 33–48; and Gillian Beer, *Darwin's Plots: Evolutionary Narrative in Darwin, George Eliot and Nineteenth-Century Fiction* (London, 1983), have dealt with aspects of Charles's reading of Wordsworth in these years.

"in the evening or on blowy days"—*CCD,* 2.440, in a note on recollections from childhood, linked with Charles's interest in the faculty of memory in Notebooks C (*Notebooks*, p. 315–C 242e) and M (*Notebooks,* p. 526–M 28).

Charles noticed Wordsworth's comments—*Notebooks,* p. 529 (M 40) "V. Wordsworth about science being sufficiently habitual to become poetical"; pp. 578–9 (N 57) "there are some notes . . . on Wordsworth's dissertation on Poetry." The passages in the "Preface" are on p. 881 of William Wordsworth, *The Poems,* ed. John Hayden (Harmondsworth, 1990). Charles's comments on landscapes and trees are on p. 529 of *Notebooks* (M 40–41).

Charles read Wordsworth's long poem *The Excursion*—*Autobiography,* p. 85.

the "innate repugnance, disgust, and abhorrence"—William Swainson, *On the Natural History and Classification of Quadrupeds* (London, 1835), p. 7.

The first chimpanzee to be exhibited—Henry Scherren, *The Zoological Society of London—A Sketch of its Foundation and Development* (London, 1905), pp. 59–60.

Mrs Lyell saw Tommy in 1835—Diary, 11 March 1838, commenting on Jenny the orang's expression.

42 The Zoological Society's veterinary surgeon—William Youatt, "Contributions to comparative pathology no. V: Intestinal fever and ulceration," *The Veterinarian,* vol. 9, no. 101, New Series, no. 41 (May 1836), pp. 271–82. The other two articles were "Account of the habits and illness of the late chimpanzee," *The Lancet,* vol. 2 (1835–36), pp. 202–6, and a piece in *The London Medical Gazette,* vol. 18, issue 440 (1836), pp. 214–16. Charles read through *The Veterinarian* in January 1842 (*CCD* 4.464).

43 "The personage who has lately arrived"—Reprinted in William Broderip, *Zoological Recreations* (London, 1847), p. 249.

43 Queen Victoria saw the second Jenny—Queen Victoria's journal for 1842, Royal Archives, quoted in Wilfrid Blunt, *The Ark in the Park: The Zoo in the Nineteenth Century* (London, 1976), p. 38.

44 Whewell praised Owen's achievements—*Proceedings of the Geological Society of London,* vol. 2 (1838), pp. 625–6, 642.

45 "pterodactyl pie"—Charles Lyell, *Life, Letters and Journals of Sir Charles Lyell,* Bart. (London, 1881), 2.39.

45 people "often talk" about the "wonderful event"—*Notebooks,* pp. 222–3 (B 207). Charles probably made this note shortly after Whewell's speech and had Whewell's comment in mind, because Whewell spoke on 16 February and the timings of an earlier and a later entry in the notebook fit closely. On B 199e Charles had referred to an article which had appeared in the *Athenaeum* of 10 February (*Notebooks,* pp. 220, 658). On B 235 he noted the title of an article which appeared in the *Athenaeum* of 24 February, possibly to read later (*Notebooks,* pp. 230, 679).

45 "speculated much about [the] existence of species"—*CCD,* 2.431.

46 "Animals whom we have made our slaves"—*Notebooks,* p. 228 (B 231).

46 David Hume—For Charles's reading of Hume in 1838 and 1839, see *Notebooks,* pp. 325, 545, 559 and 591–2 *(Enquiry Concerning Human Understanding); Notebooks,* p. 596 *(Dissertation on the Passions); Notebooks,* p. 627 *(Enquiry Concerning the Principles of Morals); CCD* 4.458 and *Notebooks,* pp. 591–2 *(Natural History of Religion and Dialogues Concerning Natural Religion).* One part of Hume's *Enquiry Concerning Human Understanding* that Charles noted for special attention was Section IX "Of the reason of animals." When Charles

wrote in *Notebooks,* p. 564, "Experience shows the problem of the mind cannot be solved by attacking the citadel itself," he may have been commenting on Hume's metaphor in the introduction to the *Treatise of Human Nature* "the only expedient [is] . . . instead of taking now and then a castle or village on the frontier, to march up directly to the capital or centre of these sciences, to human nature itself; which once being masters of, we may everywhere else hope for an easy victory" (*Treatise of Human Nature,* Harmondsworth, 1969, p. 43). Charles's page numbers for the passages he referred to in *Notebooks* pp. 591–2 correspond to those of the London, 1788 edition of Hume's *Essays and Treatises.* Charles's uncle Josiah Wedgwood II had a copy of that edition in the library at Maer, together with the *Treatise of Human Nature* (Sotheby's Auction Catalogue, 16 November 1846, p. 10). William Huntley wrote on Charles's reading of Hume in "David Hume and Charles Darwin," *Journal of the History of Ideas,* vol. 33, no. 3, (July–September 1972). Edward Manier discussed Hume and natural religion in *The Young Darwin and his Cultural Circle* (Dordrecht, 1978), pp. 86–8. For Hume's influence on Charles's later thinking, see pp. 62, 313, 323, 337 and 347 in this book.

"Such a sight has seldom been seen"—*CCD,* 2.80. In the same letter Charles wrote that Erasmus had been with Harriet Martineau "noon, morning and night" and she had "been as frisky lately [as] the rhinoceros."

"the whole fabric totters and falls"—*Notebooks,* pp. 263 (C 76), 264 (C 79), 300 (C 196).

## Chapter Three: Natural History of Babies

"Is insanity an unhealthy vividness of thought?"—*Marginalia,* p. 5.

"the accidental discovery"—John Abercrombie, *Inquiries Concerning the Intellectual Powers and the Investigation of Truth* (London, 1840), pp. 104–25.

a seven-page note—*CCD,* 2.438–42. See note on "in the evening" above.

"Therefore affections effect of organisation"—*Notebooks,* p. 525 (M 26).

"To avoid stating how far I believe"—*Notebooks,* pp. 532–3 (M 57).

"The possibility of the brain having whole trains of thoughts"—Herbert Mayo, *The Philosophy of Living* (London, 1837), p. 155; *Notebooks,* p. 538 (M80–81). Mayo's book was an essay on the "principles which contribute to the maintenance of health and the preservation of the body" (p. iii). In a chapter on diet, there is a section entitled "A first-rate dinner in England the best in the world."

51 "the religious sentiment"—*Autobiography,* p. 91.

52 "It is an argument for materialism"—*Notebooks,* p. 524 (M 19).

52 Charles questioned other historical parts—*Autobiography,* pp. 85–6.

52 Dr. Darwin suggested firmly—*Autobiography,* p. 95.

53 free will and oysters—*Notebooks,* p. 536 (M 72–3).

53 the moral sense was "an impress"—William Whewell, *Bridgewater Treatise III. On Astronomy and General Physics* (London, 1833), p. 267.

53 humans, "like deer"—*Notebooks,* p. 537 (M 76).

54 "Origin of man now proved"—*Notebooks,* p. 539 (M 84e).

54 "Do the Ourang Outang like smells"—*Notebooks,* p. 560 (M 156).

54 she "readily put it when guided"—*Notebooks,* p. 554 (M 139e).

55 she was "astonished beyond measure"—CUL DAR 119.1–2.

55 "Jenny understands, when told door open"—CUL DAR 119.1–2.

55 "The appearance of dejection"—*Expression,* p. 136.

56 "Natural history of babies."—*Notebooks,* p. 560 (M inside back cover).

57 "Pain and disease in world"—*Notebooks,* p. 636 (Macculloch 57r).

57 "analyse this out"—*Notebooks,* p. 558 (M 150).

58 If our motives were "originally mostly instinctive"—*Notebooks,* p. 608 (OUN 26).

58 his "principal object"—William Wordsworth, *The Poems,* ed. John Hayden (Harmondsworth, 1990), 1.869.

58 "the poorest poor"—Wordsworth, *The Poems* (Harmondsworth, 1990), 1.267.

"the commonplace truths of the human affections"—Letter from Wordsworth to Coleridge, 22 May 1815, *The Letters of William and Dorothy Wordsworth: The Middle Years* (Oxford, 1937), 2.669.

"probably the foundation of all that is most beautiful"—*Notebooks,* p. 409 (E 49).

her interleaved Bible, reading lists and notes for prayers—Emma's Bible is in the collection of the Reverend David Smith; her reading lists are in her diaries which are on deposit in Cambridge University Library, and her notes for prayers are in the Wedgwood/Mosley Collection in Keele University Library.

the "perfection of friendship"—Richard Whately, *A View of the Scripture Revelations Concerning a Future State* (London, 1830), p. 245.

"I think all melancholy thoughts"—*CCD,* 2.122–3.

"I believe from your account"—*CCD,* 2.169–70.

jottings in his metaphysical notebook—*Notebooks,* pp. 578–9 (N 57, 59, 60).

a long note on the moral sense—*Notebooks,* pp. 618–29 (OUN 42–55).

she was troubled again by her worries—*CCD,* 2.171–3.

"Your father's opinion"—*Autobiography,* p. 93.

"W. Erasmus Darwin born Dec 27th 1839"—*CCD* 4.410–33.

"anxious to observe accurately"—*Life and Letters,* 1.132.

a set of Wordsworth's *Poetical Works*—Stephen Gill, *Wordsworth and the Victorians* (Oxford, 1998), pp. 268–9. It was Edward Moxon's six-volume edition of 1840. Charles wrote on the title page of each volume, "Charles Darwin 1841." According to his reading notebooks, he finished the first two volumes on 13 March and the third volume on 7 May; he finished "some Wordsworth" on 26 September and the final volume on 17 February 1842 (*CCD,* 4.462–4). He marked many poems and passages and noted his judgements, ranging from "fine" and "beautiful" to "obscure," "wretchedly poor" and "stupidest." Emma also marked a number of poems. The set is in a private collection.

68 "like the beams of dawn"—"Address to My Infant Daughter, Dora on Being Reminded that She Was a Month Old that Day, September 16," William Wordsworth, *The Poems,* ed. John Hayden (Harmondsworth, 1990), 1.622. The poem is on pp. 74–6 of the second volume of the 1840 edition.

69 Annie's first smile—*CCD,* 4.412.

71 The Polytechnic and the Adelaide—Richard Altick, *The Shows of London* (London, 1978), pp. 377–89.

71 "It is a wonderful, mysterious operation."—Maria Edgeworth, *Letters from England 1813–1844,* ed. Christina Colvin (Oxford, 1971), pp. 593–4.

71 "The common remark"—Andrew Winter, "The Pencil of Nature," *The People's Journal,* vol. 2 (21 November 1846), pp. 288–9.

71 "Photographic Phenomena"—*George Cruikshank's Omnibus* (London, 1841), p. 31.

## Chapter Four: Young Crocodiles

73 a house in Woking—Jonathan Topham, "Charles Darwin of Woking? Emma Darwin's recollections of house-hunting," *Darwin College Magazine,* March 1997, pp. 50–54. The article reproduces and discusses her note in CUL DAR 251.

75 "The charm of the place to me"—"The General Aspect," English Heritage Darwin Collection at Down House. Part is transcribed in *MLCD,* pp. 33–6.

75 Dr. Mantell gave a vivid account—Gideon Mantell, *The Wonders of Geology* (London, 1838), 1.368–9.

75 Charles Lyell speculated—Charles Lyell, *Principles of Geology* (Harmondsworth, 1997), p. 67.

76 "to stand on the North Downs"—*Origin,* pp. 296–7.

77 Emma had read Thomas Carlyle's *Chartism*—*CFL* (1915), 2.52.

77 *The Times* had reported a confrontation—15 and 18 August 1842.

78 *The Illustrated London News*—27 August 1842.

*Little Robert and the Owl*—Mary Sherwood, *Little Robert and the Owl* (Wellington, 1828). *CCD,* 4.424.

"A few days later some of the copses"—"The General Aspect" as above.

"I had hoped (for experience I have none)"—*CCD,* 2.352.

All except Betty and Horace wrote accounts of their childhoods—MSS and family papers in CUL Darwin Archive, and recollections by William, Etty, Francis and Leonard in a number of books and articles.

*The Complete Servant*—Samuel and Sarah Adams, *The Complete Servant; Being a Practical Guide to the Peculiar Duties and Business of all Descriptions of Servants* (London, 1825), p. 6.

*Little Servant Maids*—Charlotte Adams, *Little Servant Maids* (London, 1851). The book was published by the Society for the Promotion of Christian Knowledge. The British Library's copy was destroyed by enemy action. There is one in the SPCK archive at Cambridge University Library.

Jessie Brodie . . . a tall, erect woman—*CFL* (1915), 2.86.

she had worked for the novelist—Lillian Shankman, Abigail Bloom and John Maynard, *Anne Thackeray Ritchie: Journals and Letters* (Columbus, 1994), Anne Thackeray Ritchie's journal for Laura Stephen, her sister's daughter, p. 197; Gordon Ray, *The Letters and Private Papers of William Makepeace Thackeray* (New York, 1945), 1.476–7, 478–9, 2.101, 2.193; Anne Thackeray Ritchie, *Biographical Introductions to the Works of William Makepeace Thackeray* (London, 1894–98), 5.xiii; Gordon Ray, *Thackeray, The Uses of Adversity* (New York, 1955), p. 303.

it was a pity Mr. Darwin had not something to do—Bernard Darwin, "On being a Darwin," in *Green Memories* (London, 1928), p. 22.

"Think how happy I was"—CUL DAR 219.8:27.

"I find fish will greedily eat"—*CCD,* 6.324.

he was now "almost convinced"—*CCD,* 4.2.

an essay, arguing each step carefully—"Essay"

94 it was "repugnant" to all our feelings—Henry Brougham, *Dissertations on Subjects of Science Connected with Natural Theology* (London, 1839), 2.62.

95 "the creation of a world so full of evil"—Henry Hallam, *Introduction to the Literature of Europe in the Fifteenth, Sixteenth and Seventeenth Centuries* (Paris, 1839), 4.94.

95 "Man acts on and is acted on"—*Notebooks,* p. 415 (E 65).

95 "To marvel at the extermination"—"Essay," pp. 167–8.

96 "Nature is cruel"—Matthew Arnold, "In Harmony with Nature," *Poetical Works,* ed. C. B. Tinker and H. F. Lowry (London, 1950), p. 5.

96 "I have just finished my sketch"—*CCD,* 3.43–5.

97 "A great assumption"—CUL DAR 113 folio 89.

98 "Loving she is, and tractable, though wild"—"Characteristics of a Child Three Years Old," William Wordsworth, *The Poems,* ed. John Hayden (Harmondsworth, 1990), 1.858–9.

98 He was very sad all the next day—Emma Wedgwood, *My First Reading Book,* CUL DAR 219.112.

98 a story book—*Cobwebs to Catch Flies* (London, 1837), pp. 22–3, 26–7.

100 The Darwins' neighbour, Louisa Nash—Louisa Nash, "Some memories of Charles Darwin," *Overland Monthly,* October 1890, p. 407.

101 "Everyone who has had much to do"—*Expression,* p. 239.

102 "Numbers of children were playing on the beach"—*Beagle Diary,* p. 367.

103 *Lamprotornis Burchellii*—Andrew Smith, *Illustrations of the Zoology of South Africa* (London, 1838–49), *Aves,* plate 47.

## Chapter Five: The Galloping Tune

105 "When we were young"—Louisa Nash, "Some memories of Charles Darwin," *Overland Monthly,* October 1890, p. 406.

the room "had a white painted floor"—Gwen Raverat, *Period Piece* (London, 1987), pp. 142–3.

"Children have an uncommon pleasure"—*Notebooks,* p. 582 (N 66).

"Jugglers and tumblers performed in the garden"—Diary of Harriet, Lady Lubbock, Lubbock family papers.

the habits of the wireworm—Autobiographical note quoted in Horace Hutchinson, *Life of Sir John Lubbock, Lord Avebury* (London, 1914), 1.23.

"No doubt the wireworm fulfils"—*The Skip-Jack or Wireworm and the Slug . . . for the Use of Parish Schools* (Edinburgh, 1858), p. 23.

"While there are boys and birds-nests"—William Howitt, *The Boy's Country Book: Being the Real Life of a Country Boy, Written by Himself* (London, 1841), p. 49.

"a most diverse kind of mortal"—Thomas Carlyle, *Reminiscences,* ed. Charles Eliot Norton (London, 1887), 1.173.

their uncle's "quaint, delicate humour"—Julia Wedgwood, *Spectator,* 3 September 1881, p. 1132.

Joseph Hooker—Ray Desmond, *Sir Joseph Dalton Hooker, Traveller and Plant Collector* (Woodbridge, 1999).

"Often I worked in the dining room"—Leonard Huxley, *Life and Letters of Sir Joseph Dalton Hooker* (London, 1918), 2.459.

reading and being read to—Gillian Beer surveyed "Darwin's reading and the fictions of development" in *The Darwinian Heritage,* ed. David Kohn (Princeton, 1985) pp. 543–88. She focused on the references in the notebooks of 1836–44, Charles's reading notebooks and his *Autobiography.* Other books and periodicals read by Charles, Emma and their children can be identified, with exceptional completeness for a Victorian household, from the surviving books known to have been in the house, a number of catalogues and inventories, the many references to Charles and Emma's reading in *CCD* and *CFL,* Emma's correspondence and pocketbooks, the London Library's surviving issue books, and the reminiscences of visitors to Down House.

117 "The Purple Jar"—The story first appeared in Maria Edgeworth's *The Parent's Assistant* (1796), but was later included in *Rosamond* (1801).

117 *The Bird Talisman*—Henry Allen Wedgwood, *The Bird Talisman, an Eastern Tale,* reprinted with illustrations by Gwen Raverat (London, 1956), pp. 1–2.

118 *The Emigrant's Manual*—*The Emigrant's Manual: British America and United States of America* (Edinburgh, 1851), pp. 104, 105, 107.

119 *Our Cousins in Ohio*—Mary Howitt, *Our Cousins in Ohio* (London, 1849), pp. 5–7, 232, 239.

120 "life in Ohio"—*CCD,* 4.479.

120 Jean-Jacques Rousseau's revolutionary ideas in *Émile*—Jean-Jacques Rousseau, *Émile or On Education* (Harmondsworth, 1991), pp. 79, 90, 93, 135.

121 a plan for a girls' boarding school—Erasmus Darwin, *A Plan for the Conduct of Female Education in Boarding Schools, Private Families and Public Seminaries* (Derby, 1797).

121 "a sympathy with the pains and pleasures"—Erasmus Darwin, *A Plan for the Conduct of Female Education,* pp. 46–7; Charles Darwin, *Erasmus Darwin* (London, 1879), p. 116.

121 Tom Wedgwood had met the radical philosopher—David Erdman, "Coleridge, Wordsworth and the Wedgwood Fund," *Bulletin of the New York Public Library,* Part I, vol. 60, no. 9 (September 1956), pp. 430–33.

121 a plan for his governess to follow—Frank Doherty, "The Wedgwood system of education," *The Wedgwoodian,* November 1983, pp. 182–7.

122 Rousseau's Swiss follower—E. Woodall, "Charles Darwin," *Transactions of the Shropshire Archaeological and Natural History Society,* vol. 8 (1884), p. 14.

122 *Levana*—Jean Paul Friedrich Richter, *Levana* (London, 1901), pp. 146, 152, 154, 162, 175, 176. The London Library's Issue Book for 1850 records that "Richter Levana" was issued to Charles on 15 May.

124 *Conversations on Optics*—Jane Marcet, "Conversation XVII," in *Conversations on Natural Philosophy* (London, 1833), p. 400.

4 *Werner's Nomenclature*—*CCD,* 5.541–2; *Werner's Nomenclature of Colours,* ed. Patrick Syme (Edinburgh, 1821), pp. 27, 29, 40, 42.

5 "A delicate girl submitted to such a discipline"—James Clark, *A Treatise on Pulmonary Consumption* (London, 1835), p. 291.

6 The position of the governess—Kathryn Hughes, *The Victorian Governess* (London, 1993), explains the role and position of the governess in Victorian families.

7 "a sliding board for the children"—Bernard Darwin, *Life is Sweet, Brother* (London, 1940), p. 22.

8 "Miss Thorley and I are doing a little botanical work"—*CCD,* 5.343, 354; *Natural Selection,* p. 230. The field "allowed to run waste for fifteen years" was Great Pucklands meadow, known to the Darwin family as "Stony Field," on the downward slope of the valley to the west of the Sand-walk copse. Miss Thorley was among the personal friends invited by the family to Charles's funeral at Westminster Abbey in 1882.

0 Mulhauser's method—M. A. Mulhauser, *A Manual of Writing; Founded on Mulhauser's Method of Teaching Writing, and Adapted to English Use* (London, 1842).

1 *The Wide, Wide World*—Elizabeth Wetherell, *The Wide, Wide World* (London, 1880), pp. 25–9.

4 a book of recipes—CUL DAR 214.

4 A book of medical preparations—English Heritage Darwin Collection at Down House.

## Chapter Six: Faith, Cricket, and Barnacles

5 Walking in the village—James Moore has written about Darwin's life as a country gentleman in Downe in his essay, "Darwin of Down: The Evolutionist as Squarson-Naturalist," in *The Darwinian Heritage,* ed. David Kohn (Princeton, 1985), pp. 435–82.

5 In one year his gifts to charities—The year was September 1850 to August 1851. The details are in Charles's account books, on deposit by English Heritage in Cambridge University Library.

135 to form a Friendly Club—*CCD*, 4.304, *Life and Letters*, 1.142. The club was registered with the Registrar of Friendly Societies as No. 3043, "The Down Friendly Society." The documents are in the Public Record Office.

136 "Frugality and providence"—Charles Ansell, *A Treatise on Friendly Societies* (London, 1835), p. 2.

136 "The game is free"—James Pycroft, *The Cricket Field: or The History and Science of the Game of Cricket* (London, 1854), pp. 33, 34.

136 a gin cordial—"Receipts and Memoranda" book, in Ralph Colp, *To be an Invalid* (Chicago, 1977), Appendix A, pp. 155, 163.

136 The treatment of animals—James Turner, *Reckoning with the Beast: Animals, Pain and Humanity in the Victorian Mind* (Baltimore, 1980).

137 Francis remembered his father—*Life and Letters*, 3.200.

137 a gentleman farmer—Speech by William Darwin, *Darwin Celebration, Cambridge, June, 1909. Speeches Delivered at the Banquet Held on June 23rd* (Cambridge, 1909), pp. 11–12.

138 their mother "was not only sincerely religious"—*CFL* (1915), 2.175.

138 A prayer book used by Unitarians—Theophilus Lindsey, *The Book of Common Prayer Reformed . . . for the Use of the Congregation in Essex Street Chapel* (London, 1836), pp. 65–6.

138 "Jesus Christ was a person"—Joseph Priestley, *A Catechism for Children and Young Persons* (London, 1817), pp. 15, 19–20.

138 "God looks down from heaven on high"—Ann and Jane Taylor, *Hymns for Infant Minds* (London, 1852), p. 44.

139 "Tell me, Mama, if I must die"—*Hymns for Infant Minds*, p. 48.

139 "Let not young persons"—Jane Taylor, "Revelation XIV.13" in *The Contributions of Q.Q.* (London, 1838), p. 103.

140 a small choir—Diary of Harriet, Lady Lubbock, Lubbock family papers.

0 Their book of chants—*The Hymns of the Church Pointed for Chanting* (London, 1849), p. 15. There is a copy among the Downe parish papers in Bromley Public Library's Local History Collection.

1 Charles once went to a vestry meeting—Wedgwood/Mosley Collection 193, Letter from Emma Darwin to Jessie Sismondi, 1848.

1 A book of sermons—John Innes, *Five Sermons Preached in Advent and on the Festival of the Holy Innocents, 1851, in the Parish Church of Downe, Kent* (London, 1852), pp. 62, 12–13, 3–4. Later the Darwins became good friends with Mr. Innes. After Charles's death, Innes wrote to Francis that when they last met, Charles remarked that they had "been fast friends for thirty years. We never thoroughly agreed on any subject but once and then we looked hard at each other and thought one of us must be very ill." Robert Stecher, "The Darwin–Innes letters: The correspondence of an evolutionist with his vicar, 1848–1884," *Annals of Science,* vol. 17, no. 4 (December 1961), p. 256.

1 a National Ecclesiastical Census—The returns are in the Public Record Office.

2 James Carter—"The late Mr James Carter, Baptist minister, of Down, Kent," *The Earthen Vessel,* 2 December 1861, pp. 295–7.

3 They set out their beliefs in a "declaration"—Kindly shown to me by the church secretary of Downe Baptist Church.

3 An inspector of schools—Canterbury Cathedral Archives, Inspection of Dartford Deanery Schools, no. 80, 2 July 1851.

4 Charles thought carefully about his own beliefs—James Moore showed the gradual change through the 1840s in "Of love and death: Why Darwin 'gave up Christianity' " in *History, Humanity and Evolution,* ed. James Moore (Cambridge, 1989), pp. 195–229.

4 *Enquiry Concerning Human Understanding*—Hume dealt with miracles in Section X, "Of miracles." Charles had noted in 1838 that he found the *Enquiry* "well worth reading" (*Notebooks,* p. 559).

4 *Rationale of Religious Enquiry*—James Martineau, *Rationale of Religious Enquiry* (London, 1845).

4 "the clearest evidence would be requisite"—*Autobiography,* p. 86.

145 *The Evidences of the Genuineness of the Gospels*—Andrews Norton, *The Evidences of the Genuineness of the Gospels* (Boston, 1837–44); *CCD,* 4.476; annotations in Emma's interleaved Bible.

145 "to invent evidence"—*Autobiography,* pp. 86–7.

146 a memoir of John Sterling—John Sterling, *Essays and Tales Collected and Edited with a Memoir of his Life by J. C. Hare* (London, 1848). Sterling's comments on Paley are on p. lxxxviii of Hare's memoir.

146 "how good God was"—Letter from Erasmus Alvey Darwin to Fanny Wedgwood, Wedgwood/Mosley Collection.

146 *Eastern Life*—Harriet Martineau, *Eastern Life, Present and Past* (London, 1848); Vera Wheatley, *The Life and Work of Harriet Martineau* (London, 1957), p. 264; *CCD,* 4.478.

147 "disbelief crept over me"—*Autobiography,* pp. 86–7.

147 "I never gave up Christianity"—Edward Aveling, *The Religious Views of Charles Darwin* (London, 1883), pp. 5, 6.

147 Francis Newman—Francis Newman, *A History of the Hebrew Monarchy from the Administration of Samuel to the Babylonish Captivity* (London, 1847), and *The Soul, her Sorrows and her Aspirations: An Essay towards the Natural History of the Soul* (London, 1849); *CCD,* 4.451, 479.

148 a small barnacle—*CCD,* 4.388–409.

151 a character called Professor Long—Edward Bulwer-Lytton, *What Will He Do with It?* (Novels, Knebworth Edition, London, n.d.), 1.162–4.

157 "And this is a most wonderful process"—Philip Gosse, *Evenings at the Microscope* (London, 1884), p. 201.

## Chapter Seven: Worlds Away from Home

160 Sophy's music—Ursula Vaughan Williams, *R.V.W.: A Biography of Ralph Vaughan Williams* (London, 1964), p. 13.

162 a parliamentary committee—Select Committee on the State of the Children Employed

in the Manufactories of the United Kingdom, 1816; Report of the Minutes of Evidence, 25 April and 18 June 1816.

"We turned and threw and blunged"—Josiah Wedgwood, *Essays and Adventures of a Labour M.P.* (London, 1924), p. 31.

"In the labyrinthine stairs and corridors"— C. V. Wedgwood, "Out on a limb," *The Listener,* 20 September 1956.

Dr. Gully met Dr. James Wilson—John Harcup, *The Malvern Water Cure* (Malvern, 1992), and Janet Browne, "Spas and sensibilities: Darwin at Malvern," in *The Medical History of Waters and Spas,* ed. Roy Porter, *Medical History,* Supplement no. 10 (1990).

"How often have I observed the undertaker's house"—James Wilson, *The Water Cure, A Practical Treatise on the Cure of Diseases by Water, Air, Exercise and Diet* (London, 1842), pp. xvi–xviii.

"It is, I think, in medicine as in religion"—Joseph Leech, *Three Weeks in Wet Sheets* (London, 1851), p. 118. The book was published anonymously, but a copy owned by John Harcup has a contemporary inscription attributing it to Leech.

Thomas Carlyle caught an element of the appeal—Charles Norton, *The Correspondence of Thomas Carlyle and R. W. Emerson 1834–1872* (London, 1883), 2.206.

"The three hydropathic doctors"—Joseph Leech, *Three Weeks in Wet Sheets,* p. 16.

*Willy's Travels on the Railroad*—Jane Marcet, *Willy's Travels on the Railroad* (London, 1847), pp. 6–7.

When Dickens passed through the city—Charles Dickens, "Malvern water," *Household Words,* 11 October 1851, p. 67.

The Lodge—The house is now called Hill House.

"The water itself, which dribbles away"— Joseph Leech, *Three Weeks in Wet Sheets,* p. 56.

*The Young Lady's Book—The Young Lady's Book: A Manual of Elegant Recreations, Arts, Sciences, and Accomplishments* (London, 1859), pp. 473, 486.

174 "he bothered my father"—In George's account of the episode (CUL DAR 112.9), he wrote that it took place in 1851. Francis (CUL DAR 140) placed it during the family's visit in 1849. Francis's date is to be preferred, as Charles was not at Malvern "for some time" in 1851.

176 *Old Stones*—W. S. Symonds, *Old Stones: Notes of Lectures on the Plutonic, Silurian, and Devonian Rocks in the Neighbourhood of Malvern* (Malvern, 1855), pp. 7–8, 17, 68. Symonds also knew Charles Lyell.

176 Edwin Lees—Edwin Lees, *Pictures of Nature in the Silurian Region Around the Malvern Hills* (Malvern, 1856), pp. 326–7.

## Chapter Eight: The Fretfulness of a Child

180 "Nothing comes up to the misery"—*CCD,* 4.155.

180 Obaysch—Wilfrid Blunt, *The Ark in the Park: The Zoo in the Nineteenth Century* (London, 1976), pp. 106–21.

181 A guide published in 1839—John Brady, *The Visitor's Guide to Knole, in the County of Kent* (Sevenoaks, 1839), pp. 69–70.

183 Charles watched with his own interests—*Notebooks,* p. 371 (C 249e); annotations in Johann Bechstein, *Naturgeschichte der Stubenvögel* (Halle, 1840), *Marginalia,* pp. 44–7; B. P. Brent, *The Canary, British Finches, and Some Other Birds* (London, 1864), p. 15; *Descent,* 2.59.

183 A medical encyclopaedia—Robert Hooper, *Lexicon Medicum; or Medical Dictionary* (London, 1839), p. 243.

184 "one of the most elegant resorts"—*Davidson's Ramsgate and Margate Guide* (London, n.d.), pp. 50–51.

184 "As most of the company prefer the morning"—*Picture of Margate, Being a Complete Guide to All Persons Visiting Margate, Ramsgate, and Broadstairs* (London, 1809), pp. 50–51, 52–3.

186 A booklet of the time—*The Shells of Margate, Ramsgate, and Broadstairs* (Margate, n.d.).

186 "Now 'tis high water"—Charles Williams, *Pickings on the Sea-shore; or Cliffs, Sands, Plants and Animals* (London, 1857), p. 46.

7 "a subject which makes me more wrath"—*CCD,* 4.354.

8 "It is my misfortune to be not of an affectionate disposition"—*CFL* (1915), 1.108.

9 Annie and Etty arranged their shells—The shells are now in a child's box which belonged to Annie and was given to Etty after Annie's death. Etty added to the collection and made labels using pieces of paper cut from her father's notes on barnacles. Solene Morris, former curator of Down House, and her husband Noel kindly identified all the shells which could have been found at Ramsgate and on neighbouring beaches. Kathie Way of the Natural History Museum kindly identified the foreign shells for me and pointed out those which corresponded with entries in Syms Covington's list for Charles of the shells in the *Beagle* collection (CUL DAR 39, ii).

1 Dr. Gully explained the methods—James Gully, *The Water Cure in Chronic Disease* (London, 1846), pp. 564–660. John Harcup pointed out to me that Dr. Gully first described the spinal wash in print in the ninth edition of 1863 (p. 386), but wrote then that he had been using it in his practice since 1846.

2 "For two or three minutes"—Joseph Leech, *Three Weeks in Wet Sheets* (London, 1851), pp. 42–3.

2 "The youngest nurse or nursery maid"—Samuel and Sarah Adams, *The Complete Servant* (London, 1825), p. 255.

3 In his daily notes on the treatment—Family papers, on deposit in Cambridge University Library, CUL DAR 185:125.

3 some books to read—London Library Issue Book no. 3 for February 1850 to January 1851.

4 "No, I have no pity"—William Howitt, *The Boy's Country Book: Being the Real Life of a Country Boy, Written by Himself* (London, 1841), pp. 112–13.

4 *Phases of Faith*—Francis Newman, *Phases of Faith; or Passages from the History of my Creed* (London, 1851).

4 "excellent"—*CCD,* 4.479.

4 "the fretfulness of a child"—Newman, *Phases of Faith,* p. 78.

195 snow fell that night—J. H. Belville's journal of weather at Blackheath, National Meteorological Archive, Bracknell.

195 "lying on the bed with me"—CUL DAR 210.13.

## Chapter Nine: The Last Weeks in Malvern

197 Wombwell's Royal Menagerie—*Berrow's Worcester Journal,* 27 March 1851.

199 an eight-year-old watercress girl—Henry Mayhew, *London Labour and the London Poor* (Harmondsworth, 1985), pp. 64–5.

200 *Letters on the Laws*—Henry Atkinson and Harriet Martineau, *Letters on the Laws of Man's Nature and Development* (London, 1851), p. 173. In her will, Harriet Martineau left her head to Atkinson for further phrenological research.

200 phrenology and mesmerism—Janet Oppenheim, *The Other World: Spiritualism and Psychical Research in England, 1850–1914* (Cambridge, 1985), pp. 207–17.

200 Charlotte Brontë wrote to a friend—Elizabeth Gaskell, *The Life of Charlotte Brontë* (Harmondsworth, 1997), p. 353.

200 "Man has his place in natural history"—Atkinson and Martineau, *Letters on the Laws of Man's Nature and Development,* pp. 16, 28.

201 "Safety of Dr Hooker"—*Berrow's Worcester Journal,* 3 April 1851.

203 That day in Malvern—John Rashdall's diary, Bodleian Library MS Eng. Misc. e. 356.

203 The next day—Charles and Emma's letters to each other and others during Annie's illness are in *CCD,* 5.13–24. Other letters are in *CFL* (1915), pp. 132–40, the Darwin Archive at Cambridge University Library and the Wedgwood/Mosley Collection in Keele University Library.

207 "It is a curious thing to observe"—Florence Nightingale, *Notes on Nursing* (London, 1859), p. 35.

208 "Compelled by her dress"—Florence Nightingale, *Notes on Nursing,* p. 26.

216 a jonquil—The fold of paper with the fragments of the flower is in DAR 210.13 in the

Darwin Archive at Cambridge University Library. The flower was identified with expert advice from Dr. Chris Preston of the Institute for Terrestrial Ecology at Monkswood, and Sally Kington, the Royal Horticultural Society's Daffodil Registrar.

## Chapter Ten: Loss and Remembering

Cox & Co.—Charles's accounts, English Heritage Darwin Collection at Down House.

"Seated on one of the wooden benches"—Joseph Leech, *Three Weeks in Wet Sheets; Being the Diary and Doings of a Moist Visitor to Malvern* (London, 1851), p. 10.

The details of a funeral—Julien Litten, *The English Way of Death: The Common Funeral since 1450* (London, 1991).

Charles found the ceremony—*CFL* (1915), 2.161. Charles was writing to Willy about the funeral of Aunt Sarah at Downe. "We walked down to Petleys, and there put on black cloaks and crape to our hats, and followed the [coffin], which was carried by six men."

Fanny returned to her life with Hensleigh—*CFL* (1915), 2.143.

For parents in the 1850s—Pat Jalland, *Death in the Victorian Family* (Oxford, 1996). Jalland has a chapter on agnostics which deals with Darwin, Hooker and Huxley. Laurence Lerner, *Angels and Absences* (Nashville, 1997) deals with child deaths in Victorian literature.

"Being troubled with continued retchings"—"The dying experiences of Mrs Carter, wife of Mr Carter, Baptist minister, Down, Kent," in *The Earthen Vessel,* vol. 7, no. 83 (December 1851), pp. 295–6.

"I consider children so entirely from Heaven"—Diary of Harriet, Lady Lubbock, Lubbock family papers.

*Christian Aspects of Faith and Duty*—John James Tayler, *Christian Aspects of Faith and Duty* (London, 1851), pp. 216, 214.

"It was, I must own, a heavy trial"—Gladstone papers in the British Library.

Dean Tait of Chester—A. C. Tait's journal in Lambeth Palace Library, quoted by Jalland in *Death in the Victorian Family,* p. 138.

226 "the brief, dry, unpretending, uncircumstantial manner"—Richard Whately, *A View of the Scripture Revelations Concerning a Future State* (London, 1830), pp. 211–12.

227 "where my living soul would go"—"Upon Death," in Ann and Jane Taylor, *Hymns for Infant Minds* (London, 1852), p. 119.

227 the "venerable and consolatory" belief—James Mackintosh, *Memoirs of the Life of Sir James Mackintosh* (London, 1835), 2.12.

227 "Talk as we will of immortality"—Frederick W. Robertson, *Sermons Preached at Brighton* (London, 1868), pp. 215–16, 223, 228.

228 People took their wish for a life after death—Henry Atkinson and Harriet Martineau, *Letters on the Laws of Man's Nature and Development* (London, 1851), p. 164.

228 "I did not know the parting would be such a pang"—Elizabeth Birks quoted by Pat Jalland in *Death in the Victorian Family,* pp. 119–20.

230 "the great roll of the organ"—Letter to Edmund Lushington in Alfred Tennyson, *The Letters of Alfred Lord Tennyson,* ed. Cecil Lang and Edgar Shannon (Oxford, 1987), 2.14.

230 "Little bosom not yet cold"—Alfred Tennyson, *The Poems of Tennyson in Three Volumes,* ed. Christopher Ricks (London, 1987), 2.464–5.

232 "amidst tears, punishments, threats, and slavery"—Jean-Jacques Rousseau, *Émile or On Education* (Harmondsworth, 1991), p. 79.

233 "There is something strangely pathetic"—*CFL* (1904), 1.81–2.

233 "She passed away, like morning dew"—Hartley Coleridge, *Poems,* 1833, ed. Jonathan Wordsworth (Oxford, 1990), pp. 93–6. In the second of the verses Emma copied, the original "Admired" was changed to "Beloved."

235 chose instead to write a piece about her—Ralph Colp, "Charles Darwin's 'insufferable grief,'" *Free Associations* 9 (London, 1987), pp. 7–44.

236 "If I could have been left alone"—*Life and Letters,* 1.11.

236 "I remember the faces of persons formerly well-known"—*Life and Letters,* 3.239.

its "most sweet expression"—letter to Caroline Darwin of 20 September 1881, CUL DAR 153.

"Our poor child, Annie"—*CCD,* 5.540–42.

## Chapter Eleven: The Destroying Angel

In the medical language of the time—Ordinary people's understanding of the medical terms is shown in Elizabeth Gaskell's account of Ben Davenport's fever in *Mary Barton,* her "tale of Manchester life" in the "hungry forties" published in 1848. "'The fever' was . . . of a low, putrid, typhoid kind." Davenport was delirious, and when John Barton described the symptoms to the druggist, he "concluded it was typhus fever." Elizabeth Gaskell, *Mary Barton,* ed. Stephen Gill (Harmondsworth, 1985), pp. 99, 102.

no one in the Darwin family said anything more—In her letter to her daughter Effie about Annie's death (Wedgwood/Mosley Collection 310), Fanny Wedgwood wrote: "It was at last a bad fever that she died of. Typhous it is called, and there are more children ill of it here." As noted on p. 204, there is no trace of any other deaths from any kind of fever in the parish records or local newspapers for the month.

"We can scarcely indeed touch on this subject of fever"—Henry Holland, *Medical Notes and Reflections* (London, 1840), p. 101.

I gave all the clues—My thanks to Dr. Denis Gibbs, Dr. John Ford, Dr. Gordon Cook and Dr. John Brown for their opinions.

Tuberculosis is caused by a slow-working bacillus—René and Jean Dubos, *The White Plague: Tuberculosis, Man and Society* (London, 1953), gave a detailed medical and social history of the disease in the nineteenth and twentieth centuries. Many other books have been published since; among the most recent is Thomas Dormandy, *The White Death: A History of Tuberculosis* (London, 1999).

"Consumption, Decline or Phthisis"—Thomas Yeoman, *Consumption of the Lungs, or Decline: The Causes, Symptoms and Rational Treatment* (London, 1848), pp. i, 8.

*Treatise on Pulmonary Consumption*—James Clark, *A Treatise on Pulmonary Consumption* (London, 1835).

"so obscure or doubtful"—Yeoman, *Consumption of the Lungs,* pp. 19–20.

244 "In childhood . . . the child is peevish"—Richard Cotton, *The Nature, Symptoms, and Treatment of Consumption* (London, 1852), p. 102.

244 "Many persons acquire a predisposition"—Yeoman, *Consumption of the Lungs,* p. 6.

245 "No physician acquainted with the morbid anatomy"—Clark, *A Treatise on Pulmonary Consumption,* p. 6.

245 "I romanced internally about early death"—Harriet Martineau, *Harriet Martineau's Autobiography* (London, 1877), 2.435.

245 "From some cause"—Cotton, *The Nature, Symptoms, and Treatment of Consumption,* p. 100.

246 "The extreme prevalence of consumption"—Henry Hillier, "Preface" in *A Popular Treatise on Diseases Resembling Consumption* (London, 1854).

246 Royal Brompton Hospital—Charles noted a contribution in his accounts for 1841. The history of the setting up and early years of the hospital is told by Maurice Davidson and F. G. Rouvray in *The Brompton Hospital: The Story of a Great Adventure* (London, 1954).

246 "Pain and suffering must be alleviated"—W. H. Howard, *Introductory Sermon Preached in the Chapel of the Hospital for Consumption and Diseases of the Chest* (London, 1842), p. 11.

247 "There is a dread disease"—Charles Dickens, *Nicholas Nickleby,* ed. Michael Slater (Harmondsworth, 1986), pp. 731–2.

247 "We have great doubts about the propriety"—Review article on novels by Charles Dickens and Captain Marryat, *The Christian Remembrancer,* December 1842, pp. 581–611.

249 "To her mind it seemed an evil"—Elizabeth Wetherell, *The Wide, Wide World* (London, 1880), pp. 412, 450.

250 the child's mind was "quick, forward, intelligent"—Herbert Mayo, *The Philosophy of Living* (London, 1837), p. 26.

"Water is one of the best prophylactics"—Yeoman, *Consumption of the Lungs,* p. 50.

he was "convinced, that the judicious use"—James Gully, *The Water Cure in Chronic Disease* (London, 1846), pp. 260–61.

"proof of common origin of man"—*Notebooks,* p. 293 (C 174).

he "fully recognised the truth"— Charles Darwin, *Erasmus Darwin* (London, 1879), p. 110.

"excellent observations of sickly offspring"—*Notebooks,* p. 279 (C 133).

"rearing up of every hereditary tendency"—*Notebooks,* p. 415 (E 67).

he wrote to Fox about his surviving children—*CCD,* 5.84, 100.

he revealed his worries indirectly—*CCD,* 5.194.

"It is only by convincing the public"— Clark, *A Treatise on Pulmonary Consumption,* pp. vi–vii.

William Farr—John Eyler, *Victorian Social Medicine: The Ideas and Methods of William Farr* (Baltimore, 1979). Farr's writings can be surveyed most easily in the memorial volume, *Vital Statistics: A Memorial Volume of Selections from the Reports and Writings of W. F. Farr,* ed. N. A. Humphreys (London, 1885).

"Diseases are more easily prevented than cured"—*1st Annual Report of the Registrar-General of Births, Deaths, and Marriages in England,* 1839, p. 89, quoted in Farr, *Vital Statistics,* pp. 213–15.

"The great source of the misery of mankind"—Farr, *Vital Statistics,* p. 136.

In 1851, five children died—*14th Annual Report of the Registrar-General of Births, Deaths, and Marriages in England,* 1852.

they might be dwarfish or ill-formed—John Hogg, *London as It is* (London, 1837), pp. 352–3.

254 "for the most part . . . the parent's gift"—Thomas Bull, *The Maternal Management of Children in Health and Disease* (London, 1848), pp. 386–99.

255 "ignorant members of our legislature"—*Descent,* 2.438–9.

255 "natural history of infectious disease"—Macfarlane Burnet and David White, *Natural History of Infectious Disease* (Cambridge, 1972).

255 mistletoe as a parasite—*Origin,* pp. 114, 122.

256 a copy of his periodical—CUL DAR 161.2:205.

256 "I well remember saying to myself"—*Life and Letters,* 3.234.

## Chapter Twelve: The Origin of Species

257 "The only chance of forgetting"—CUL DAR 211.11.

258 "I suggest not halfway"—CUL DAR 205.5: 130–31.

259 a boy's visit to the Polytechnic—Mary Howitt, *The Children's Year* (London, 1847), p. 47.

259 "Etty nearly 8 years old"—Family papers, *CCD,* 5.542–3.

260 *Stories for Sunday Afternoons*—Susan Crompton, *Stories for Sunday Afternoons. From the Creation to the Advent of the Messiah* (London, 1845), p. 3.

262 "the sense of loss was always there unhealed"—*CFL* (1915), 2.137.

262 "Why sorrow should make us shy"—*CFL* (1915), 2.85.

262 "my very dear coadjutor"—*CFL* (1915), 2.203.

263 "The Darwin family *are* a nice family"—Family papers.

263 *Chapters on Mental Physiology*—Henry Holland, *Chapters on Mental Physiology* (London, 1852), pp. 185, 189.

263 a matter of "deep interest"—Henry Holland, Review of James Prichard, *Researches into*

*the Physical History of Mankind* (London, 1836–47), *Quarterly Review,* vol. 86, no. 171 (December 1849), p. 16.

"Sobbing in child"—*Marginalia,* p. 385.

*Confessions of an English Opium Eater*—Thomas De Quincey, *Confessions of an English Opium Eater* (Harmondsworth, 1986), pp. 104, 108, 52–3, 64–5, 112.

"very poor"—*CCD,* 4.481, 491.

"One instance . . . has fallen under my own observation"—Charles Darwin, *The Variation of Animals and Plants Under Domestication* (London, 1888), 1.450–51.

"I thank you sincerely"—*CCD,* 5.151.

"Poor dear happy little thing"—*CCD,* 6.238.

"I went up to my father"—Leonard Darwin, "Memories of Down House," *The Nineteenth Century and After,* vol. 106 (July 1929), p. 120.

"I was indeed grieved"—*CCD,* 8.365–6.

He did not attend church services—George Foote, *Darwin on God* (London, 1889), p. 20. The constable who told Foote about his conversations with Mr. Darwin may have been William Soper, who served at Downe between 1858 and the mid-1860s.

He did, though, still firmly believe—*Autobiography,* p. 93.

When he returned to the theme—James Moore, "Of love and death: Why Darwin 'gave up Christianity,'" in *History, Humanity and Evolution,* ed. James Moore (Cambridge, 1989), p. 222; David Kohn, "The aesthetic construction of Darwin's theory," in *The Elusive Synthesis: Aesthetics and Science,* ed. A. Tauber (Boston, 1996), pp. 13–48.

"The indecency of the process," "What a book a Devil's Chaplain might write"—*CCD,* 6.178.

"Can the instinct, which leads the female spider"—*Natural Selection,* p. 526.

"She cares not for mere external appearance"—*Natural Selection,* pp. 224–5.

271 "very small parts of one general law"—*Natural Selection,* p. 527.

271 "Next to the greatness of these cosmic forces"—John Stuart Mill, "Nature," in *Essays on Ethics, Religion and Society by John Stuart Mill,* ed. J. M. Robson (Toronto, 1969), pp. 384, 385.

272 Charles saw that the range of breeds—James Secord, "Darwin and the breeders," in *The Darwinian Heritage,* ed. David Kohn (Princeton, 1985), gives the background and history of Charles and Etty's experiments with pigeons. Etty's comments are in CUL DAR 246.

273 "He was small for his age"—*CCD,* 7.521.

273 I showed it recently to a consultant paediatrician—My thanks to Martin Gardiner, Professor of Paediatrics at University College London Medical School, for his opinion.

274 Dr. John Langdon Down—Conor Ward, *John Langdon Down: A Caring Pioneer* (London, 1998), gives a full account of Down's approach to his patients and his achievements. Down first described the syndrome in his paper, "Observations on an ethnic classification of idiots," in *London Hospital Reports* (London, 1866).

274 "to rescue the feeble one"—John Langdon Down, *On the Education and Training of the Feeble in Mind* (London, 1876), p. 8.

275 "the contented face of Nature"—"Essay" p. 116, *Notebooks,* p. 429 (E 114).

275 "All Nature . . . is at war"—*Natural Selection,* pp. 175–6.

276 "Nothing is easier than to admit"—*Origin,* pp. 115–16.

276 "There is a force like a hundred thousand wedges"—*Notebooks,* p. 375 (D 135e).

276 each creature "lives by a struggle"—*Origin,* p. 119.

277 we must "keep steadily in mind"—*Origin,* p. 129.

277 "It is interesting to contemplate an entangled bank"—*Origin,* p. 459.

## Chapter Thirteen: Going the Whole Orang

"the habit of looking at man as an animal"—Etty's notes on Charles's *Autobiography,* CUL DAR 199.1:2.

the "highest and most interesting problem"—*CCD,* 6.515.

Wombwell's Menagerie had a male orang—BL 1889.b.10, vol. 8, folio 85v.

the living ape "exhibits an intelligence"—Samuel Phillips, *Official General Guide to the Crystal Palace and Park* (London, 1856), p. 97.

the first complete gorilla skeleton—*Descriptive Catalogue of the Osteological Series Contained in the Museum of the Royal College of Surgeons,* vol. 2, *Mammalia Placentalia* (London, 1853), pp. 782–802.

another was displayed—*Roger Fenton, Photographer of the 1850s* (London, 1988), p. 12 and Cat. 29 and 30. It has been presumed that the display and Fenton's photograph reflected interest in the link between man and ape after the publication of *The Origin of Species* in 1859. The Minute Book of the Trustees of the British Museum records that they paid Fenton for an "additional negative photograph of the gorilla skeleton" on 26 June 1858.

the evident close links—Richard Owen, "On the characters, principles of division, and primary groups of the class mammalia," *Journal of the Proceedings of the Linnean Society,* vol. 2, no. 5 (1858), pp. 1–37.

"I wonder what a chimpanzee would say"—*CCD,* 6.419.

Huxley suggested in a lecture—Adrian Desmond, *Huxley,* p. 241.

a corpse of a young adult male—Richard Owen, "On the external characters of the gorilla," *Transactions of the Zoological Society of London,* vol. 5, part 4 (1866), pp. 243–82. Charles's copy of the issue is in Cambridge University Library.

Abraham Bartlett—Abraham Bartlett, *Bartlett's Life Among Wild Beasts in the "Zoo"* (London, 1900), pp. 3–4.

a terrifying monster—*Illustrated London News,* 9 April 1859.

283 declared his belief twice—*Origin,* pp. 451, 458; *Autobiography,* p. 130.

283 "To show how minds graduate"—*CCD,* 7.345.

284 a "villainous shifty fox of an argument"—*CCD,* 7.379.

284 "he applies his scheme"—*Quarterly Review,* vol. 108, no. 215 (1860), pp. 257–8.

284 the notorious debate on evolution—Adrian Desmond, *Huxley* (Harmondsworth, 1998), pp. 277–81.

284 Explorations—Paul Du Chaillu, *Explorations and Adventures in Equatorial Africa* (London, 1861), pp. 60, 352.

285 "One thing we may as well state"—*Bromley Record,* 1 November 1863.

286 "a grand and almost awful question"—*CCD,* 10.71.

286 "no one is more strongly convinced than I"—Thomas Huxley, *Evidence as to Man's Place in Nature* (London, 1863), p. 112.

287 The dignity of man was Lyell's worry—Charles Lyell, *Sir Charles Lyell's Scientific Journals on the Species Question,* ed. Leonard Wilson (New Haven, 1970), pp. 335–6, 332.

287 "the whole orang"—*CCD,* 11.230–31.

287 *The Geological Evidences*—Charles Lyell, *The Geological Evidences of the Antiquity of Man* (London, 1863), pp. 504–5.

288 "vomiting preceded by shivering"—Ralph Colp, *To be an Invalid* (Chicago, 1977), p. 83.

288 "She asked me a good deal about the Darwinian theory"—K. M. Lyell, *Life, Letters and Journals of Sir Charles Lyell* (London, 1881), 2.369.

288 The *Spectator* wrote: "The purpose of this tale"—Quoted in Arthur Johnston, " 'The Water-Babies': Kingsley's debt to Darwin," *English,* vol. 12, no. 72 (Autumn 1959).

289 "a poor, lean, seedy, hard-worked old giant"—Charles Kingsley, *The Water Babies* (Harmondsworth, 1995), pp. 293–9.

He read a paper—Alfred Russel Wallace, "The origin of human races and the antiquity of man deduced from the theory of 'Natural Selection,'" *Anthropological Review,* May 1864.

George Eliot had made sympathy—Thomas Noble, *George Eliot's Scenes of Clerical Life* (New Haven, 1965), Chapter 3, "The doctrine of sympathy."

He read *Adam Bede*—*CCD,* 7.300. Charles judged it "excellent" in his reading notebook (*CCD,* 4.496). In 1861, Emma wrote to William: "We have just finished reading aloud *Silas Marner* to our great sorrow. We like it better than *Mill on the Floss,* though not so well as *Adam Bede*" (CUL DAR 219.43).

an example for a scientific point—Chad Cranage's daughter in George Eliot, *Adam Bede,* ed. Stephen Gill (Harmondsworth, 1980), p. 240; *Expression,* p. 354.

"I have collected a few notes on man"—*MLCD,* 2.33–4.

a review of a new edition—Review of the tenth edition of Lyell's *Principles of Geology, Quarterly Review,* April 1869, p. 391.

"I hope you have not murdered"—*MLCD,* 2.39–40.

Professor W. B. Carpenter—William Carpenter, *Principles of Human Physiology* (London, 1855), covered "the mind and its operations" on pp. 546–633. He wrote: "It is much to be desired that a systematic study should be made . . . of that wide and almost unexplored domain, which comprehends the whole range, not only of what may be termed *Mental Physiology,* but also of *Mental Pathology,* and, in addition, the *Comparative Psychology* of the lower animals, and the *History of the Development of* [the human mind], from the earliest manifestation of its powers" (p. 547). He dealt with memory and recollection on pp. 600–03, and unconscious cerebration on pp. 608–10. He removed the sections on the human mind from later editions of the work, and in 1874 published an expanded version of them in *Principles of Mental Physiology.* He drew the distinction between memory and recollection on pp. 369 and 370. Jonathan Miller deals with Carpenter's treatment of unconscious cerebration in his essay, "Going unconscious," in *Hidden Histories of Science,* ed. Robert Silvers (London, 1997), pp. 1–35.

the power of deep-rooted prejudices—William Lecky, *History of the Rise and Influence of the Spirit of Rationalism in Europe* (London, 1870), pp. 94–5.

293 Henry Maudsley—Michael Collie, *Henry Maudsley: Victorian Psychiatrist* (Winchester, 1988).

294 "a wondrous entity"—Henry Maudsley, *The Physiology and Pathology of Mind* (London, 1867), pp. 67, 57.

294 "The beatings of the heart"—*Marginalia,* pp. 572–3.

## Chapter Fourteen: God's Sharp Knife

295 "This is always painful to me."—Letter to Asa Gray of 22 May 1860, *Life and Letters,* 2.310–12.

296 She set out the problem—Julia Wedgwood, "The boundaries of science: A second dialogue," *Macmillan's Magazine,* July 1861, pp. 237–47.

297 Charles wrote to Snow—*CCD,* 9.200.

298 a wild cucumber plant—*CCD,* 11.506.

299 "Twiners entwining twiners"—Nora Barlow, *Charles Darwin and the Voyage of the Beagle* (London, 1945), p. 162.

301 a wax flower from Queensland—CUL DAR 157.1:39.

302 "I shall be glad to hear sometime about your boy"—*CCD,* 11.682.

302 "Nothing is so dreadful in this life as fear"—*CCD,* 11.687.

304 "It has always appeared to me more satisfactory"—*Life and Letters,* 3.64.

305 "My heart has often been too full"—*CCD,* 9.155–6.

305 "The Lord of all"—William Cowper, *Poetry and Prose,* ed. Brian Spiller (London, 1968), p. 520.

306 "The groans of nature"—Cowper, *Poetry and Prose,* pp. 534–5, 537.

306 *Words of Peace*—Ashton Oxenden, *Words of Peace; or the Blessings and Trials of Sickness with Meditations, Prayers and Hymns* (London, 1863), pp. 2–3, 5.

7 *Fervent Prayer*—Ashton Oxenden, *Fervent Prayer* (London, 1860), pp. 115–16.

7 "As years went by"—*CFL* (1915), 2.175.

8 Henry James, then a young American visitor—Letter of 1 April 1869 in the Houghton Library, Harvard University, published in Ralph Colp, "'The perfect counterpart of our Cambridge luminary': Henry James meets Charles Darwin," *Clio's Psyche, Psychohistory Forum,* vol. 2, no. 3 (December 1995).

8 *The Index*—Adrian Desmond and James Moore, *Darwin* (London, 1991), p. 591.

8 pressed the newspaper's claims—Note by Francis Darwin, CUL DAR 140.3.

8 "the question whether there exists a Creator"—*Descent,* 1.143–6.

8 He made occasional tongue-in-cheek comments—Another passage in which he may not have been entirely serious is the paragraph in *The Expression of the Emotions* in which he noted the advice of "a lady who is a great blusher" that awareness of a fault before God never excites a blush. *Expression,* p. 331.

9 "There is said to be 'gnashing of teeth' in hell"—*Expression,* p. 73.

9 spiritual forces and life beyond death—Janet Oppenheim, *The Other World: Spiritualism and Psychical Research in England, 1850–1914* (Cambridge, 1985).

) Robert Chambers—Milton Millhauser, *Just Before Darwin: Robert Chambers and Vestiges* (Middletown, 1959), pp. 174–86; Oppenheim, *The Other World,* pp. 272–8; Daniel Dunglas Home, *Incidents in my Life* (London, 1872), pp. 140–43.

1 "My idea is that the term 'supernatural' "—Alfred Russel Wallace, *My Life: A Record of Events and Opinions* (London, 1905), 1.286.

1 Home held a séance for Crookes—R. G. Medhurst, *Crookes and the Spirit World* (London, 1972), p. 158.

2 a séance with a paid medium—*Life and Letters,* 3.186–8; *CFL* (1915), 2.216–17.

2 Charles had "quite made up his mind"—Letter from Snow Wedgwood to Emily Gurney of 9 July 1874, Wedgwood/Mosley Collection 438 (IV).

## Chapter Fifteen: The Descent of Man

313 "I am thinking of writing a little essay"—Letter to Fritz Muller of 22 February 1869, *Life and Letters,* 3.112.

313 *The Descent of Man*—Robert Richards, *Darwin and the Emergence of Evolutionary Theories of Mind and Behavior* (Chicago, 1987).

313 He took up Hume's suggestion—CUL DAR 80B:117. Charles's note is headed "D. Hume, An Enquiry concerning the Principles of Morals—edit 1751." It consists of transcripts of four passages from the book including the one about the social virtues—David Hume, *Enquiries Concerning Human Understanding and Concerning the Principles of Morals,* ed. P. H. Nidditch (Oxford, 1979), p. 214. Another of the passages, *Enquiries* pp. 243–4, Charles quoted in *The Descent of Man,* 1.166.

314 The subjects are still as bedevilled—Most recent writing on the evolutionary view of human nature has dwelt on instinctive elements which may have developed by natural selection in our recent past. One of the most interesting lines of thought has been the work on altruism in which W. D. Hamilton and others have shown how the "selfish gene" can adapt for cooperative life. Matt Ridley has explained their findings about animal and human behaviour in *The Origins of Virtue* (Harmondsworth, 1997). Charles saw an element of the problem of altruism in his puzzling over the evolution of neuter insects (Robert Richards, "Instinct and intelligence in British natural theology: Some contributions to Darwin's theory of the evolution of behavior," *Journal of the History of Biology,* Fall 1981, vol. 14, no. 2); he worked out the first part of the answer to the problem with his notion of family selection in *The Origin of Species* (pp. 258–9), and he mentioned the selective value of "higher" morality in *The Descent of Man* (1.203). He wrote little, though, about the evolution by natural selection of particular human instincts. In his treatment of humans as social animals, he focused instead on aspects of self-awareness and understanding, on the interplay between thought and feeling and the tangle of conscious and unconscious factors that determine human behaviour. I have set out Charles's comments on "thinking about feeling" on pp. 57–9, 62–3, 318–322 and 325–6. His suggestions beg many obvious questions, but the philosopher Mary Midgley has drawn intriguing ideas from them in her book *The Ethical Primate: Humans, Freedom and Morality* (London, 1994), pp. 139–45 and 177–83.

314 "I think it will be very interesting"—*CFL* (1915), 2.196.

314 She was to repeat that phrase—Letter to Frances Power Cobbe, CB 390 in the Huntington Library, San Marino.

"that arrogance which made our forefathers declare"—*Descent,* 1.36–7.

"It is notorious that man is constructed"—*Descent,* 1.7–8.

"As some of my readers may never have seen"—*Descent,* 1.12–13.

"a pedigree of prodigious length"—*Descent,* 1.255.

"self-preservation, sexual love"—*Descent,* 1.100.

"Parental affection, or some feeling which replaces it"—*Descent,* 1.162.

It was Etty who had suggested the point—Note by Charles in CUL DAR 88.

"the same senses, intuitions and sensations"—*Descent,* 1.120.

One aspect of his approach—G. H. Lewes, *Problems of Life and Mind, Third Series, The Study of Psychology* (London, 1879), pp. 118–58.

"Who can say what cows feel"—*Descent,* 1.156.

"A man cannot prevent past impressions"—*Descent,* 1.173.

"Man, from the activity of his mental faculties"—*Descent,* 1.171.

A man's "early knowledge"—*Descent,* 1.173.

"Even when we are quite alone"—*Descent,* 1.172.

"We recognise the same influence"—*Descent,* 1.186.

"a mother may passionately love"—*Descent,* 1.162.

"by far the most important of all the differences"—*Descent,* 1.148.

Charles believed that "The social instincts"—*Descent,* 1.190–91. He wrote that the social instincts "no doubt were acquired by man as by the lower animals for the good of the community." He went on to comment that "As a struggle may sometimes be seen going on between the various instincts of the lower animals, it is not surprising that there should be a struggle in man between his social instincts, with their derived virtues, and his lower, though momentarily stronger impulses or desires." He looked forward to a future in which "virtual habits will grow stronger, becoming perhaps fixed by inheritance."

322 "The moral faculties"—*Descent,* 2.429.

322 George Eliot saw memory and feeling—George Eliot, *Scenes of Clerical Life,* ed. David Lodge (Harmondsworth, 1985), p. 358; *Adam Bede,* ed. Stephen Gill (Harmondsworth, 1980), pp. 203, 531; *The Mill on the Floss,* ed. A. S. Byatt (Harmondsworth, 1985), p. 571.

323 "If . . . men were reared"—*Descent,* 1.151–2.

323 "The same high mental faculties"—*Descent,* 1.146.

324 Marian Marsden—Mary Hartman, "Child-abuse and self-abuse: Two Victorian cases," *History of Childhood Quarterly, Journal of Psychohistory,* vol. 2, no. 2 (1974), pp. 221–48.

324 Mademoiselle Doudet was tried in Paris—Armand Fouquier, *Les Causes célèbres de tous les peuples* (Paris, 1858–74), 3.1–56.

325 "Ultimately our moral sense or conscience"—*Descent,* 1.203.

326 "With savages, the weak in body and mind"—*Descent,* 1.205–6.

327 Charles noted his suggestion—*Marginalia,* pp. 571–2.

328 "an echo from a far-distant past"—Henry Maudsley, *Body and Mind: An Inquiry into their Connection and Mutual Influence* (London, 1873), pp. 59–60.

328 "Most persons who have suffered from the malady of thought"—Henry Maudsley, *Responsibility in Mental Disease* (London, 1874), p. 268.

328 *The Expression of the Emotions*—Janet Browne, "Darwin and *The Expression of the Emotions*" in *The Darwinian Heritage,* ed. David Kohn, pp. 307–26. Robert Richards, *Darwin and the Emergence of Evolutionary Theories of Mind and Behavior* (Chicago, 1987), pp. 230–4, emphasises how Charles "did not invoke natural selection" to explain the features of emotional expression in which he was interested.

9 "With mankind some expressions"—*Expression,* p. 19.

9 "A strong desire to touch the beloved person"—*Expression,* p. 212.

9 "I used to like to hear him admire the beauty of a flower"—*Life and Letters,* 1.117.

0 "No emotion is stronger than maternal love"—*Expression,* p. 82.

0 Charles's comments on infants and young children—*Expression,* pp. 209, 221, 329, 80, 210.

1 "The feelings which are called tender"—*Expression,* p. 214.

2 The prospectus declared—Letter from George Croom Robertson of 17 February 1875 with the prospectus, CUL DAR 176; Charles's reply of 19 February 1875, Croom Robertson papers, University College London.

2 "We are not content with saying"—Frederick Pollock, "Evolution and ethics," *Mind: A Quarterly Review of Psychology and Philosophy,* vol. 1, no. 3 (July 1876), pp. 335–7.

3 "I hope that you will read it"—Charles's letter to George Croom Robertson of 27 April 1877 in Croom Robertson papers, University College London.

3 Thirty-five years after Annie's first smile—"Biographical Sketch," pp. 288, 290.

4 "At what age does the new-born infant"—*Descent,* 1.194.

## Chapter Sixteen: Touching Humble Things

5 Charles wrote a memoir—*Autobiography,* pp. 21, 97–8.

5 "And so it is with many other facts"—Charles Darwin, *The Effects of Cross and Self Fertilisation in the Vegetable Kingdom* (London, 1876), p. 457.

6 "more of misery or of happiness"—*Autobiography,* pp. 88–90.

7 "Epicurus's old questions"—David Hume, *Dialogues Concerning Natural Religion,* ed. Martin Bell (Harmondsworth, 1990), pp. 108–9, 121.

338 an "infinite spider"—Hume, *Dialogues,* p. 91.

338 "Can the mind of man"—*Autobiography,* p. 93.

339 "The building was crammed"—CUL Add MS 7831(2).

341 The tremor in his signature—CUL University Archives, Subscriptions, vol. 11, p. 222.

341 "Science has nothing to do with Christ"—Letter to N. A. von Mengden of 5 June 1879, CUL DAR 139.12.

341 "My judgement often fluctuates."—Letter to John Fordyce of 9 May 1879, *Life and Letters,* 1.304.

342 "I dare say I should find my soul"—*Life and Letters,* 3.92.

342 his "curious and lamentable loss"—*Autobiography,* p. 139.

342 "His life had reduced itself"—George Eliot, *Silas Marner,* ed. David Carroll (Harmondsworth, 1996), p. 20.

343 "The place will be propitious"—Letter to E. R. Lankester of 9 July 1879, American Philosophical Society, Philadelphia.

343 He still had his old Wordsworth—Etty's notes for *Life and Letters,* CUL DAR 262.23. She wrote: "He reread most of *The Excursion* during this visit after an interval of I suppose thirty or more years. I think it was a disappointment. He found parts of it preachy and it did not give him much pleasure. Parts of it he had admired extremely and his old Wordsworth—a funny series edition in one volume—is marked with his notes as to what to skip and what he cared for." I have found no other information about Charles's copy of the one-volume edition.

343 "Surprised by joy"—William Wordsworth, *The Poems,* ed. John Hayden (Harmondsworth, 1990), 1.863. The poem was one of the "Miscellaneous Sonnets" in the third volume of the six-volume edition of 1840 (see p. 304 above). Charles marked it in his copy and wrote on the first page of contents: "All the sonnets except for five which I have marked appear to me wretchedly poor." See Stephen Gill, *Wordsworth and the Victorians* (Oxford, 1998), pp. 269–70.

4 *Souvenirs entomologiques*—Jean-Henri Fabre, *Souvenirs entomologiques: Etudes sur l'instinct et les moeurs des insectes* (Paris, 1879), p. 323.

4 "Permit me to add"—*Life and Letters,* 3.220–21.

4 "wormograph"—CUL DAR 262 DH/MS* 10.

5 "taming earthworms"—CUL DAR 239 LD/23/2/9.

5 "I am rather despondent about myself"—*MLCD,* 2.433.

5 "He looked at me very hard"—*Life and Letters,* 1.316.

6 "the horrid doubt"—Letter to W. S. Graham, *Life and Letters,* 3.316.

6 he came up to her in the dining room—Letter from Snow Wedgwood to Francis, CUL DAR 139(12):17. In Snow's account, Charles spoke of a link between human and animal "feelings" rather than mind. She went on to say that she found later that she had misunderstood his precise meaning, but believed it was linked with what he had written to Graham. Her comments suggest that Charles's point was the one he explained in his letter to Graham.

7 "a voice like a euphonium"—Michael Holroyd, *Bernard Shaw: The Search for Love* (London, 1988), p. 154.

7 Charles asked Büchner and Aveling—Edward Aveling, *The Religious Views of Charles Darwin* (London, 1883), p. 5.

8 Anne Thackeray Ritchie—Winifred Gérin, *Anne Thackeray Ritchie, A Biography* (Oxford, 1981).

8 "We drove to the door"—Anne Thackeray Ritchie's diary for 1882, published in Hester Ritchie, *Letters of Anne Thackeray Ritchie* (London, 1924), pp. 183–4.

0 "I can remember the tortoise"—Anne Thackeray Ritchie, *Chapters from Some Memoirs* (London, 1894), p. 73.

0 "I used to dream a great deal"—Ritchie, *Chapters from Some Memoirs,* p. 34.

351 "I am in love with Madame de Sévigné"—*CCD,* 4.146.

351 "If we deny the derivation of life"—Letter of 22 February from Daniel MacKintosh, 1882, CUL DAR 171.

351 "I hardly know what to say"—Letter of 28 February 1882 to Daniel MacKintosh, CUL DAR 146.

352 Etty wrote that "He was utterly ill"—Account of her father's last weeks and death, CUL DAR 262 DH/MS* 23:1–13.

352 Laura Forster—CUL DAR 112 A38–44.

352 "You will perhaps remember"—Letter to Joseph Fayrer of 30 March 1882, CUL DAR 144.

353 "The facts which you relate"—Letter to Henry Groves of 3 April 1882, BL Add MS 46917:66.

354 "No one can have any words"—Family papers.

355 "I know that this was Life"—*In Memoriam,* XXV, CFL (1915), 2.49.

356 "Is it, then, regret for buried time"—*In Memoriam,* CXVI, CFL (1915), 2.307.

357 "As soon as the door was opened"—Gwen Raverat, *Period Piece* (London, 1952), pp. 142, 144.

358 Margaret was called into the garden—Note by Margaret Keynes in Annie's writing case.

359 She played the "galloping tune"—*CFL* (1904), 2.463.

359 Emma died early in the morning—*CFL* (1915), 2.313; *Bromley Record,* November 1896.

359 "strange vividness"—*CFL* (1904), 2.149–50.

# ACKNOWLEDGEMENTS

I am grateful to the late George Darwin for permission to quote from Charles Darwin's letters and manuscripts, and to the Syndics of Cambridge University Library for permission to quote from the material owned by the Library. Passages from the documents in English Heritage's Darwin Collection at Down House are reproduced with the kind permission of English Heritage. Material from the Wedgwood/Mosley Collection at Keele University Library is quoted by courtesy of the Trustees of the Wedgwood Museum, Barlaston, Stoke-on-Trent, Staffordshire. The passage from Queen Victoria's Journal of 1842 is quoted with the permission of Her Majesty Queen Elizabeth II. I am grateful to Mrs. Belinda Norman-Butler for showing me Anne Thackeray Ritchie's diary, and for her permission to quote from it. Likewise to Gordon Hoppé, Church Secretary of Downe Baptist Church, for the Declaration of the congregation of Downe Chapel in 1851, and to Lyulph Lubbock for the diary of Harriet, Lady Lubbock.

I am grateful to the Syndics of Cambridge University Press for permission to quote from *The Correspondence of Charles Darwin,* and to Sophie Gurney for permission to quote from Gwen Raverat's *Period Piece.*

My thanks to the staff of the London Library, the British Library, English Heritage, Cambridge University Library, the Darwin Correspondence Project and Keele University Library for all their kind help. Also to the archivists and library staff at Aberdeen City Archives, the Bethnal Green Museum of Childhood (London), the Bodleian Library (Oxford), the British Museum (London), Bromley Public Library, Cambridge University Museum of Zoology, Canterbury Cathedral Archives, Cheshire Record Office (Chester), Christ's College Library (Cambridge), Dr. Williams's Library (London), Eton College Library, the Evangelical Library (London), the Family Records Centre (London), the Geological Society of London, Guildhall Library (London), Hanley Library, Holborn Library (London), Islington Central Library (London), Kent

Record Office (Maidstone), Lambeth Palace Library (London), the London Metropolitan Archives, Malvern Library, Malvern Registration District Office, the Metropolitan Police Museum (London), the National Meteorological Archive (Bracknell), the Natural History Museum (London), the Pierpont Morgan Library (New York), the Public Record Office (Kew), the Royal College of Surgeons (London), the Royal Horticultural Society (London), the Royal Photographic Society (Bath), the Royal Society (London), the Royal Society for the Prevention of Cruelty to Animals (London), St. Bride Printing Library (London), Staffordshire Record Office (Stafford), Stoke-on-Trent City Archives, University College London Library, University College London Medical School Library, University of Birmingham Library, University of Reading Library, the Wellcome Institute Library (London), Hereford and Worcester County Record Office (Worcester), the Wordsworth Trust (Grasmere), and the Zoological Society of London.

Darwin scholars and other experts have given me information and much-needed advice on many points. My first debt is to Jim Moore for his warm encouragement and guidance at every stage, and for his great generosity in providing details from his research and sharing his findings and ideas. Janet Browne, Ralph Colp, Adrian Desmond, Nick Gill, John Harcup and David Kohn were equally generous, each in their areas of special knowledge and authority. I owe special thanks also to Edna Healey for her wise advice in our many conversations about Emma Darwin. My thanks for help on particular points to Dave Annal, Gordon Baldwin, Jeremy Barlow, Paul Betz, Anne Bissell, Geoffrey Breed, Chris Brooks, John Brown, Helen Burton, Tessa Chester, Gordon Cook, Geoffrey Copus, Shirley and Emma Corke, Rosemary Dinnage, Kenneth Dix, John Ford, Adrian Friday, Martin Gardiner, Marilyn Gaull, Denis Gibbs, Stephen Gill, Gary Hatfield, Andrew Hill, Kathryn Hughes, Nick Humphrey, Michael Jaye, Peter Kay, Terry Kidner, Desmond King-Hele, Sally Kington, Eric Korn, Julien Litten, Noreen Marshall, Lyn Martin, Jonathan Miller, Solene and Noel Morris, Irene Palmer, Halina Pasierbska, Duncan Porter, Richard Preece, Chris Preston, Henry Quinn, Martin Rudwick, Robert Ryan, Jim Secord, Eoin Shalloo, Michael Twyman, Derek Wallis, Kathie Way and Cora Weaver.

My thanks to Mrs. Connie May of Downe and Mrs. Nell Davis of Malvern for their recollections, Mrs. Doreen Speare for showing me Montreal House and Hill House in Malvern, the Reverend Tim Hatwell, vicar of Downe, for information about the Darwin tomb in the parish churchyard, and the Reverend David Smith for showing me Emma Darwin's interleaved Bible.

Friends and family have helped greatly. My special thanks to Robin Holloway, Paul Martin and Michael Phillips for their advice; to Angelo Hornak for his photographs; to Jeremy Barlow, Andrew Cornford, George and Angela Darwin, Sophie Gurney, Philip and Nellie Trevelyan, and Barbara, John and Martin Wedgwood; and to Laura, Milo, Richard, Roger, Simon and Stephen Keynes.

Warmest thanks to Virginia Bonham-Carter and Julie Grau, my editors in London and New York, for their guidance and help through the planning and shaping of the book.

Edna Healey, Robin Holloway, Eric Korn and Jim Moore kindly read the book in draft and provided many helpful comments.

Anne, Cecil, Skandar, Soumaya and Zelfa helped in all the ways they know. My deepest thanks to them.

# PICTURE CREDITS

The British Library, London, pl. 1; The British Museum, London, p. 6; The Syndics of Cambridge University Library, pl. 9, 14, pp. 36, 131, 341; English Heritage Photographic Library, London, pl. 2, 3, 5, 7, 8, 15, 16, 20, pp. iv, 106; The Provost of Fellows of Eton College, Windsor Berkshire, England, pl. 10; Guildhall Library, Corporation of London, p. 340; Dr. John Harcup, pl. 13; Professor Richard Keynes, pp. xviii, 183, pl. 20, 21, 22 (Angelo Hornak). pl. 17, 18, 19 (English Heritage Photographic Library, London). pl. 6, p. 14 (The Syndics of Cambridge University Library); The Linnean Society of London, p. 93; Philip and Nellie Trevelyan, pl. 12 (English Heritage Photographic Library, London); Mrs. Barbara Wedgwood, pl. 4; The Trustees of the Wedgwood Museum, Barlaston, Staffordshire, England, p. 116; Worcestershire County Council Cultural Services, Malvern, Worcestershire, England, pl. 11.

### Published Sources

Darwin, Charles, *The Descent of Man* (London 1888), Vol. I, p. 14.

Darwin, Charles, *On the Movements and Habits of Climbing Plants* (London 1865), p. 79.

Darwin, Charles, *Monograph on the Sub-Class Cirripedia* (London 1851–54), Vol. I, pp. 384–5, pl. 5, fig. 9; Vol. II, pp. 659–60, pl. 23, fig. 5.

Darwin, Francis, *The Life and Letters of Charles Darwin,* John Murray, London 1887, Vol. I, facing p. 320.

Fitzroy, Robert, *Narrative of the Surveying Voyages of His Majesty's Ships* Adventure *and* Beagle, Vol. II, Henry Colburn, London 1839, Frontispiece; p. 324.

Fouquier, A, "Mademoiselle Doudet," *Causes célèbres de tous les peuples,* Paris 1858–74, Cahier 13, 59e Livre, p. 9.

*Illustrated London News,* 20 August 1842, p. 232.

Leech, Joseph, *Three Weeks in Wet Sheets* (London 1851), p. 72.

Owen, Richard, "Contributions to the Natural History of the Anthropoid Apes. No. III. On the External Characters of the Gorilla," *Transactions of the Zoological Society of London,* Vol. 5, part 4, pl. 45.

*Punch*'s Almanack for 1882, *Punch,* 6 December 1881.

Ritchie, Hester Thackeray, *Thackeray and his Daughter, the Letters and Journals of Anne Thackeray Ritchie* (London 1924), p. 62.

# INDEX

# INDEX

INDEX

INDEX

# INDEX